FRACTALS
Concepts and Applications in Geosciences

FRACTALS
Concepts and Applications in Geosciences

Editors

Behzad Ghanbarian

Department of Petroleum and Geosystems Engineering
University of Texas at Austin, Texas, USA

Allen G. Hunt

Department of Physics, Wright State University
Dayton Ohio, USA

CRC Press is an imprint of the
Taylor & Francis Group, an **informa** business
A SCIENCE PUBLISHERS BOOK

CRC Press
Taylor & Francis Group
6000 Broken Sound Parkway NW, Suite 300
Boca Raton, FL 33487-2742

First issued in paperback 2021

© 2017 by Taylor & Francis Group, LLC
CRC Press is an imprint of Taylor & Francis Group, an Informa business

No claim to original U.S. Government works

Version Date: 20170808

ISBN-13: 978-0-367-78170-5 (pbk)
ISBN-13: 978-1-4987-4871-1 (hbk)

This book contains information obtained from authentic and highly regarded sources. Reasonable efforts have been made to publish reliable data and information, but the author and publisher cannot assume responsibility for the validity of all materials or the consequences of their use. The authors and publishers have attempted to trace the copyright holders of all material reproduced in this publication and apologize to copyright holders if permission to publish in this form has not been obtained. If any copyright material has not been acknowledged please write and let us know so we may rectify in any future reprint.

Except as permitted under U.S. Copyright Law, no part of this book may be reprinted, reproduced, transmitted, or utilized in any form by any electronic, mechanical, or other means, now known or hereafter invented, including photocopying, microfilming, and recording, or in any information storage or retrieval system, without written permission from the publishers.

For permission to photocopy or use material electronically from this work, please access www.copyright.com (http://www.copyright.com/) or contact the Copyright Clearance Center, Inc. (CCC), 222 Rosewood Drive, Danvers, MA 01923, 978-750-8400. CCC is a not-for-profit organization that provides licenses and registration for a variety of users. For organizations that have been granted a photocopy license by the CCC, a separate system of payment has been arranged.

Trademark Notice: Product or corporate names may be trademarks or registered trademarks, and are used only for identification and explanation without intent to infringe.

Library of Congress Cataloging-in-Publication Data

Names: Ghanbarian, Behzad, editor. | Hunt, Allen G. (Allen Gerhard), editor.
Title: Fractals : concepts and applications in geosciences / editors, Behzad Ghanbarian, Department of Petroleum and Geosystems Engineering University of Texas at Austin, Austin, TX, USA, Allen Hunt, Department of Physics Wright State University, Dayton, OH, USA.
Description: Boca Raton, FL : CRC Press, Taylor & Francis Group, 2017.
| "A science publishers book." | Includes bibliographical references and index.
Identifiers: LCCN 2017034326 | ISBN 9781498748711 (hardback : alk. paper)
Subjects: LCSH: Fractals. | Geology--Mathematics. | Earth sciences--Mathematics.
Classification: LCC QE33.2.F73 F6845 2017 | DDC 550.1/514742--dc23
LC record available at https://lccn.loc.gov/2017034326

Visit the Taylor & Francis Web site at
http://www.taylorandfrancis.com

and the CRC Press Web site at
http://www.crcpress.com

Dedicated to
our loving families
for their continuous, countless, and unconditional support,
love, understanding, and inspiration

Preface

It is now half a century since Benoit Mandelbrot in 1967 published his article, How Long is the Coastline of Britain? Statistical Self-Similarity and Fractional Dimension, in the journal Science. Although not immediately perceived by the community as being of fundamental relevance, the fractional dimension, subsequently named by Mandelbrot as a fractal, has become recognized as a concept with pervasive relevance. This journey has been long and arduous, as many novel directions have been. One of the challenges has been to turn the subject of fractals from being descriptive in nature to a predictive enterprise. Related to this question is whether fractals are enough, or are there other measures that are necessary for understanding? Finally, it is essential to address the overall structure, or role, of fractals in statistical mechanics and optimization theory. For example, hierarchical structures are known to provide optimal dissipation of energy, access to information, etc., but how do they develop in the real world? Are they built from the small scale up (optimization, as in constructal theory)? Or are they generated from a kind of fragmentation, as traditionally entertained in fractal models (fractal fragmentation, for example)? Does it matter? Aside from the first chapter, this book is more of a collection of practical applications of the concept of fractals than a treatise addressing the more fundamental questions posed in this paragraph. But an examination of such specific applications supports development of an intuition that helps us in our own understanding. So, specifically, what does this particular book do?

It provides theoretical concepts and applications of fractals and multifractals to a broad range of audiences from various scientific communities, such as petroleum, chemical, civil and environmental engineering, atmospheric research, and hydrology. In the first chapter, Schertzer and Tchiguirinskaia introduce fractals and multifractals, and replace scale symmetries in a general framework of symmetry groups, which provides a generalized scale invariance framework that have a much wider applicability than the too restricted framework of the classical self-similarity. The next chapters address mainly implications of fractal medium geometry to measurable properties. In chapter 2, Ghanbarian and Millán review and discuss various fractal capillary pressure curve models. They show how the incorporation of surface roughness effect affects capillary pressure curve. In chapter 3, Perrier addresses modeling of a porous medium either as a bi-phase (e.g., solid and pore) or a tri-phase material. In the latter, a third phase is explicitly introduced to account for the porous matrix. She also addresses upscaling permeability by mixing discrete pore network modeling and continuous approaches at matrix level. In chapter 4, Martín studies the mathematics of complexity in soils and granular media. He shows that particle-size distribution of such media can be characterized by an entropic or heterogeneity level. Chapter 5 by Hunt and Yu discusses applications of fractals developed from percolation theory to fluid flow and solute transport in porous media. The main point of this chapter is the realization that it is the fractal characteristics of the dominant flow paths, not necessarily the structure of the medium itself, that generate predictive scaling relationships for relevant rates. As applications, Hunt and Yu demonstrate applications to chemical weathering, non-Gaussian solute transport, soil production, vegetation growth, etc. In chapter 6, nonlocal transport in fractal porous media is addressed by O'Malley and Cushman using two approaches: one is based on thermodynamics and statistical mechanics, and another on a continuum mechanics approach. O'Malley and Cushman provide an overview of the two approaches, and then compare them with each other. The last chapters focus more on spatial and temporal series analyses, which have been under investigations for several decades in physics, hydrology, atmospheric research, civil engineering, and water resources. Chapter 7, by Tarquis and her coworkers, addresses applications from multifractals to geostatistics. They show that multifractal analysis can be used as one- or two-dimensional spatial statistics based on several orders of moments, in contradiction to classical geostatistical techniques, which are based on the second-order moment. An important application is to the normalized difference vegetation index, NDVI, which is highly correlated with the important ecohydrological quantity, net primary productivity, NPP. In chapter 8, Lovejoy provides a detailed explanation of the nonlinear geophysics of climate closure. The most important implication of this research is the clear inference that recent climate change is human-induced, and the so-called "pause" in warming after 1998 is merely a cool fluctuation on top of anthropogenic warming. In chapter 9, Bogachev and his coauthors address applications from fractals and multifractals to climate and hydrometeorological time series and their relation to long-term memory. They show how occurrence of rare events may be quantified in long-term correlated records. Chapter 10 by Jafari and his collaborators discusses applications of the multifractal random walk method to analyze petrophysical (spatial) time series. In chapter 11, Puente and his coauthors introduce application of a fractal geometric technique called fractal-multifractal (FM) to model geophysical records, such as rainfall, streamflow and temperature. In the last chapter, Hunt discusses the use of constructal theory in river basins. By using percolation concepts for smaller scales, but results of Adrian Bejan's constructal research for continental scales, it is possible to present a picture of river basins that start with linear, random fractal organization at shorter time and smaller spatial scales, and evolve towards non-linear optimization, in which the river becomes an active agent organizing the landscape.

May, 2017 **Behzad Ghanbarian**
Allen G. Hunt

Contents

CHAPTER

1

An Introduction to Multifractals and Scale Symmetry Groups

Daniel Schertzer* **and Ioulia Tchiguirinskaia**

Hydrology Meteorology and Complexity (HMCo), Ecole des Ponts ParisTech, U. Paris-Est, 6-8 av. B. Pascal, Cité Descartes, 77455 Marne-la-Vallée, France

1. Introduction and Motivations

The complexity of geosciences has inspired the development of many innovative concepts and techniques. For instance, the extreme variability of geophysical fields over wide ranges of space-time scales, particularly their intermittency, has been a key driver towards the development of multifractals. In turn, this development has significantly improved our understanding and modeling capacities of our complex environment as a whole, i.e. not only its physico-ecological component from its smallest scales (e.g. micro-turbulence) to its largest scales (e.g. climate (Lovejoy and Schertzer 2013), astrophysics (Sylos Labini et al. 1998)), but also its socio-economical component (e.g. city dynamics (Murcio et al. 2015, Dupuy 2016)). This introduction intends to demonstrate that recent theoretical developments on multifractals, particularly on their formalism, will make them even more useful and indispensable for geophysics.

In general, a multifractal is not a geometric set, but a map from a space-time domain X onto a codomain $\tilde{X}$ that is a vector space or a manifold, displaying structures at all scales. Furthermore, a multifractal is not necessarily a pointwise function, but is often a (mathematical) measure or a generalized function. In fact, multifractal formalisms have made manageable, by physicists and engineers, mathematical objects like stochastic multi-singular measures, which were earlier mathematical curiosities. Multifractals are indeed very broad generalizations of the

*Corresponding author: Daniel.Schertzer@enpc.fr

geometrical fractals whose field is a set indicator function with a binary codomain $\tilde{X} = \{0, 1\}$, therefore a kind of degenerate case for multifractal fields.

The property to have structures at all scales is trivially scale invariant since it does not depend on the scale of observation. This points out that a multifractal field can also be defined as being invariant under a given scale transform, which defines a symmetry of this field. A precise definition of this invariance and of the corresponding scale transform, which could either be deterministic or stochastic (e.g. involves only equality in probability distribution or other statistical equivalences), isotropic or not, will be given below. However, the above rather intuitive definition already shows that multifractals are not only quite general, but also quite fundamental. Indeed, symmetry principles are the building blocks of physics and of many other disciplines (Weyl 1952, Zee 1986). Scale symmetry is in fact an element of the extended Galilean invariance. Unfortunately, the maximum attention in mechanics, especially in point mechanics, has been given to the space shifts between two (Galilean) frameworks that differ only by a constant relative velocity or by a given rotation that defines the pure Galilean group. But with continuous mechanics, it has broadened to other transforms such as scale dilations. Following the Buckingam Π-theorem (Buckingham 1914, Buckingham 1915, Sonin 2004), the key role of the scale symmetry in fluid mechanics was demonstrated by Sedov (1972), using many applications. More recently, Galilean invariance was used to assess subgrid models in turbulence (Speziale 1985). Scale symmetry has been widely used under the denomination of self-similarity, but with unnecessary limitations. Indeed, the main goal was to find a unique, deterministic, scale transformation under which the nonlinear dynamical equations remain invariant, in particular the Navier-Stokes equations associated with others, such as the advection-dissipation equation. This was mainly achieved by a dimensionalisation of these equations with the help of various so-called characteristic quantities, e.g. characteristic space and time scales. Multifractals are in fact invariant with respect to multiple scale symmetry and therefore correspond once again to a broad generalization of properties that were once perceived in a more restrictive framework. We indeed have to go from scale analysis to scaling analysis, as illustrated by Schertzer et al. (2012) on the seminal example of the quasi-geostrophic approximation and resulting model of atmospheric turbulence, both derived by Charney (Charney 1948, Charney 1971).

Reviewing multifractal formalisms shows that beyond their strong communality of being defined as a hierarchy of fractals, hence a common complexity, they have fundamental differences due to basic assumptions on the nature of the process and more or less stringent hypotheses (e.g. geometric or analytic). The formalisms yield rather different properties; for instance they can be either deterministic or statistical with various meanings (e.g. weakly, in probability, almost surely, surely or in a given norm).

2. From Geometry to Analytics

2.1 Multifractals, Scale Transforms and Symmetries

Although it is usual to introduce multifractals after introducing fractal (geometric) sets, since this corresponds partly—but not totally—to historical developments, we believe it is more fruitful to proceed in an opposite manner, i.e. to see fractals as a very special case with a number of limitations. In fact, these limiting features of fractal sets have impeded the initial development of multifractals. The most obvious case is presumably the question of the uniqueness of the Hausdorff dimension or the Hurst exponent. This uniqueness is a mere mathematical fact for geometric sets (Falconer 1985, Falconer 1990) and respectively for fractal processes, e.g. the celebrated Brownian motion (Mandelbrot and Van Ness 1968). It was unfortunately believed to be a rather general feature and thus has led to a kind of dogma. As a consequence, the differences in estimates of these exponents obtained by different algorithms were unfortunately considered as unessential (Mandelbrot 1983, Mandelbrot 1989), whereas these algorithms were in fact capturing different features of a process that could not be reduced to a geometric object.

Here, we go a step further with respect to what was done (Schertzer et al. 2012, Schertzer and Tchiguirinskaia 2015) towards a presentation of multifractals based on symmetry groups. This is illustrated by the commutative diagram of Fig. 1, where the field φ that maps the domain X into the codomain $\tilde{X}$, is composed with a scale transform T_λ of the domain X. In the simplest case (see Section "Generalized Scale Invariance (GSI)" for extensions), T_λ, is the usual (isotropic) contraction/dilation for any

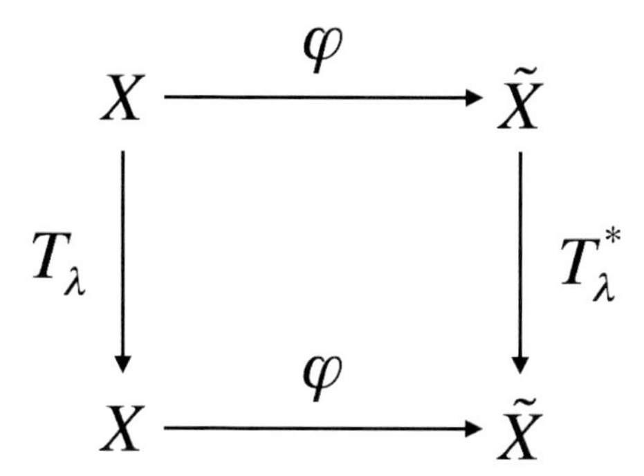

Figure 1: Commutative diagram illustrating how the analytical pullback transform T_λ^* is generated on the codomain $\tilde{X}$ of the field φ by the geometric transform T_λ on the domain X.

given positive scale ratio λ ($\lambda > 1$ for a contraction, $0 < \lambda < 1$ for a dilation):

$$\forall x \in X : T_\lambda x = x/\lambda \tag{1}$$

The commutative diagram of Fig. 1 shows that, independently of its precise definition, the map T_λ of X into itself defines a map T_λ^* of $\tilde{X}$ into itself:

$$\forall x \in X : T_\lambda^* \circ \varphi(x) = \varphi \circ T_\lambda(x) \tag{2}$$

which merely means that the action of T_λ^* on $\tilde{X}$ counterbalances the action of T_λ on X, and corresponds to a transfer of a geometric change of scale (T_λ) into an analytic transform (T_λ^*) of the field (φ). The couple (T_λ, T_λ^*)[1] therefore defines a symmetry that is satisfied by the field φ. If T_λ^* is furthermore invertible, this symmetry can be written under the form:

$$S(\varphi) = \varphi \tag{3}$$

with:

$$S(\varphi) = T_\lambda^{*-1} \circ \varphi \circ T_\lambda \tag{4}$$

A (geometric) fractal set A, of indicator function 1_A, corresponds to the simplest and somehow degenerate case of scale symmetry (*Id* denotes the identity transform):

$$\varphi = 1_A;\ T_\lambda(x) = x/\lambda;\ T_\lambda^* = Id \tag{5}$$

which corresponds to the more familiar expression:

$$A = T_\lambda(A) \tag{6}$$

One may note that if A is compact, i.e. has an outer scale, one needs to make this equation local by considering the intersection of the two sides of the equation with a small enough sphere. The definition of a fractal process $B(t)$ with stationary increments (Lamperti 1962):

$$|\Delta B(\Delta t/\lambda)| = \lambda^{-H} |\Delta B(\Delta t)| \tag{7}$$

corresponds to:

$$T_\lambda(x) = x/\lambda;\ T_\lambda^*(\tilde{x})/\lambda^H \tag{8}$$

The classical choice of the letter H for the exponent refers to the hydrologist Hurst (Hurst 1951, Klemes 1974, Koutsoyiannis 2002), who empirically uncovered this exponent in analyses of long time hydrological series, particularly that of the Nile floods. The classical value $H = 1/2$ corresponds to the Brownian motion, whereas other values corresponds to fractional Brownian motions (Mandelbrot and Van Ness 1968, Kahane

1 Which obviously is not unique.

1997) or other "anomalous" diffusion processes (Painter 1996, Yanovsky et al. 2000, Lévy 1965) (see Fig. 2 for illustrations). Unfortunately, the existence of a unique exponent H in these examples gave some credence that this uniqueness was somewhat a general property of scale symmetries.

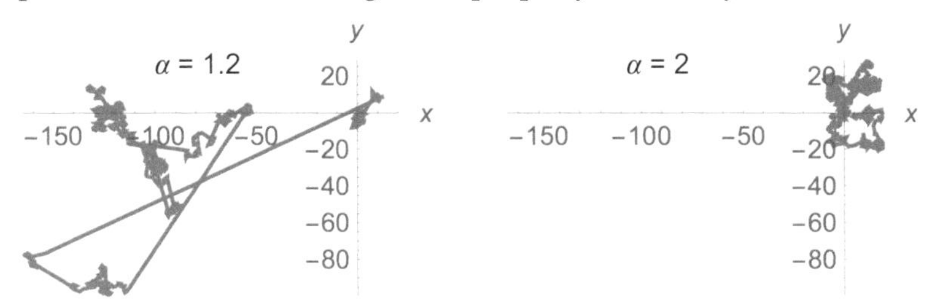

Figure 2: Trails of two classical monofractal processes: a Levy flight with jumps (left, with a stable index $\alpha = 1.2$) and a Brownian (right, with the upper limit case $\alpha = 2$), which is much more compact. In both cases, the Hurst exponent is: $H = 1/\alpha$.

2.2 Pullback, Duality and Push Forward Transforms

The transform T_λ^* (Eq.(2) and Fig. 1) is known as the "pullback transform" of the field φ (or "composition operator" (Shapiro 1993)): it corresponds to a generalization to (infinite dimensional) functional spaces of the contravariant change of coordinates of (possibly finite dimensional) vector spaces. Indeed, the $T_\lambda^*(\varphi)$ pulls back the map φ from the new coordinates $y = T_\lambda x$ to the old coordinates x, with the "change of coordinates" T_λ. The superscript '*' is in agreement with the usual notation of contravariant tensors.

In fact, the pullback transform generates by duality (see Fig. 3 for illustration) another transform on a more complex functional space, as required by the fact that multifractal fields are often not (pointwise) functions, but (mathematical) measures or generalized functions μ's. The latter belongs to a dual space $C(X,\tilde{X})'$ of a given set $C(X,\tilde{X})$ of test functions φ's. $C(X,\tilde{X})'$ is a set of linear transforms over the functional space $C(X,\tilde{X})$. It is therefore needed to define the effect of the transforms T_λ's not only on functions φ's, but also on measures or generalized functions μ's. The definition of the pullback transform T_λ^* (Eq. (2)) can be rewritten as:

$$\forall\varphi \in C(\tilde{X},X): T_\lambda^*(\varphi) = \varphi(T_\lambda) \tag{9}$$

which shows that it plays a similar role on $C(X,\tilde{X})$ as the scale transform T_λ on X and can therefore define by duality (see the corresponding commutative diagram of Fig. 3) a "push forward" transform $T_{*,\lambda}$ on $C(X,\tilde{X})'$:

$$\forall\mu \in C(X,\tilde{X})', \forall\varphi \in C(X,\tilde{X}): \int T_\lambda^*(\varphi)d\mu = \int \varphi dT_{*,\lambda}(\mu) \tag{10}$$

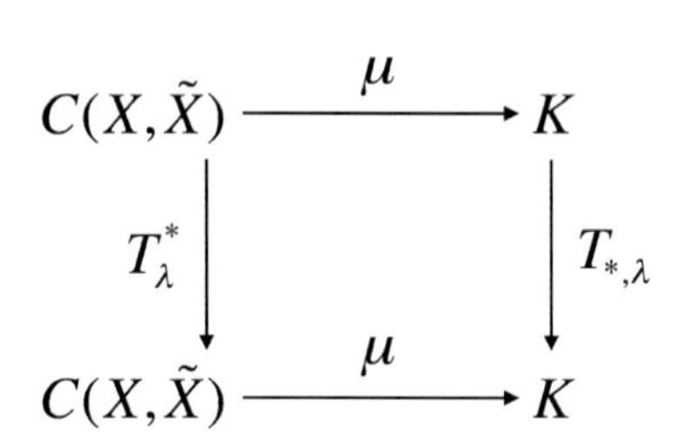

Figure 3: Commutative diagram, similar to that of Fig. 1, illustrating how the analytical pullback transform T_λ^* generates in turn the push forward $T_{*,\lambda}$ for measures or generalized functions μ's.

Although diagrams of Figs 1 and 3 look very similar, an important difference is that the "push forward" transform $T_{*,\lambda}$ indeed pushes forward the measure μ from the old coordinate x to the new coordinate $y = T_\lambda x$ and thus generalizes the covariant coordinate transform of the (finite dimensional) dual vector spaces. We will use the generic notation $\tilde{T}_\lambda$ to represent both transforms T_λ^* and $T_{*,\lambda}$ on the spaces they map, when their differences are not relevant. It is worthwhile to note that the renormalization group approach (Wilson 1971) and its variants defines transformations similar to T_λ^* by a decimation process (i.e. reducing the degrees of freedom of a system) rather than by a change of scale. Loosely speaking, the starting point is not the same, but both approaches converge.

2.3 Generalized Scale Invariance (GSI)

The definitions of "pullback" and "push forward" transforms are extremely broad and we need to specialize them with respect to a given concept of scale. We are going to use for the domain and the codomain a generalization of the Euclidean metrics, still constituting a building block of the Fractal Geometry of Nature (Mandelbrot 1977, Mandelbrot 1983) in spite of its claim to be fundamentally non-Euclidean. This is at first required by the ubiquitous evidence of anisotropic, multiscale fields and patterns, and the resulting necessity to avoid the usual hypothesis of rotational symmetry prior to a scale symmetry (Schertzer and Lovejoy 1985a, Schertzer and Lovejoy 2011). For instance, this hypothesis had very unfortunate consequences in atmospheric turbulence research, which had been blocked for decades by considering only 2D turbulence and 3D turbulence, whereas none of this regime can be truly relevant (Schertzer and Lovejoy 1983, Schertzer and Lovejoy 1985b, Schertzer and Lovejoy 1988, Schertzer et al. 2012). Figure 4 displays examples of strongly anisotropic (closed) balls B_t defined by the generalized scale $\|.\|$:

$$\forall \ell \in R^+ : B_t = \{x \,|\, \|x\| \leq \ell\} \tag{11}$$

Figure 4: Contrary to the isotropic self-similar spheres defined by a classical norm, the balls corresponding to a given generalized scale can be strongly anisotropic due to rotation and stratification. This anisotropy increases with the nonlinearity of the involved scale transformation, here from top to bottom and left to right, starting with the linear case.

contrary to the isotropic self-similar spheres defined by a classical norm $|.|$ that satisfies the scalability relation for any isotropic scale transformation T_λ (Eq.(1)):

$$\forall x \in X : |T_\lambda x| = |x| / \lambda \tag{12}$$

The isotropic scale transformations obviously satisfy a multiplicative group property with respect to the scale ratio λ:

$$T_{\lambda\lambda} = T_\lambda \circ T_\lambda \tag{13}$$

Both properties are preserved for a generalized scale change operator T_λ, but with respect to a generalized scale $||.||$ that satisfies both these conditions, as well as the non-degeneracy condition:

$$||x|| = 0 \Leftrightarrow x = 0 \tag{14}$$

However, the triangular inequality is no longer required for the generalized scale $||.||$, only the weaker property of ball embedding (Schertzer et al. 1999):

$$\forall L \in R^+, \lambda' \geq \lambda \geq 1 : B_{L/\lambda'} \subset B_{L/\lambda} \tag{15}$$

Figure 5 illustrates that the group property of T_λ propagates in a straightforward manner to the "pullback" transform T_λ^* and then to "push forward" transform $T_{*,\lambda}$. Nonlinear GSI balls are not necessarily convex as shown in Fig. 4, although they conform with the aforementioned properties, in particular that of embedding. The simplest case of Generalized Scale

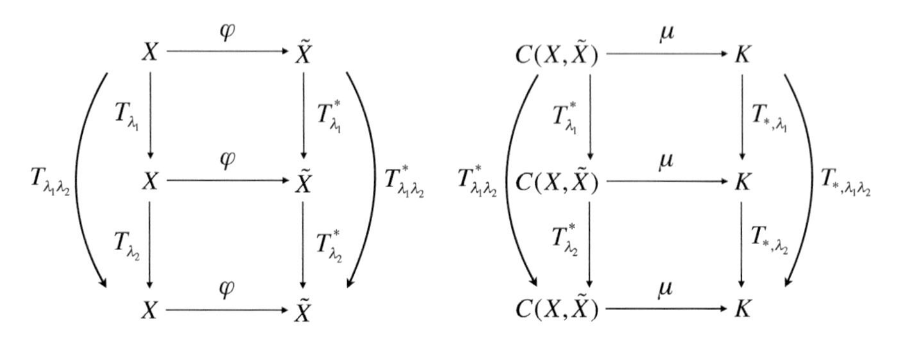

Figure 5: These diagrams show how the group property of T_λ propagates in a straightforward manner to the "pullback" transform T_λ^* (left) and then (by duality) to the "push forward" transform $T_{*,\lambda}$ (right).

Invariance corresponds to that of linear GSI, where the scale transform is a multiplicative group generated by a given matrix G:

$$T_\lambda = \lambda^G \equiv \exp(\mathrm{Log}(\lambda)G) \tag{16}$$

It is straightforward to explicitly obtain an associated generalized scale for any given $\alpha > 0$ generalizing the classic L_α norm, if this generator has a positive spectrum Spec(G) = $\{\mathrm{v}_i\}$, with corresponding eigenvectors e^i (the classical L_α norm corresponds to $v_i \equiv 1$):

$$\left\|\sum_i x_i e^i\right\|_\alpha - \left(\sum_i \left\|x_i e^i\right\|_\alpha^\alpha\right)^{1/\alpha} ; \left\|x_i e^i\right\|_\alpha = \left|x_i\right|^{1/\mathrm{v}_i}\left|e^i\right| \tag{17}$$

This illustrates the fact that, like classical norms, several generalized scales could correspond to the same scale transform. Although not necessarily, they may have some strong common properties. There is a similar definition of equivalence between two generalized scales $||.||_1$ and $||.||_2$:

$$\exists A, B > 0, \forall x \in X : A\|x\|_1 \le \|x\|_2 \le B\,\|x\|_1 \tag{18}$$

3. Scalar-valued Multifractals

3.1 Scaling of the Statistical Moments and Singularities, the Mellin and Legendre Transforms

Before addressing the more recent and more complex case of vector-valued multifractals, we consider the somewhat classical case of scalar-valued multifractals. Although this took some time to be understood, the scalar-valued multifractals are in many respects a straightforward generalization of the uni/mono-fractal case. Indeed, instead of having a unique Hurst exponent H like in Eq.(7) that fully defines the scale change operator $\tilde{T}_\lambda$ —of either the codomain $\tilde{X}(\tilde{T}_\lambda = T^*_\lambda)$ or of the field $K(\tilde{T}_\lambda = T^*_{*,\lambda})$—a given distribution of singularities γ's is required. They define in fact a stochastic generator Γ_λ (Schertzer and Lovejoy 1987):

$$\tilde{T}_\lambda = \lambda^\gamma \equiv \exp(\Gamma_\lambda) \tag{19}$$

The statistical moments of $\tilde{T}_\lambda$ are therefore defined by the cumulant function (or second characteristic function) K_λ of the generator Γ_λ:

$$E(\tilde{T}_\lambda^q) = E[\exp(q\Gamma_\lambda)] \equiv \exp[K_\lambda(q)] \approx \lambda^{K(q)} \tag{20}$$

where $E(.)$ denotes the mathematical expectation or ensemble average. The last equation corresponding to the scaling behavior is obtained as soon as K_λ has a $\mathrm{Log}(\lambda)$ divergence:

$$K_\lambda(q) = \mathrm{Log}[\exp(q\Gamma_\lambda) \cong \mathrm{Log}(\lambda)K(q) \tag{21}$$

$K(q)$ is called the scaling moment function and is the generator of the multiplicative group of the statistical moments $E(\tilde{T}_\lambda^q)$. The Mellin transform can be used to relate this multiplicative group to that of the probability distribution of the singularities $\Pr_\lambda(\gamma' > \gamma)$ at a given scale ratio λ (Eq. (24)) (Schertzer et al. 1997). Indeed, under fairly general conditions, the Mellin transform $M(.)$ relates the probability distribution function $p(x)$ of a non-negative random variable X to its statistical moments $E(X^{q-1})$:

$$E(X^{q-1}) = M_X(q) = \int_0^\infty x^{q-1} p_X(x)dx \tag{22}$$

$$p_X(x) = M_X^{-1}(E(X^{q-1})) = \frac{1}{2i\pi}\int_{c-j\infty}^{c+i\infty} E(X^{q-1})x^{-q}dq \tag{23}$$

The direct and inverse Mellin transforms correspond to the Laplace and inverse Laplace transforms for $\mathrm{Log}(x)$. In the present case, the distribution of interest is:

$$\Pr_\lambda(\gamma' > \gamma) = \Pr(\tilde{T}_\lambda > \lambda^\gamma) \approx \lambda^{-c(\gamma)} \tag{24}$$

Here the last equality, representing the scaling or multiplicative group property of the probability distribution, is obtained by the Mellin transform of the corresponding equation (Eq. (20)) for the statistical moments. This Mellin duality is very general and $\lambda \to \infty$ yields the Legendre duality between the corresponding exponents $K(q)$[2] and $c(\gamma)$:

$$K(q) = \sup\{q\gamma - c(\gamma)\} \Leftrightarrow c(\gamma) = \sup_q\{q\gamma - K(q)\} \tag{25}$$

2 The field K and the scaling function $K(q)$ have only a common letter (due to respective usages), nothing else.

The Legendre duality was first pointed out by Parisi and Frisch (1985) in a different context and with restrictive geometric hypotheses that are discussed below. The codimension function $c(\gamma)$ has been called Cramer function by Mandelbrot (1991) due to its relation to the large deviation theory (Oono 1989, Varadhan 1984). Both $K(q)$ and $c(\gamma)$ are convex. The convexity of $K(q)$ is based on its cumulant function definition. The convexity of $c(\gamma)$ results from the property that the Legendre transform yields a convex function. When $K(q)$ and $c(\gamma)$ are derivable, the Legendre duality (Eq.(25)) takes a more explicit form:

$$q_\gamma = \frac{dc(\gamma)}{d\gamma}; \gamma_q = \frac{dK(q)}{dq} \tag{26}$$

i.e. one curve is the envelop of the tangencies of the other one. Hence, there is an explicit one-to-one correspondence between moment orders q's and orders of singularities γ's.

3.2 Comparison of Multifractal Formalisms

We have discussed above a codimension multifractal formalism (Schertzer and Lovejoy 1987, Schertzer and Lovejoy 1989), which is more general than the dimension formalism used by Parisi and Frisch (1985) and Halsey et al. (1986). These choices resulted from different motivations and theoretical frameworks. The dimension formalism was developed in a deterministic and geometric framework, either to explain (Parisi and Frisch 1985) the nonlinearity of the scaling exponents of the velocity structure functions (the statistical moments of the velocity increments) empirically observed by Anselmet et al. (1984), or the fact that strange attractors (Halsey et al. 1986) could not be characterized by a unique dimension (Grassberger and Procaccia 1983). It turned out that both cases correspond to characterize the singularities α_D as local Holder exponents (Fraysse and Jaffard 2006) supported by embedded fractal sets of dimensions $f(\alpha_D)$. In a finite D-dimensional embedding space, which is required only by the dimension formalism (hence, the D-dimensional dependence and corresponding sub-index D), the respective notations correspond through a codimension/dimension transform:

$$\alpha_D = D - \gamma;\ f(\alpha_D) = D - c(\gamma) \tag{27}$$

For a stochastic process, the underlying probability space is usually infinite-dimensional, whereas realizations are (often) finite D-dimensional cuts. The codimension formalism enables to explore various types of cuts, including those with negative dimensions:

$$D - c(\gamma) > D \Leftrightarrow f_D(\alpha_D) < 0 \tag{28}$$

and therefore solve in a straightforward manner the problem of the so-called latent negative dimensions (Mandelbrot 1991). The scaling exponent $\tau_D(q)$ of the (deterministic) partition function (Procaccia 1983) have a similar D-dimensional dependence and relation to the moment scaling function $K(q)$:

$$\tau_D(q) = (q-1)\,D - K(q) \tag{29}$$

It is worth noting that the notations $\tau_D(q)$ and $f(\alpha_D)$ are frequently used instead of $K(q)$ and $c(\gamma)$ for stochastic multifractals in a very confusing manner, e.g. $K(q)$ is not understood as being a cumulant function of a stochastic generator and the properties of the latter are not used. Let us mention that some early concepts of multifractals were related to the so-called Renyi dimensions $D(q)$ algebraically defined by Grassberger (1983):

$$D(q) = \tau_D(q)/(q-1);\; D(1) = \tau_D'(1) \tag{30}$$

Unfortunately, the physical meaning of $D(q)$, including the claim that they were dimensions, was not immediately obvious and raised doubts on the interest of the notion of multifractals (Mandelbrot 1984). It finally turns out that $D(q)$ corresponds to the dimension of the cuts where a divergence of moments of order q occurs (Schertzer and Lovejoy 1984, Schertzer and Lovejoy 1985b), as discussed further in the Section "Multifractal extremes: phase transitions and self-organized criticality".

The difference of formalisms bears also on the hierarchy of fractals supporting the singularities. Note that this question being a bit technical can be skipped in a first reading. In the geometric and deterministic approach, these supports are defined as the sets where the action of T_λ boosts the multifractal value of $\varphi(\underline{x})$ exactly by the singularity γ:

$$S_\lambda(\gamma) = \{x \in X \mid T_\lambda^{*}\varphi(x) = \lambda^{\gamma}\varphi(x)\} \tag{31}$$

which corresponds to the (deterministic) boundary $\partial A_\lambda(\gamma)$ of the event:

$$A_\lambda(\gamma) = \{(x,\omega) \in X x \Omega \mid T_\lambda^{*}\varphi(x) \geq \lambda^{\gamma}\varphi(x)\} \tag{32}$$

that is involved in the exceedance probability distribution, whereas $S_\lambda(\gamma)$ would rather correspond to the probability distribution function if defined, which is not necessarily the case. A subsequent difference occurs in the definition of the support in the limit $\lambda \to \infty$. In the geometric approach, one considers the limit inferior, i.e. the set of points that has the given singularity γ for eventually all resolutions λ's (i.e. for large enough λ):

$$\underline{S}(\gamma) = \underline{\lim}_{\Lambda\to\infty} S_\Lambda(\gamma) \equiv \bigcup_{\lambda}\bigcap_{\Lambda>\lambda} S_\Lambda(\gamma) \tag{33}$$

whereas the asymptotic scaling of the probability (Eq.(24)) is related to the limit superior, i.e. have the singularity γ for infinitely often resolutions λ's (Schertzer et al. 2002):

$$\bar{A}(\gamma) = \overline{\lim}_{\Lambda\to\infty} A_\Lambda(\gamma) \equiv \bigcap_{\lambda}\bigcup_{\Lambda>\lambda} A_\Lambda(\gamma) \tag{34}$$

Similar properties hold for measures μ's and $T_{*,\lambda}$.

This choice is in agreement with the mathematical approach of the "multiplicative chaos" initiated by Kahane (Kahane 1974, Kahane 1985b, Fan 1989, Kahane and Peyriere 1976) and based on discrete cascades. The limit superior $\overline{A}(\gamma)$ is less stringent than the limit inferior $\underline{A}(\gamma)$, i.e. $\underline{A}(\gamma) \subset \overline{A}(\gamma)$, and this difference corresponds to the fact that stochastic multifractal singularities are non-local (Schertzer and Lovejoy 1992), because they undergo a random walk. Tracking in a pointwise manner the singularities, i.e. looking for a limit $\gamma(x) = \lim_{\lambda \to x} \gamma_\lambda(x)$, can therefore be misleading, e.g. the notion of so-called local singularities (Cheng 2006).

4. Multifractal Extremes: Phase Transitions and Self-organized Criticality

4.1 Sample Size Limitations and Second Order Multifractal Phase Transition

An important difference between various multifractal processes, and corresponding formalisms, is the type of their extremes. For instance, singularities of geometric multifractals, embedded in a finite d-dimensional set, are necessarily bounded above by the singularity γ_s whose support dimension is the lowest possible one, i.e. it cannot be negative:

$$D(\gamma_s) = 0 \Leftrightarrow c(\gamma_s) = d \tag{35}$$

Because the topological dimension d_T of a set cannot exceed its fractal dimension, the support of the singularity γ_s is at best a set of isolated points ($d_T = 0$), the field can therefore correspond to sum of Dirac measures. But this singularity is not necessarily an upper bound for stochastic multifractals, it only corresponds to the maximal probable singularity on a d-dimensional sample (Hubert et al. 1993, Douglas and Barros 2003), hence the sub-index s for sampling singularity. If $N_s \geq 1$, independent samples of resolution λ are analyzed and it is convenient to introduce the sampling dimension D_s (Schertzer and Lovejoy 1989, Lavallée et al. 1991):

$$D_s = \mathrm{Log}_\lambda\,(N_s) \tag{36}$$

that enables to increase the sampling singularity γ_s and its codimension $c(\gamma_s)$ by somewhat generalizing Eq.(28) into:

$$c(\gamma_s) = d + D_s;\ D(\gamma_s) = d - c(\gamma_s) = -D_s \leq 0 \tag{37}$$

where $d + D_s$ is the overall sampling dimension of the sample. It explains how a negative dimension $D(\gamma_s)$, formally defined from the codimension $c(\gamma_s)$ (see Eq.(37)), becomes negative as soon as $D_s > 0$, i.e. $N_s > 1$. This resolves the somewhat paradoxical problem of latent dimensions (Mandelbrot 1991). Furthermore, because γ_s is almost surely the highest singularity observed on N_s samples it yields the following effective codimension $c_d(\gamma_s)$

$$\gamma < \gamma_s : c_d(\gamma) = c(\gamma_s); \gamma \geq \gamma_s : c_d(\gamma) = \infty \tag{38}$$

whose Legendre transform is the corresponding effective moment scaling function $K_s(q)$

$$q < q_s : K_s(q) = K(q); q \geq q_s : K_s(q) = K(q_s) + (q - q_s)\, c(\gamma_s) \tag{39}$$

i.e. $K_s(q)$ follows its tangency for order $q \geq q_s$ (see Fig. 6). This corresponds to a second order multifractal phase transition, where $K(q)$ is the analog of a thermodynamic potential and $1/q$ of the temperature (Feder 1988, Tel 1988, Schertzer and Lovejoy 1989); see Table 1 that displays the analogy between multifractal parameters and those of thermodynamics.

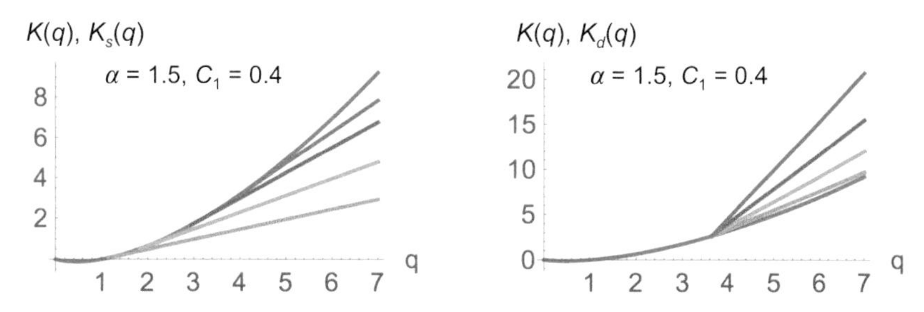

Figure 6: Second order (left) and first order (right) multifractal phase transitions of the (multifractal) potential analogue $K(q)$ for $\alpha = 1.5$ and $C_1 = 0.4$ (and the embedding dimension $d = 1$). For second order phase transitions (left), the effective moment scaling functions $K_s(q)$ (from yellow to dark blue curves) follow the tangency of the theoretical $K(q)$ (blue curve) for $q \geq q_s$ (the analogue of the inverse of a critical temperature) for the overall sampling dimensions $d+D_s = 0.5$, 1, 2 and 3. For first order phase transitions (right), the effective moment scaling functions $K_d(q)$ (from yellow to dark blue curves) corresponds to inner chords of the theoretical $K(q)$ (blue curve) for $q \geq q_d$ (the critical order of the divergence of moments) for the overall sampling dimensions $d+D_s = 5$, 7.5, 11.25 and 16.87, with $q_d \approx 3.6$.

4.2 First Order Multifractal Phase Transition and Self-organized Criticality

A more drastic behavior may occur for large samples. Indeed, they are able to have singularities γ's larger than the embedding dimension d. An observable at a given resolution λ is usually obtained by d-dimensional integration from the same observable at a much larger resolution Λ, possibly infinite, that corresponds to the inner scale of the process. For $\gamma < d$, this averaging succeeds to smooth out the sub-scale activity and to impose the resolution of observation λ as the effective resolution, because the integration shifts the singularities by $-d$, the scaling exponent of the elementary volume of integration.

Table 1: Correspondence between (stochastic) multifractals and thermodynamics (setting for notation simplicity $k_B = 1$ for the Boltzman's constant k_B): $\Sigma(\beta)$ being the thermodynamic potential, $F(\beta)$ the Helmhotz free energy

(Stochastic) multifractals	*Thermodynamics*
Probability space	Phase space
Moment order: q	(Reciprocal) temperature: $\beta = T^{-1}$
Singularity order: γ	(Negative) energy: $-E$
Generator	(Negative) Hamiltonian
Singularity codimension: $c(\gamma)$	Codimension of entropy: $D - S(E)$
Scaling moment function: $K(q) = \sup_{\gamma}(q\gamma - c(\gamma))$	(Negative) thermodynamic potential: $-\Sigma(\beta) = -\inf_{E}(\beta E - S(E))$
Dual codimension function: $C(q) = K(q)/(q-1)$	(Negative) free energy: $-F(\beta) = -\Sigma(\beta)/\beta$
Dimension of integration: D	External field: h
Ratio of scales: λ	Correlation length: ξ

Adapted from: Schertzer and Lovejoy (1989)

This is no longer the case for $\gamma \geq d$: the scale of observation is no longer relevant, whereas the inner scale (L/Λ) remains relevant. Therefore statistics may diverge as $\Lambda \to \infty$. This generates a drastic difference between a "bare" process that is stopped at the observation resolution λ and a "dressed" process that proceeds up to a very large resolution Λ and is then averaged down to the observation resolution. This renormalization jargon was introduced for multifractals, because of the similarity between taking into account more and more small scale activity and taking higher and higher order interactions in renormalization theory (Schertzer and Lovejoy 1987). Nevertheless, this singular behavior may be statistically negligible up to a critical singularity $\gamma_d > d$. For $\gamma > \gamma_d$, the observed dressed codimension function $c_d(\gamma)$ no longer corresponds to the bare codimension $c(\gamma)$: the dressed fluctuations will be much larger than the bare fluctuations and $c_d(\gamma)$ can therefore be estimated as maximizing the occurrences of high singularities, while nevertheless respecting the convexity constraint. This requires that $c_d(\gamma)$ follows the tangency of $c(\gamma)$ at γ_d, whose slope is $q_d = \left.\frac{c_d(\gamma)}{d\gamma}\right|_{\gamma_d}$ (Schertzer and Lovejoy 1992):

$$\gamma < \gamma_d : c_d(\gamma) = c(\gamma);\ \gamma \geq \gamma_d : c_d(\gamma) = c(\gamma_d) + q_d(\gamma - \gamma_d) \tag{40}$$

i.e. a similar behavior to second phase transition of $K_s(q)$ (see Eq.(39)); however $c(\gamma)$ is rather a local entropy analog, not a thermodynamic potential analog. It is worthwhile to note that the linear behavior of $c_d(\gamma)$

for $\gamma \geq \gamma_d$ corresponds to a power-law falloff of the probability distribution. This feature has often been taken as a hallmark of self-organized criticality (Bak et al. 1987, Bak and Chen 1991), but this occurs here for a dissipative system with non-vanishing input (Schertzer et al. 1993, Schertzer and Lovejoy 1994).

The Legendre transform yields a first order phase transition for the corresponding moment scaling function $K_d(q)$ for a finite number $N_s > 1$ of samples with a sampling singularity $\gamma_s > \gamma_d > d$:

$$q \leq q_d : K_d(q);\ K(q);\ q \geq q_d : K_d(q) = k(q_d) + \gamma_s(q - q_d) \tag{41}$$

Contrary to $K_s(q)$ (Eq. (39)), $K_d(q)$ does not follow the tangent, but an inner chord of the theoretical $K_d(q)$ (see Fig. 6). This yields a singularity for its derivative with a jump from γ_s to γ_d at the statistical order γ_d [see Chigirinskaya et al. (1994), Schmitt et al. (1994) for studies of the multifractal phase transitions of atmospheric turbulence, Hoang et al. (2012) for precipitations and Hooge et al. (1994) for seismicity].

5. Universal Multifractals

5.1 Universal Parameters

The only mathematical constraint on the functions $K(q)$ and $c(\gamma)$ is that they should both be convex, $c(\gamma)$ should furthermore be non-decreasing. It means that these functions *a priori* depend on an infinite number of parameters. For obvious theoretical and empirical reasons physics abhors infinity. Fortunately, the codimension C_1 of the mean intermittency and the multifractality index α (Schertzer and Lovejoy 1987) play a key role and define up to the second order any conservative field ($K(1) = 0$) at least around its mean:

$$K(q) = C_1[(q-1) + \alpha(q-1)^2 / 2] + o((q-1)^2) \tag{42}$$

$$C_1 = \left.\frac{dK(q)}{dq}\right|_{q=1};\ \alpha = \left.\frac{d^2K(q)}{C_1 dq^2}\right|_{q=1} \tag{43}$$

- $C_1 > 0$ is both the singularity of the mean field and its codimension ($c(C_1) = C_1$). It measures the mean intermittency of the field. Due to the Legendre transform, C_1 is the slope of the tangency of $K(q)$ at $q = 1$.
- $0 \leq \alpha \leq 2$ measures, on the contrary, how the intermittency varies with statistical order. It is related to the curvature of $K(q)$, still at $q = 1$.
- For non-conservative multifractals, the value $H = K(1) \neq 0$ adds a third parameter to be discussed in the Section "Simulations of Universal Multifractals".

5.2 Universality and Lévy Stable Variables

This local characterization of $K(q)$ around $q = 1$ already provides some universality to the parameters C_1 and α, but this universality is strongly reinforced by the fact that this characterization becomes global (up to some critical order) in the case of Universal multifractals (UMs). This large class of multifractals results from a broad generalization of the central limit theorem (CLT) for multiplicative processes. Because the CLT for additive processes is often reduced to the Gaussian case, i.e. for variables with finite variance, we first recall its generalization to strongly non-Gaussian variables, i.e. for variables with finite invariance.

A random variable X is said to be a Lévy stable variable (Lévy 1937, Lévy 1965, Gnedenko and Kolmogorov 1954, Kahane 1974, Feller 1971, Kahane 1985a, Zolotarev 1986) if and only if it is stable under renormalized sums, i.e. it is a fixed point, with the rescaling factor $a(n)$ and centering term $b(n)$, of any n of its independent realizations X_i $(i = 1, n)$. This corresponds to ($\overset{d}{=}$ denotes equality in distribution):

$$\forall n \in N, \exists a(n), b(n) \in R : \sum_{i=1}^{n} X_i \overset{d}{=} a(n)X + b(n) \tag{44}$$

Furthermore, any Lévy stable variable X is attractive for renormalized sum of independent realizations Y_i $(i = 1, n)$ of a random variable Y having similar distribution tails:

$$\lim_{n\to\infty} \frac{\sum_{i=1}^{n} Y_i - b(n)}{a(n)} \overset{d}{=} X \tag{45}$$

i.e., power-law tail whose exponent is the Lévy stability index $\alpha \in [0,2]$:

$$\forall s >> 1 : \Pr(|X| > s) \approx s^{-\alpha} \tag{46}$$

It is worthwhile to note that the finiteness of this critical exponent α is equivalent to the divergence of all statistical moments of order larger or equal to α:

$$\forall q \geq \alpha : E\left[|X|^q\right] = \infty \tag{47}$$

Another important property of the Lévy stability index is that its inverse $1/\alpha$ is the generator of the multiplicative group of the renormalizing group $a(n)$:

$$a(n) = n^{1/\alpha} \tag{48}$$

A consequence of the stability (Eq. (44)) is that the second Fourier characteristic function of the Lévy stable variable is:

$$K_F(q) = imq - D|q|^{\alpha}(1 - i\beta\,\text{sign}(q)\omega(q,\alpha));$$
$$\omega(q,1) = \frac{\pi}{2}\text{Log}|q|;\ \alpha \neq 1 : \omega(q,\alpha) = \tan\frac{\pi\alpha}{2} \tag{49}$$

where D is the scale parameter, β the skewness parameter and m the centering term. The Gaussian case ($\alpha = 2$) is necessarily symmetric because $\omega = 0$ and the skewness value β is therefore undefined, but rather corresponds to 0 because of the symmetry.

5.3 Lévy Stable Generators

The general idea is that the linear stability and attractivity of Lévy generators are transformed by exponentiation into a multiplicative stability and attractivity (Schertzer and Lovejoy 1987, Fan 1989, Brax and Pechanski 1991, Schertzer and Lovejoy 1997). Nevertheless, this requires to be done cautiously. Indeed, the following inequality:

$$\forall n \in N, \forall X, q > 0 : \exp(qX) \geq (qX)^n / n! \tag{50}$$

demonstrates that the exponentiation of a positive variable with a power-law tail exponent α (Eq. 46) generates a new random variable having no finite moment of positive order q. It suffices to take the ensemble average of Eq. (50) with $n > \alpha$ to obtain the divergence (Eq. (47)) of any moment $E[\exp(qX)]$ of statistical order $q > 0$ (Eq. (47)). Therefore, the cumulant generating function $K(q)$ is finite for $q > 0$ only for fully asymmetrical Lévy stable variables ($\beta = -1$, see Fig. 7) and then has the following expression:

$$q \geq 0 : K(q) = mq + D\text{sign}(\alpha - 1)q^{\alpha};\ q < 0 : K(q) = \infty \tag{51}$$

where m and D have the same parameters as for the Fourier characteristic function $K_F(q)$ (Eq. (49)). It is no surprise that the finiteness of the Laplace characteristic function requires more constraints than its Fourier counterpart, but it is, on the contrary, surprising that it has been scarcely used and almost exclusively (Feller 1971) for the case $\alpha < 1$, i.e. stable Lévy variables being only negative and therefore upper bounded. For the scaling moment function $K(q)$ of the conservative ($K(1) = 0$) universal multifractals, Eq. (51) yields:

$$q \geq 0 : K(q) = \frac{C_1}{\alpha - 1}(q^{\alpha} - q);\ q < 0 : K(q) = \infty \tag{52}$$

5.4 Simulations of Universal Multifractals

We show in this section that universal multifractals are obtained along a 4-step procedure, see Figs 7 and 8 for illustration:

(i) Create a sub-generator $\gamma_0^{(\alpha)}$, i.e. an extremely asymmetric Lévy white noise $\gamma_0^{(\alpha)}$ of index α

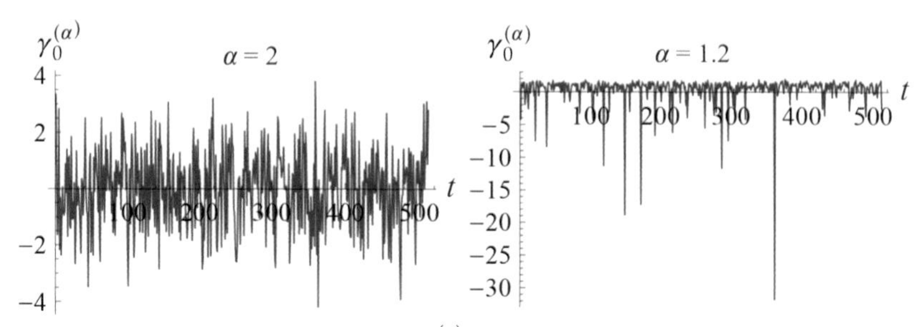

Figure 7: Examples of white noises $\gamma_0^{(\alpha)}$ with λ = 512 (the horizontal axis is the time $t \in [0, \lambda]$), admitting a finite second Laplace characteristic function $K(q)$ for $q > 0$, respectively, for the Gaussian case $\alpha = 2$ (left), which is symmetrical (positive and negative fluctuations have same amplitudes), and a Lévy case $\alpha = 1.2$ (right), which is extremely asymmetrical with huge negative fluctuations, but moderate positive fluctuations. Both have a fast probability falloff for positive extremes, as required (Schertzer and Tchiguirinskaia 2015).

(ii) Perform a fractional integration of order D/α' on the sub-generator $\gamma_0^{(\alpha)}$ to obtain the generator Γ_λ

(iii) Take the exponential of the generator Γ_λ to obtain the (normalized) flux ε_λ of universal multifractal parameters C_1 and α

(iv) Perform a fractional integration of order H on the forcing $f_R = \varepsilon_\lambda^a$ to obtain the non-conservative field u_λ.

Generalizing what we reviewed on Lévy stable variables, the generator of a conservative multifractal, which is scalar valued and defined on a D-dimensional domain, can be obtained by a fractional integration of extremely asymmetric Lévy white noise $\gamma_0^{(\alpha)}$ of index α (i.e. independently identically distributed Lévy variables on each pixel/voxel):

$$\Gamma_\lambda(x) = \left|\frac{\mathrm{var}(\alpha)}{m_{D-1}(\partial B_L)}\right|^{1/\alpha} \int_{B_L \setminus B_{L/\lambda}(x)} G(x-x')d^D\gamma_0^{(\alpha)}(x') - \mathrm{var}(\alpha)\mathrm{Log}\lambda;$$

$$\mathrm{var}(\alpha) = \frac{C_1}{\alpha - 1} \tag{53}$$

where $B_L(x)$ is the ball of center x and size L (which could be arbitrarily chosen), $m_{D-1}(\partial B_L)$ is the $(D-1)$-dimensional measure of its (hyper-) surface ∂B_L. For isotropic cases:

$$B_L(x) = \{x'[|x-x'|] \leq L/2\} \text{ and } m_{D-1}(\partial B_L) = 2(L/2)^{D-1}\pi^{D/2}/\Gamma_E(D/2)$$

where Γ_E is the Euler Gamma function. var(α) corresponds to a generalization of the (quadratic) variation of a (semi-) martingale (Metivier 1982) and $G(x)$ is the Green function of a fractional Laplace operator $(\Delta = \nabla^2)$:

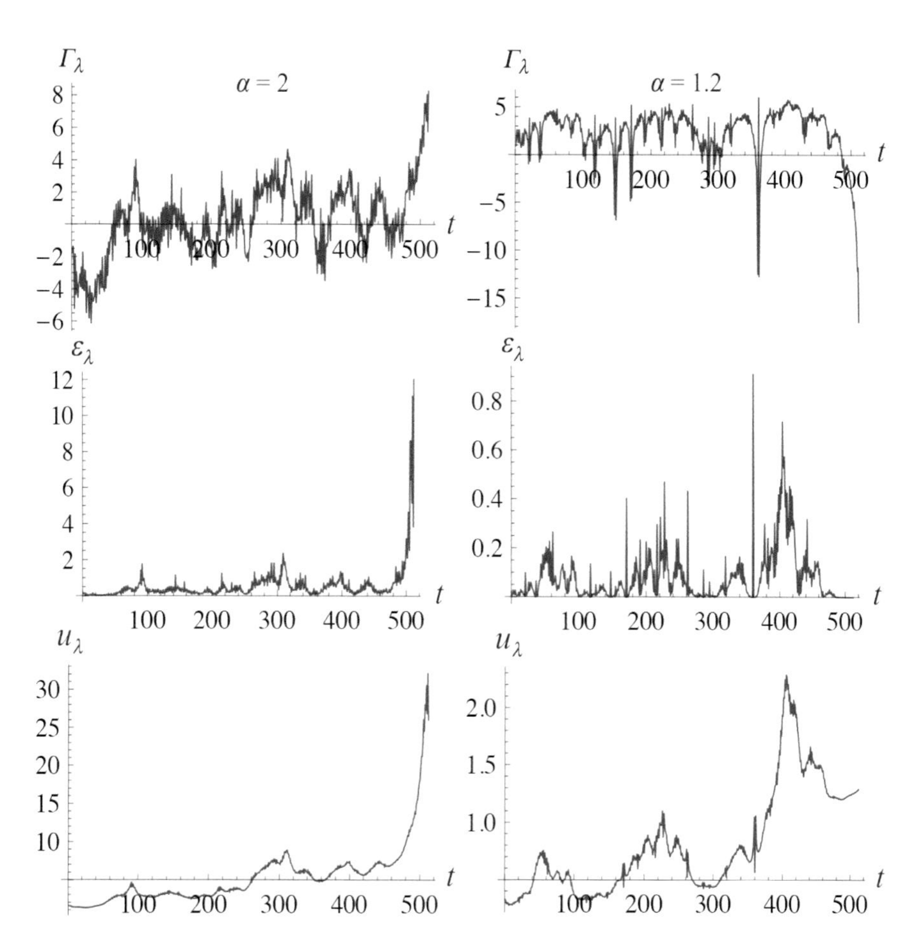

Figure 8: Illustration of the three last steps (see text) to obtain multifractal fields from sub-generators (white noises) $\gamma_0^{(\alpha)}$ of Fig. 7 with parameters $C_1 = 0.2, a = 1, H = 1/9, \lambda = 512$ (the horizontal axis is the time $t \in [0, \lambda]$), respectively, for $\alpha = 2$ (left column) and $\alpha = 1.2$ (right column), from top to bottom: generator Γ_λ; (conservative) flux ε_λ; multifractal field u_λ obtained by a fractional integration of ε_λ (Schertzer and Tchiguirinskaia 2015).

$$-(-\Delta)^{D/2\alpha'} G(x - x') = \delta(x - x'); \; G(x) \propto |x|^{-D/\alpha} \tag{54}$$

with: $1/\alpha + 1/\alpha' = 1$.

Applying $\tilde{T}_\lambda^{-1} = \exp(\Gamma_\lambda)$ to a homogeneous large scale flux ε_1, which could be taken as unity without loss of generality, yields a highly inhomogeneous flux ε_λ. Nevertheless, ε_λ is conservative $(\forall\lambda : E(\varepsilon_\lambda) = E(\varepsilon_1))$ and has in fact the stronger property to be a martingale:

$$\forall \lambda : E_\lambda(\varepsilon_\Lambda) = \varepsilon_\lambda \tag{55}$$

i.e. the conditional expectation $E_\lambda(.)$ at resolution λ of ε_Λ is simply ε_λ. However, many geophysical fields are not conservative, like the wind velocity u_λ. These fields can result from a fractional diffusion equation forced by a (renormalized) forcing $f_R = \varepsilon_\lambda^a$, which is a given a^{th} power of the flux (Schertzer et al. 1997). Such a field corresponds to fractional integration of order H of the forcing f_R:

$$u_\lambda(x) \propto \int_{B_L \setminus B_{L/\lambda(\underline{x})}} G_R(x-x')f_R(x')d^D x';\ f_R = \varepsilon_\lambda^a \tag{56}$$

where the renormalized propagator G_R satisfies:

$$-(-\Delta)^{H/2} G_R(x-x') = \delta(x-x');\ G_R(x) \propto |x|^{-(D-H)} \tag{57}$$

Figures 7 and 8 display the time series corresponding to each of the four steps to obtain a 1D scalar-valued universal multifractal with parameters: $C_1 = 0.2$, $a = 1$, $H = 1/9$ and respectively for $\alpha = 2$ and $\alpha = 1.2$.

6. Vector Valued Multifractals

Pioneering works on multiplicative cascades (Yaglom 1966, Mandelbrot 1974), whose generic outcome was later on recognized as multifractal, were obtained by products of identically independently distributed variables along a dyadic (more generally a p-adic) tree, i.e. these cascades were discrete in scale. To obtain continuous in scale processes, the products were replaced by exponentials of additive processes (Schertzer and Lovejoy 1987). In spite of the logarithmic divergence of these generators that yields some technical problems, this straightforward substitution nevertheless opens the road to broad generalizations needed to obtain multifractals that are vector or manifold valued (Schertzer and Lovejoy 1995, Schertzer and Tchiguirinskaia 2015). Indeed, as underlined in the present paper, similar concepts should be used for domain X and the codomain $\tilde{X}$; therefore the dimension of the codomain cannot be an obstacle to define multifractality, as already established for the domain. In fact, in the case of scalar domain and codomain, the (usual) exponential maps the additive group R into the multiplicative group R^+ of positive real numbers. This is readily extended to complex numbers, as well as to square matrices of dimension n into the general linear group $GL(n, R)$ of non-singular matrices. These generalizations correspond, nevertheless, to particular cases of mapping of a Lie algebra into an associated Lie group, with the help of a (generalized) exponential. The Lie algebra is in fact the tangent vector space to the Lie group at its unity (Gilmore 1941, Sattinger and Weaver 1986). It therefore has a simpler structure than the group that in general is a manifold. The interest of the wide and generic case of Clifford algebra, more precisely real Clifford algebra $Cl_{p.q}(R)$[3], was pointed out by Schertzer and Tchiguirinskaia (2015). Clifford algebra is indeed generated by symmetry generators e^i that are either square roots of plus or minus unity, which generate by exponentation respectively rotations or deformations. These generators form an orthogonal basis of a non-degenerate, but possibly defective, quadratic form Q of signature (p, q) that defines the square of each generator v, of coordinates v_i's on this basis (1 denotes the identity operator):

$$v^2 = Q(v)1, \; Q(v) = v_1^2 + v_2^2 + \cdots + v_p^2 - v_{p+1}^2 - v_{p+2}^2 - \cdots - v_{p+q}^2 \tag{58}$$

This provides a very broad generalization of the complex numbers, which corresponds to the signature (0,1). For instance, the classical Euler-Moivre identity for any (spherical) angle θ:

$$(\exp(i\theta))^\alpha \equiv \cos(\alpha\theta) + i \sin(\alpha\theta) \tag{59}$$

becomes:

$$(\exp(u\theta))^\alpha \equiv \cosh(\alpha\theta)1 + u \sinh(\alpha\theta) \tag{60}$$

where $\alpha\theta$ is an "hyperbolic" angle when the unitary direction u is one of the square roots of plus unity ($Q(u) = 1$): it is the curvilinear coordinate along a geodesic in a hyperbolic geometry framework (Milnor 1982). When u is a square root of minus unity ($Q(u) = -1$), we are back to the spherical geometry:

$$(\exp(u\theta))^\alpha \equiv \cos(\alpha\theta)1 + u \sin(\alpha\theta) \tag{61}$$

but, contrary to the exceptional case of the complex numbers, there is no longer uniqueness of this square root. Clifford algebra can be used to generate both change groups, respectively on the domain (T_λ) and the codomain ($\tilde{T}_\lambda$). This has been practiced for a while for the domain, notably with the special case of the pseudo-quaternions $Cl_{2.0}(R) = Cl_{1.1}(R)$ (Schertzer and Lovejoy 1985a), whereas it is only recent for the codomain, including for the quaternions $H \equiv Cl_{0.2}(R)$ (Schertzer and Tchiguirinskaia 2015). In the latter case, supplementary technical difficulties occur when dealing with the stochastic nature of the codomain scale change group ($\tilde{T}_\lambda$), especially to generalize universal multifractals.

At first, similar to the scalar case, the exponential of too large fluctuations may prevent the existence of any finite statistical moment. But, non-scalar cases are far more involved because Lévy stable vectors are non-parametric contrary to Gaussian vectors. Nevertheless, precise

3 The moment order q and the index q of a Clifford algebra have nothing else in common, except to be the same alphabetical letter due to respective usages.

conditions of finiteness of the statistical moments $E\left[\tilde{T}_\lambda^q\right]$, where the (vector) order q belongs to the same Lie algebra, can be derived from the fact that these moments are the Clifford Laplace transform of the probability distribution of the generator Γ_λ (Schertzer and Tchiguirinskaia 2015):

$$E\left[\tilde{T}_\lambda^q\right] = E\left[\exp\left(\langle q, \Gamma_\lambda \rangle\right)\right] = \exp(K_\lambda(q)) \tag{62}$$

where $\langle .,. \rangle$ denotes the scalar product generated by the quadratic form ($Q(u) = \langle u, u \rangle$) of the Lie algebra and all the involved quantities are asymptotically power laws as soon as $K_\lambda(q)$ has a Log(λ) divergence. This enables to generalize the 4-step simulation procedure described in the Section "Simulations of Universal Multifractals" in a more or less straightforward manner and Fig. 9 displays a $3D$ snapshot of a $4D$+1 simulation of a multifractal quaternion velocity field, i.e., the domain is R^4 and the codomain is $H \equiv Cl_{0,2}(R)$, with parameters $\alpha = 1.5$, $H = a = 1/3$. It appears to be in visually good agreement with the atmospheric turbulence fingerprint on the wall of the Randall museum of San Francisco that is observable with the help of the architectural scale instrument "Windswept" (Sowers 2011).

7. Conclusions and Prospects

Multifractals have been on the scene for three decades and increasingly understood as a basic and interdisciplinary framework for analyzing and simulating various processes that are extremely variable over wide ranges of space-time scales. Nevertheless, there is still a large gap between their potential and their actual use. Some strong obstacles were inherited from older, pre-existing and highly restrictive concepts of scaling: for example that they are only geometrical descriptors (with bounded singularities) or that they are inherently isotropic. The later had been a particularly important obstacle for the development of multifractals in geophysics due to privileged directions generated by gravity and Earth's rotation: a generalization of the scale notion was necessary to overcome it. Another obstacle, is that multifractal properties have been often restricted to "bare" properties, i.e. those of a process that is supposed to be homogeneous below the scale of observation, whereas small scale fluctuations generate "dressed" properties that are often much more exotic and extreme, e.g. they correspond to a non-classical self-organized criticality.

This introduction puts a particular emphasis on the need to have a formalism that will be strongly operational. This requires one to go back to the fundamentals, i.e. to scale symmetry groups acting on complex fields and the Lie algebra of their generators. This enables one

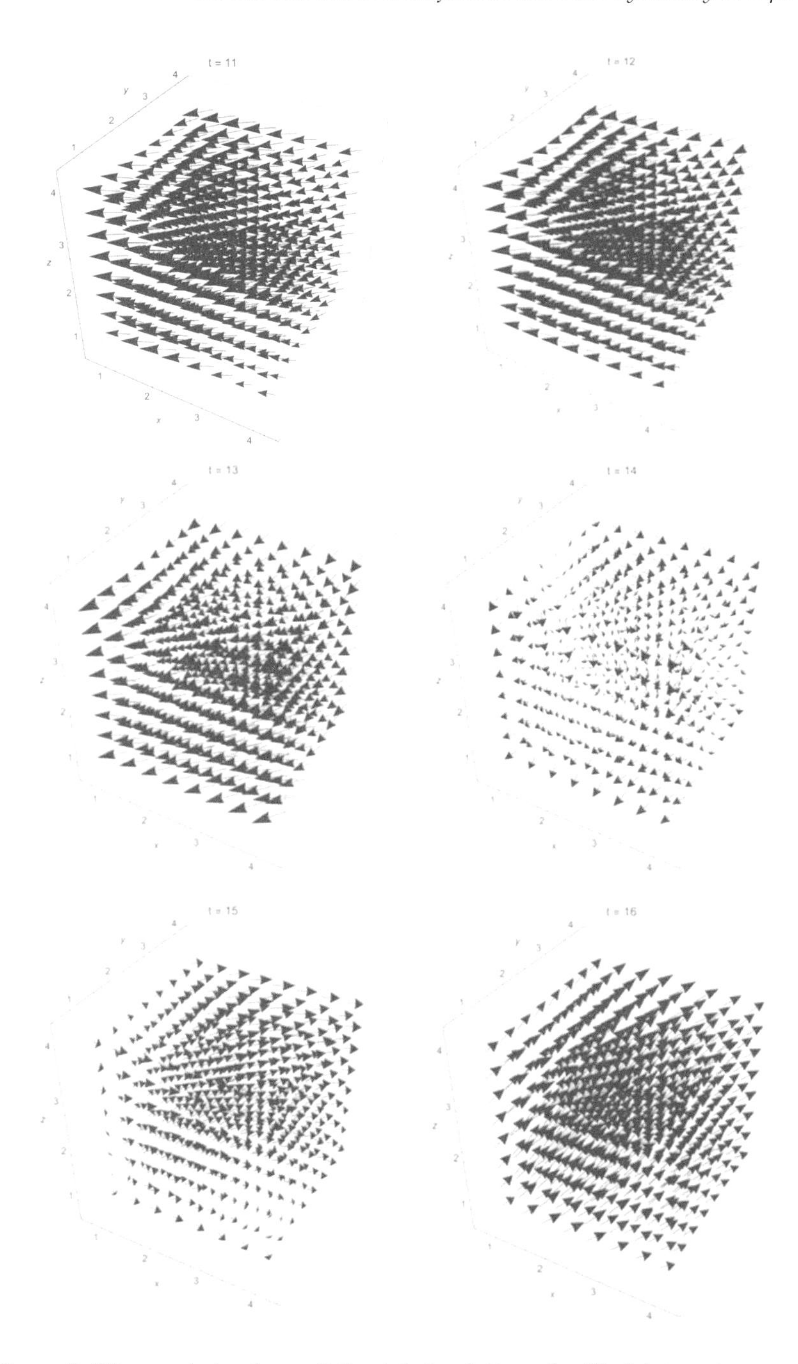

Figure 9: 3D snapshots of a multifractal simulation of a 4D+1 intermittent vector field obtained by a quaternion cascade, i.e., with values on $H \equiv Cl_{0,2}(R)$ (Schertzer and Tchiguirinskaia 2015).

to generalize in a straightforward manner multifractals to vector and manifold frameworks, whereas multifractals have been restricted for too long to scalar-valued fields. This extension is particularly indispensable in developing multifractal forecasting methods, whose development is expected to be fast. The already large class of Lévy-Clifford algebra of multifractal generators is a particularly important contribution in this direction. Indeed, these algebra combine a number of seductive properties, including universal statistical and robust algebraic properties, both defining the basic symmetries of the corresponding fields. More fundamentally, these multifractals may yield direct connections to the deterministic-like nonlinear equations that are supposed to generate them, in particular the Navier-Stokes equations.

Acknowledgements

The authors warmly thank Dr. Behzad Ghanbarian and Prof. Allen Hunt for their kind invitation to write this chapter and stimulating discussions. Part of this work was achieved during a summer workshop in the Aspen Center for Physics (sponsored by NSF). The authors acknowledge partial support from the Chair "Hydrology for Resilient Cities", endowed by Veolia.

REFERENCES

Anselmet, F. et al. 1984. High-order velocity structure functions in turbulent shear flows. *Journal of Fluid Mechanics*, 140: 63.

Bak, P. and K. Chen. 1991. Self-Organized Criticality. *Scientific American* (Jan.), pp. 46–53.

Bak, P., C. Tang and K. Weiessenfeld. 1987. Self-Organized Criticality: An explanation of 1/f noise. *Physical Review Letter* 59: 381–384.

Brax, P. and R. Pechanski. 1991. Levy stable law description on intermittent behaviour and quark-gluon phase transitions. *Physics Letter B* 253: 225–230.

Buckingham, E. 1914. Physically Similar Systems: Illustrations of the Use of Dimensional Equations. *Phys Rev.* 4: 345–376.

Buckingham, E. 1915. The principle of similitude. *Nature* 96: 396–397.

Charney, J.G. 1971. Geostrophic Turbulence. *J. Atmos. Sci* 28: 1087.

Charney, J.G. 1948. On the scale of atmospheric motions. *Geophys. Publ. Oslo* 17: 1–17.

Cheng, Q. 2006. Singularity-generalized self-similarity-fractal spectrum (3S) model. *Earth Science, Journal of China University of Geosciences* 31: 337–348.

Chigirinskaya, Y. et al. 1994. Unified multifractal atmospheric dynamics tested in the tropics, part I: Horizontal scaling and self organized criticality. *Nonlinear Processes in Geophysics* 1(2/3): 105–114.

Douglas, E.M. and A.P. Barros. 2003. Probable Maximum Precipitation Estimation Using Multifractals: Application in the Eastern United States. *Journal of Hydrometeorology* 4: 1012–1024.
Dupuy, G. (ed.). 2016. Villes, réseaux et transport, le défi fractal, Paris: Economica.
Falconer, K. 1990. Fractal Geometry: Mathematical Foundations and Applications. John Wiley and Sons.
Falconer, K. 1985. The Geometry of Fractal Sets. Cambridge: Cambridge University Press.
Fan, A.H. 1989. Chaos additif et multiplicatif de Levy. *Comptes Rendus de l'Académie des Sciences de Paris* I(308): 151–154.
Feder, J. 1988. Fractals. New York, NY: Plenum Press.
Feller, W. 1971. An Introduction to Probability Theory and Its Applications, vol. 2. New York: Wiley.
Fraysse, A. and S. Jaffard. 2006. How smooth is almost every function in a Sobolev space? *Revista Matematica Iberoamericana* 22(2): 663–682.
Gilmore. 1941. Lie Groups. New York: Wiley.
Gnedenko, B.V. and A.N. Kolmogorov. 1954. Limit Distribution for Sums of Independent Random Variables. Addison-Wesley.
Grassberger, P. 1983. Generalized dimensions of strange attractors. *Physics Letters A* 97(6): 227–230.
Grassberger, P. and I. Procaccia. 1983. Measuring the strangeness of Strange attractors. *Physica* 9D: 189–208.
Halsey, T.C. et al. 1986. Fractal measures and their singularities: The characterization of strange sets. *Physical Review A* 33: 1141–1151.
Hoang, C.T. et al. 2012. Assessing the high frequency quality of long rainfall series. *Journal of Hydrology* 438–439: 39–51.
Hooge, C. et al. 1994. Universal Multifractals in Seismicity. *Fractals* 2(3): 445–449.
Hubert, P. et al. 1993. Multifractals and extreme rainfall events. *Geophysical Research Letters* 20(10): 931.
Hurst, H.E. 1951. Long-term storage capacity of reservoirs. *Transactions of the American Society of Civil Engineers* 116: 770–808.
Kahane, J.-P. 1997. A Century of Interplay Between Taylor Series, Fourier Series and Brownian Motion. *Bulletin of the London Mathematical Society* 29 (May 1995): 257–279.
Kahane, J.P. 1985a. Definition of stable laws, infinitely divisible laws, and Lévy processes. In: G.Z. and U.F.M. Shlesinger (ed.). Lévy flights and Related Phenomena in Physics. Berlin: Springer-Verlag, pp. 99–109.
Kahane, J.P. 1985b. Sur le Chaos Multiplicatif. *Annales des Siences mathématique du Québec* 9: 435.
Kahane, J.P. 1974. Sur le modèle de turbulence de Benoit Mandelbrot. *Comptes Rendus (Paris)* 278A: 621–623.
Kahane, J.P. and J. Peyriere. 1976. Sur certaines martingales de B. Mandelbrot. *Advances in Mathematics* 22: 131–145.
Klemes, V. 1974. The Hurst phenomenon: A puzzle? *Water Resour. Res.* 10(4): 657–688.
Koutsoyiannis, D. 2002. The Hurst phonomenon and fractional Gaussian noise made easy. *Journal of Hydrological Sciences* 47(4): 573–594.

Lamperti, J. 1962. Semi-stable stochastic processes. *Transactions of the American Mathematical Society* 104: 62–78.

Lavallée, D., S. Lovejoy and D. Schertzer. 1991. On the determination of the codimension function. In: D. Schertzer and S. Lovejoy (eds). Non-linear variability in geophysics: Scaling and Fractals. Kluwer, pp. 99–110.

Lévy, P. 1965. Processus Stochastiques et Mouvement Brownien, Paris: Gauthiers-Villars.

Lévy, P. 1937. Théorie de l'addition des variables aléatoires, Paris: Gauthiers Villars.

Lovejoy, S. and D. Schertzer. 2013. The Weather and Climate: Emergent Laws and Multifractal Cascades. Cambridge: Cambridge Univeristy Press, U.K. Available at: http://www.cambridge.org/fr/academic/subjects/earth-and-environmental-science/atmospheric-science-and-meteorology/weather-and-climate-emergent-laws-and-multifractal-cascades.

Mandelbrot, B. 1989. Fractal geometry: What is it and what does it do? In: M. Fleischman, R.C. Ball, and D. Tildesley (eds). Fractals in the Natural Sciences. Princeton: Princeton University Press, pp. 3–16.

Mandelbrot, B.B. 1977. Fractals, form, chance and dimension. San Francisco: Freeman.

Mandelbrot, B.B. 1984. Fractals in physics: Squig clusters, diffusions, fractal measures, and the unicity of fractal dimensionality. *Journal of Statistical Physics* 34: 895–930.

Mandelbrot, B.B. 1974. Intermittent turbulence in self-similar cascades: Divergence of high moments and dimension of the carrier. *Journal of Fluid Mechanics* 62: 331–350.

Mandelbrot, B.B. 1991. Random multifractals negative dimensions and the resulting limitations of the thermodynamic formalism. *Proc. R. Soc. Lond. A* 434: 79–88.

Mandelbrot, B.B. 1983. The Fractal Geometry of Nature. San Francisco: Freeman.

Mandelbrot, B.B. and J.W. Van Ness. 1968. Fractional Brownian motions, fractional noises and applications. *SIAM Review* 10: 422–450.

Metivier, M. 1982. Semimartingales: A Course on Stochastic Processes. Berlin. New York: Walter de Gruyter.

Milnor, J.W. 1982. Hyperbolic geometry: The first 150 years. *Bulletin of the American Mathematical Society* 6(1): 9–25.

Murcio, R. et al. 2015. Multifractal to monofractal evolution of the London street network. *Phys. Rev. E.* 92(6): 62130.

Oono, Y. 1989. Large Deviation and Statistical Physics. *Progr. Theor. Phys. Suppl.* 99: 165–205. Available at: http://ptps.oxfordjournals.org/content/99/165.abstract.

Painter, S. 1996. Stochastic interpolation of aquifer properties using fractional Levy motion. *Water Resour. Res.* 32(5): 1323–1332.

Parisi, G. and U. Frisch. 1985. On the singularity structure of fully developed turbulence. In: M. Ghil, R. Benz, and G. Parisi (eds). Turbulence and predictability in geophysical fluid dynamics and climate dynamics. Amsterdam: North Holland, pp. 84–88.

Procaccia, I. 1983. The infinite number of generalized dimensions of fractals and strange attractors. *Physica D* 8: 435–444.

Sattinger, D.H. and O.L. Weaver. 1986. Lie groups and algebras with applications to physics, geometry and mechanics. New-York, Berlin: Springer-Verlag.

Schertzer, D. et al. 1999. Generalized Stable Multivariate Distribution and Anisotropic Dilations. Minneapolis: IMA, U. of Minnesota.

Schertzer, D. et al. 1997. Multifractal cascade dynamics and turbulent intermittency. *Fractals* 5(3): 427–471.

Schertzer, D. et al. 2012. Quasi-geostrophic turbulence and generalized scale invariance, a theoretical reply. *Atmospheric Chemistry and Physics* 12(i): 327–336.

Schertzer, D. and S. Lovejoy. 1995. From scalar cascades to Lie cascades: Joint multifractal analysis of rain and cloud processes. In: R.A. Feddes (ed.). Space/time Variability and Interdependance for various hydrological processes. Cambridge: Cambridge University Press, U.K. pp. 153–173.

Schertzer, D. and S. Lovejoy. 1985a. Generalised scale invariance in turbulent phenomena. *Physico-Chemical Hydrodynamics Journal* 6: 623–635.

Schertzer, D. and S. Lovejoy. 1992. Hard and soft multifractal processes. *Physica A* 185(1–4): 187–194. Available at: http://www.sciencedirect.com/science/article/B6TVG-46DFR3J-CD/2/fc9e5dadd7f2415df70cbca9e2b09877.

Schertzer, D. and S. Lovejoy. 1994. Multifractal Generation of Self-Organized Criticality. In: M.M. Novak (ed.). Fractals in the Natural and Applied Sciences. Amsterdam: Elsevier, North-Holland, pp. 325–339.

Schertzer, D. and S. Lovejoy. 1988. Multifractal simulations and analysis of clouds by multiplicative processes. *Atmospheric Research* 21(1): 337–361.

Schertzer, D. and S. Lovejoy. 2011. Multifractals, generalized scale invariance and complexity in geophysics. *International Journal of Bifurcation and Chaos* 21: 3417–3456.

Schertzer, D. and S. Lovejoy. 1989. Nonlinear variability in geophysics: Multifractal analysis and simulation. In: L. Pietronero (ed.). Fractals: Physical Origin and Consequences. New York: Plenum, p. 49.

Schertzer, D. and S. Lovejoy. 1984. On the dimension of atmospheric motions. In: T. Tatsumi (ed.). Turbulence and Chaotic Phenomena in Fluids. Amsterdam: North Holland, pp. 505–508.

Schertzer, D. and S. Lovejoy. 1983. On the Dimension of Atmospheric motions. In: T. Tatsumi (ed.). 6th Symposium Turbulence and Diffusion. Amsterdam: Elsevier Science Publishers B.V., pp. 69–72.

Schertzer, D. and S. Lovejoy. 1987. Physical Modeling and Analysis of Rain and Clouds by Anisotropic Scaling Multiplicative Processes. *Journal of Geophysical Research* D8(8): 9693–9714. Available at: http://www.physics.mcgill.ca/~gang/eprints/eprintLovejoy/neweprint/JGR.SL.1987.good.pdf.

Schertzer, D. and S. Lovejoy. 1985b. The dimension and intermittency of atmospheric dynamics. In: B. Launder (ed.). Turbulent Shear Flow 4. Springer-Verlag, pp. 7–33.

Schertzer, D. and S. Lovejoy. 1997. Universal Multifractals do Exist! *Journal of Applied Meteorology* 36: 1296–1303.

Schertzer, D., S. Lovejoy and P. Hubert. 2002. An Introduction to Stochastic Multifractal Fields. In: A. Ern and W. Liu (eds). ISFMA Symposium on

Environmental Science and Engineering with related Mathematical Problems. Beijing: High Education Press, pp. 106–179.

Schertzer, D., S. Lovejoy and D. Lavallée. 1993. Generic Multifractal phase transitions and self-organized criticality. In: J.M. Perdang and A. Lejeune (eds). Fractals 93. World Scientific, pp. 216–227.

Schertzer, D. and I. Tchiguirinskaia. 2015. Multifractal vector fields and stochastic Clifford algebra. *Chaos: An Interdisciplinary Journal of Nonlinear Science* 25(12): 123-127. Available at: http://scitation.aip.org/content/aip/journal/chaos/25/12/10.1063/1.4937364.

Schmitt, F. et al. 1994. Empirical study of multifractal phase transitions in atmospheric turbulence. *Nonlinear Processes in Geophysics* 1: 95–104.

Sedov, L. 1972. Similitudes et Dimensions en Mécanique, Moscow: MIR.

Shapiro, J.H. 1993. Composition operators and classical function theory. New York: Springer-Verlag.

Sonin, A.A. 2004. A generalization of P-theorem and dimensional analysis. *Proc. Nat. Acad. Sci.*

Sowers, C. 2011. Windswept. Available at: http://charlessowers.com/windswept.

Speziale, C.G. 1985. Galilean invariance of subgrid scale stress models in the large-eddy simulation of turbulence. *Journal of Fluid Mechanics* 156: 55–62.

Sylos Labini, F., F. Montuori and L. Pietronero. 1998. Scale-Invariance of galaxy clustering. *Physics Reports* 293: 61–226.

Tel, T. 1988. Fractals and multifractals. *Zeit. Naturforsch* 43a: 1154–1174.

Varadhan, S.R.S. 1984. Large Deviations and Applications, Philadelphia: SIAM. Available at: http://dx.doi.org/10.1137/1.9781611970241.

Weyl, H. 1952. Symmetry, Princeton, NJ: Princeton University Press.

Wilson, K.G. 1971. Renormalization Group and Critical Phenomena. I. Renormalization Group and the Kadanoff Scaling Picture. *Phys. Rev. B* 4(9): 3174–3183.

Yaglom, A.M. 1966. The influence on the fluctuation in energy dissipation on the shape of turbulent characteristics in the inertial interval. *Sov. Phys. Dokl.* 2: 26–30.

Yanovsky, V.V. et al. 2000. Lévy diffusion and Fractional Fokker Plank Equation. *Physica A* 282(1–2): 13–34.

Zee, A. 1986. Fearful Symmetry: The Search for Beauty in Modern Physics. New York: Macmillan Publishing Company.

Zolotarev, V.M. 1986. One-dimensional Stable Distributions. Providence RI: American Mathematical Society.

CHAPTER
2

Fractal Capillary Pressure Curve Models

Behzad Ghanbarian[1]* and Humberto Millán[2]

[1] Department of Petroleum and Geosystems Engineering University of Texas at Austin, Austin, TX 78712, USA

[2] Department of Physics, Center for Higher Studies, Tefé Amazonian State University, Manaus, Amazonas, Brazil

1. Introduction

Fractal geometry provides a promising mathematical framework to model the complexity of pore space and solid phase in disordered and hierarchical porous media. Although natural systems such as soils, rocks and fracture networks can at best only be *approximated* by fractal models (Crawford et al. 1995), concepts from fractal geometry have been frequently used to characterize hierarchical structure of pore space, solid matrix and pore-solid interface roughness.

Fractal objects have two main characteristics: (1) self-similarity (or self-affinity) and (2) fractional (fractal) dimension D_f (degree of homogeneity, Stanley 1984). The former implies that a fractal object above lower l_{min} and below upper l_{max} cutoff scales looks the same, which leads to scale-invariant property of fractals between certain scales ($l_{min} < l < l_{max}$). The latter, a non-integer value, is an index of the complexity of a fractal object, which measures the space-filling capacity of a pattern.

Figure 1 shows a Koch curve with exactly self-similar property at various scales. In nature, however, natural physical objects rarely are *exactly* self-similar. Many objects in the real world are statistically self-similar instead meaning that parts of them show the same statistical properties at various scales. Figure 2 shows an exactly self-similar Koch

*Corresponding author: b.ghanbarian@gmail.com

curve compared to a statistically self-similar one. Although they seem different, both curves have the same length and fractal dimension D_f = 1.262. We compare exactly and statistically self-similar Sierpinski carpets – frequently used to model pore space and capillary pressure curve of porous media – later in this chapter.

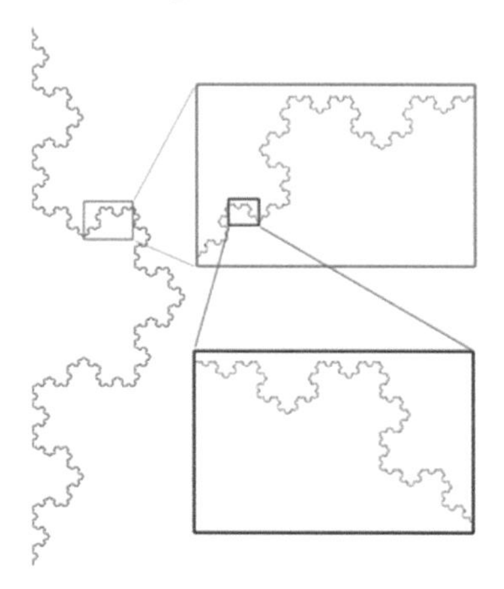

Figure 1: Exactly self-similar von Koch curve looking the same at various scales (adapted from Paul Bourke's web page http://paulbourke.net/fractals/fracdim/).

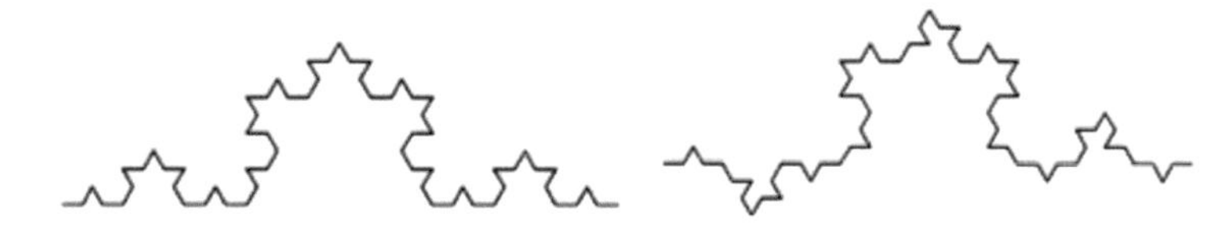

Figure 2: Exactly self-similar (left) and statistically self-similar (right) Koch curves after two iterations with fractal dimension D_f = 1.262.

If a fractal object is rescaled in all directions with the same scaling factor, an exactly similar object is reproduced and called self-similar. Self-affinity (Mandelbrot 1985), however, means that a fractal object has different scaling factors in different directions (Sahimi 2011), such as time series (see e.g., Malamud and Turcotte 1999, Pelletier and Turcotte 1999, Kantelhardt 2012) or fracture surface roughness (Poon et al. 1992, Schmittbuhl et al. 1994, Drazer and Koplik 2000, Madadi and Sahimi 2003). One should note that most natural objects that are self-similar or self-affine have that property only in a statistical sense. The concepts of self-similarity and self-affinity have widely been used to model physical and geometrical properties of complex networks, such as soils, rocks and fracture networks. See, for example, Barton et al. (1995), Barton and La Pointe (2012), Pachepsky et al. (2000), Sahimi (2011), Bunde and Havlin (2012, 2013), Feder (2013) or Hunt et al. (2014) for comprehensive reviews in different research areas.

In addition to self-similarity and self-affinity, fractal objects are characterized with a fractional (non-integer) dimension, D_f, less than the Euclidean dimension it is embedded in. Within Euclidean geometry, dimensions are limited to integer values e.g., $E = 0$ (representing a point), $E = 1$ (representing a line), $E = 2$ (denoting a plane), and $E = 3$ (corresponding to a space). In fractal geometry, however, the fractal dimension – characterizing the chaotic behavior of a pattern – is not necessarily an integer value and typically ranges between 0 and 3 ($0 < D_f < 3$). Mandelbrot (1990) showed that the degree of emptiness of empty sets is quantified with a negative fractal dimension. According to Mandelbrot (1990), negative dimensions have applications to turbulence and Diffusion Limited Aggregation (DLA) data within the context of multifractal analysis.

In Fig. 3, we compare Euclidean and fractal dimensions of a straight line, three fractal curves and a plane schematically. As can be observed, Euclidean and fractal dimensions of a straight line are identical and equal to 1. As the chaotic and space-filling patterns of a line increase its fractal dimension D_f increases as well, while its Euclidean dimension E remains the same (see Fig. 3). For a plane both Euclidean and fractal dimensions would be the same and equal to 2 ($D_f = E = 2$).

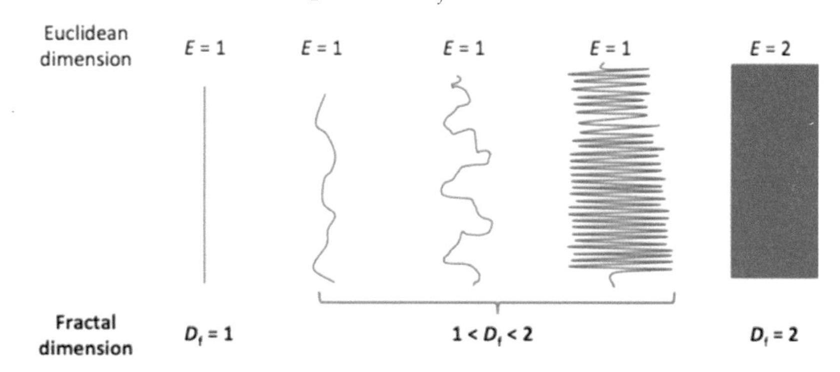

Figure 3: *Schematic* comparison of Euclidean dimension E with fractal dimension D_f. Within Euclidean geometry a straight line has the same dimension of a curve ($E = 1$), while in fractal geometry a fractal curve has a higher dimension as compared to a straight line ($D_f > E = 1$). Note that these are schematic curves and not necessarily self-similar or self-affine.

2. Sierpinski Carpet

It is well documented in the literature that pores in many natural porous media e.g., soils and rocks possess a hierarchical structure that can be represented as a fractal object. This has been used by many researchers to develop various models of porous media (see e.g., Stanley 1984, Katz and Thompson 1985, Turcotte 1986, Crawford et al. 1993, Rieu and Perrier, 1997, Perrier and Bird 2002, Filgueira et al. 2003, Guber et al. 2005, Yu 2008).

Among those fractal objects proposed in the literature, the Sierpinski carpet and its three-dimensional version i.e., Menger sponge have been frequently used to model pore space and solid matrix of porous materials. The Sierpinski carpet is constructed by starting with a solid square of size L (initiator shown in Fig. 4a). For the first iteration, the original length L is divided by the scaling factor b (= 3 in the traditional Sierpinski carpet) creating $(L/b)^2$ smaller squares. Then, the central sub-square is removed, which creates a square pore of size L/b (generator given in Fig. 4b). For the next iteration, squares of size L/b are divided by b, and accordingly those central sub-squares of size L/b^2 are removed (Fig. 4c). As the number of iteration increases, the porosity φ of the carpet also increases (see Fig. 4b-d). If the carpet is iterated infinite times ($i \rightarrow \infty$), what remains is a porous carpet with a unit value of porosity ($\varphi = 1$).

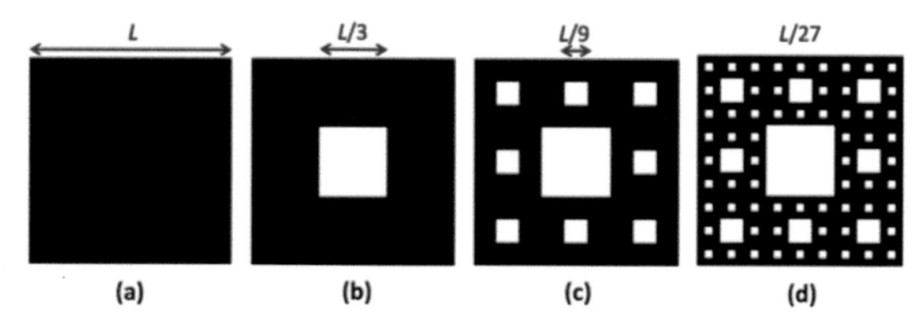

Figure 4: The traditional exactly self-similar Sierpinski carpet with fractal dimension D_f = 1.893 and scaling factor b = 3 for different iterations (a) i = 0 (initiator), (b) i = 1 (generator), (c) i = 2 and (d) i = 3. The solid matrix is in black and pores are in white.

Sierpinski carpet has a key characteristic: the carpet density varies with the length scale l. Consider the traditional Sierpinski carpet given in Fig. 4d. Assume that the initiator in Fig. 4a has a unit value of length (L = 1) and the smallest black square shown in Fig. 4d with length $l = 1/27$ has a mass $M = 1$. The density of this square therefore would be $\rho = M/l^2 = 27^2$. The next smallest black square in Fig. 4d has $l = 1/9$ but $M = 8$ and thus $\rho = 8 \times 9^2$. The next larger black square have $l = 1/3$, $M = 64$ and $\rho = 64 \times 3^2$. The largest black square shown in Fig. 4d has $l = L = 1$ and $M = 8 \times 64$ and consequently $\rho = 8 \times 64$. As can be seen, by increasing length scale, density decreases while mass increases. For irregular fractal objects, one expects the mass and accordingly the density to scale with length as

$$M \propto l^{D_f} \tag{1}$$

and

$$\rho \propto l^{D_f - E} \tag{2}$$

respectively. Recall that E is the Euclidean dimension and equal to 2 in two dimensions and 3 in three dimensions.

By fitting Eqs (1) and (2) to the density-length (ρ-l) and mass-length (M-l) data determined above for the traditional Sierpinski carpet, one easily obtains that the fractal dimension D_f is 1.893. As Stanley (1984) pointed out, it is important to have pores of all sizes in the fractal object to observe the fractal scaling i.e., power-law behavior in density-length and mass-length data. If the Sierpinski carpet only includes voids of smallest two sizes, then the density decreases up to the second smallest length scale above which the density would not change and the carpet would be macroscopically homogeneous (Stanley 1984). In fact, most porous media in the nature that show self-similarity and fractal behavior typically lose their fractal characteristics below the lower cutoff and above the upper cutoff scales (Sahimi 2003).

In addition to mass-length and density-length power laws, the number-size relationship of fractal objects conforms to a power-law function that may be distinguished from all other functions since the power law appears linear when plotted on a log-log scale. The power-law number-size relationship is (Mandelbrot 1983)

$$N(l) \propto l^{D_f}, \tag{3}$$

where $N(l)$ is the number of fractal objects whose size is equal to l and the fractal dimension D_f typically ranges between 0 and 3 in natural porous media. As a statistical representation of a natural phenomenon, Eq. (3) will be only approximately applicable, with both upper and lower bounds (i.e., l_{max} and l_{min}) to the range of applicability (Turcotte 1992). Note that Eq. (3) is sometimes used in its cumulative form, which represents number-size distribution, by replacing $N(l)$ by the number of cumulative objects of size greater than or equal to l, $N(\geq l)$ (Giménez et al. 1997).

Equation (3) provides an alternative way to determine the Sierpinski carpet fractal dimension D_f. Again assume that the initiator in Fig. 4a has a unit value of length ($L = 1$). Accordingly, for the generator shown in Fig. 4b, $N(l) = 1$ and $l = 1$. In Figs 4c and 4d, the number of objects that exactly look like the generator is 8 and 64 but $l = 1/3$ and $1/9$, respectively. Fitting Eq. (3) to the number-size data results in $D_f = 1.893$.

Tyler and Wheatcraft (1989) applied Eq. (3) in the cumulative form and determined D_f from the number-size distribution of soil particles inferred from the grain-size distribution. In the Tyler and Wheatcraft (1989) approach, calculation of the number of particles between the lower and upper sieve sizes requires dividing the mass of particles retained on the lower sieve size by the mass of a particle with size equal to the arithmetic mean of the two sieve sizes (the characteristic particle size). Tyler and Wheatcraft (1989) found that $D_f > 3$ for 9 of the 10 grain-size distributions, casting doubt on the results because one should expect D_f to be less than 3 (Turcotte, 1986). The inconsistency of the arithmetic mean of the upper and lower sieve sizes, as a characteristic particle size, was pointed out by Tyler and Wheatcraft (1992). Kozak et al. (1996) also stated that the choice of arithmetic mean to represent the characteristic size of a fraction (two successive sieves) is arbitrary and may yield errors, as anticipated by Tyler and Wheatcraft (1989). Other inconsistencies within the number-based method of Tyler and Wheatcraft (1989) were addressed by Tyler and Wheatcraft (1992), Kozak et al. (1996), and Anderson et al. (1997). We should emphasize that in fact to avoid such difficulties (e.g., arithmetic averaging) with the number-based method, Tyler and Wheatcraft (1992) proposed the mass-based approach i.e., $M(\leq l) \propto l^{3-D_f}$. Kozak et al. (1996) also pointed out that the lower sieve size should be used instead of the arithmetic mean in the number-based method when $N(\leq l) \propto l^{-D_f}$ is applied to compute D_f. By giving an example, Kozak et al. (1996) illustrated that the arithmetic averaging results in a significantly different fractal dimension in the number-based method. While the lower sieve size is appropriate to use in the number-based approach, in the mass-based model the upper sieve size is the appropriate parameter to use in determining the cumulative mass of particles of size l and less and thereby D_f.

Theoretically, there are two types of Sierpinski carpets: a solid fractal shown in Fig. 5a and a pore fractal presented in Fig. 5b (Rieu and Perrier 1997, Ghanbarian-Alavijeh et al. 2011, Hunt et al. 2014). The non-trivial property of the solid Sierpinski carpet is that, within each iteration of its

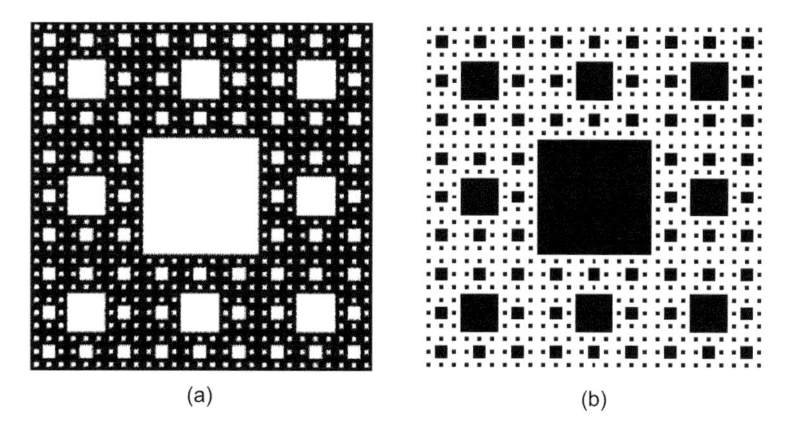

Figure 5: Exactly self-similar Sierpinski carpets after three iterations with fractal dimension $D_f = 1.893$ and $b = 3$: (a) a solid fractal model constructed of particles of the same size but pores of different sizes, (b) a pore fractal model built up of particles of different sizes but pores of the same size. The solid matrix is shown in black, while pores are white.

construction, the carpet is composed of identical-size particles denoted by black squares but pores of different sizes represented by white squares (Fig. 5a). In such a model, just the solid matrix is fractal whose number-size distribution (and consequently probability density function) follows a power-law function (or the Pareto type function named after the Italian-born Swiss professor of economics, Vilfredo Pareto, 1948-1923).

Although the pore phase in the solid Sierpinski carpet is not geometrically fractal, its number-size distribution is given by a power-law function (Eq. 3) and it has the same fractal dimensionality as the solid phase (Rieu and Perrier 1997). So, in fact, one fractal dimension scales both solid and pore phases, albeit in different ways. The same argument applies to the pore Sierpinski carpet (Fig. 5b).

The Sierpinski carpets presented in Fig. 5 are exactly self-similar. Natural porous media are instead randomly (statistically) self-similar, but it seems that the same power-law number-size distribution applies. In Fig. 6, we show randomly (statistically) self-similar Sierpinski carpets after three iterations for solid (Fig. 6a) and pore (Fig. 6b) fractal models. Albeit the fractal dimension is identically 1.893 for carpets in both Figs. 5 and 6, the random carpets shown in Fig. 6 are more heterogeneous and complex than the deterministic carpets presented in Fig. 5, and appear more similar to real natural porous media.

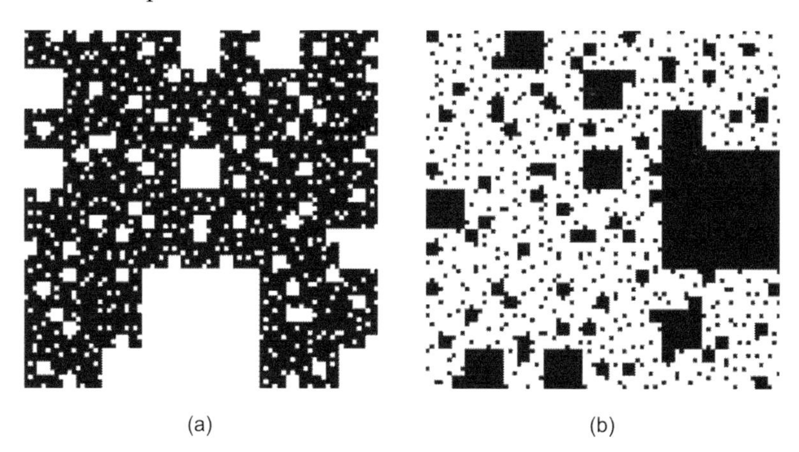

Figure 6: Randomly (statistically) self-similar Sierpinski carpets after three iterations with the same fractal dimension $D_f = 1.893$ and scale factor $b = 3$ used in Fig. 1: (a) a solid fractal model, (b) a pore fractal model. Black and white squares represent solid particles and pores, respectively.

3. Fractal Porous Media

In fractal porous media, the number-size distribution of pores follows (Mandelbrot 1983)

$$N(\geq r) = kr^{-D_p}, r_{\min} \leq r \leq r_{\max} \tag{4}$$

where $N(\geq r)$ is the number of pores whose size is equal or greater than r, k is a constant coefficient, and D_p is the pore space fractal dimension quantifying the breadth of the number-size distribution. As we mentioned earlier, Eq. (4) is a truncated power-law function meaning that it only applies between lower $l_{\min}$ and upper $l_{\max}$ cutoffs of the fractal scaling.

The total number of pores of size $r_{\min}$ and greater is given by

$$N_t = N(\geq r_{\min}) = kr_{\min}^{-D_p} \tag{5}$$

Dividing Eq. (4) by (5) yields the normalized number-size distribution of pores $N_n(\geq r)$

$$N_n(\geq r) = \left(\frac{r}{r_{\min}}\right)^{-D_p} \tag{6}$$

The number of pores whose size is within the r and $r + dr$ range is equal to the first derivative of Eq. (6), which gives

$$-dN_n(\geq r) = D_p r_{\min}^{D_p} r^{-1-D_p} dr \tag{7}$$

The probability density function of pores $f(r)$ is equal to $-dN_n(\geq r)/dr$. Therefore, we have

$$f(r) = D_p r_{\min}^{D_p} r^{-1-D_p} \quad r_{\min} \leq r \leq r_{\max} \tag{8}$$

There exists well-documented experimental evidence in the literature indicating that the size distribution of grains (Tyler and Wheatcraft 1992, Wu et al. 1993, Kozak et al. 1996, Bittelli et al. 1999, Millán et al. 2003, Ghanbarian and Daigle 2015), fragments (Turcotte 1986, Rieu and Sposito 1991, Giménez et al. 1998, Perfect et al. 2002, Perrier and Bird 2002), aggregates (Filgueiera et al. 1999, Millán and Orellana 2001, Giménez et al. 2002, Hirmas et al. 2013), and pores (Katz and Thompson 1985, Hansen and Skjeltorp 1988, Krohn 1988a, b, Anderson et al. 1996, Dathe and Thullner 2005, Bird et al. 2006, Tarquis et al. 2012, Giri et al. 2015, Ghanbarian et al. 2015) conform to power-law scaling. Nonetheless, sometimes the lognormal distribution (see e.g., Kosugi 1994, Tuli et al. 2001) i.e.,

$$f(r) = \frac{1}{\sqrt{2\pi}\sigma r} \exp\left[-\left(\frac{\ln(r) - \mu}{\sqrt{2}\sigma}\right)^2\right] \quad r \geq 0, \tag{9}$$

exponential distribution (see e.g., Lindquist et al. 2000, Omuto 2009) i.e.,

$$f(r) = \delta e^{-\delta r}, r \geq 0 \tag{10}$$

or Gamma distribution (see e.g., Heiba et al. 1992, Fredlund and Xing 1994) i.e.,

$$f(r) = \frac{1}{\Gamma(h)q^h} r^{h-1} e^{-r/q}, \; r \geq 0 \tag{11}$$

is assumed instead. In Eq. (9), σ is the standard deviation and μ is the mean, in Eq. (10) δ is a shape parameter, and in Eq. (11) h is a shape factor, q is a scale parameter and Γ is the complete Gamma function.

We should point out that with appropriate parameters, the lognormal distribution is similar in shape to the power-law (Pareto) distribution over much of its range. In particular, where variance of the lognormal distribution is large, its cumulative density function may appear linear on a log-log plot for several orders of magnitude (Mitzenmacher 2004). Baveye et al. (2008) also stated that Weibull, lognormal or exponential distribution may resemble the Pareto distribution over certain ranges. So, one may ask (Baveye et al. 2008), "Is power law equivalent to fractal?" A similar question was raised earlier by Avnir et al. (1998) as "Do power laws that are limited in range represent fractals?" Different researchers have different perspectives on these types of questions as they involve some sort of subjectivity and the interplay between two main definitions of fractals: probabilistic versus geometric. As Turcotte (1992) stated, a single mathematical definition of fractal geometry cannot encompass all its applications. Falconer (1990) also pointed out that the term 'fractal' does not have a rigorous definition. Nonetheless, Turcotte (1992) emphasized on the statistical representation i.e., a fractal is a set whose statistical distribution function conforms to power law (e.g., Pareto distribution). Crovelli and Barton (1995) referred to this type of definition of fractals as 'probabilistic'. Turcotte (1992) also stated that, "Some power-law distributions fall within the limits associated with fractional [fractal] dimensions, i.e., $0 < D < 3$, but others do not. … Such distributions are clearly scale invariant, even if not directly associated with a fractal dimension. This choice eliminates an ambiguity that can lead to considerable confusion when addressing measured data sets."

According to Falconer (1990), however, a fractal (e.g., a fractal set) has to be consistent with some geometrical properties. For example, a geometric fractal is too irregular to be described by the traditional Euclidean geometry and thus a new geometric language is required both locally and globally, it has some form of self-similarity and its fractal dimension is greater than its topological dimension (Falconer 1990). The most important aspect of Falconer (1990) and other definitions is that all consider geometric fractals as sets of points in R^n i.e., geometric constructs (Baveye et al. 2008). In brief, whereas a geometric fractal would be also probabilistic, a probabilistic fractal would not necessarily be geometric (Baveye et al. 2008).

Avnir et al. (1998) answered the question raised above "Do power laws that are limited in range represent fractals?" as, "If by "fractal" one means an object that obeys Eq. 1 [Eq. (4) here] over a limited range, then the use of this label may be acceptable, not only because of its usefulness, but because of the following additional reasons: (i) Interestingly, the sense of self-similarity in irregular objects is comprehended visually even for a limited range. (ii) In some cases, experimentally derived objects resemble simulated objects obtained from fractal models. (iii) The empirical values of D for spatial objects fall in the fractal regime of $0 < D < 3$."

Another key point for the application of fractal analysis to natural systems is the scaling range i.e., the number of orders of magnitude that a variable e.g., particle, pore or aggregate size spans. More than four decades ago Brock (1971) pointed out that experimental data points showing linearity on log-log scale should span at least two or three orders of magnitude before one tries to fit the Pareto distribution. Avnir et al. (1998) stated that, "A fractal object, in the purely mathematical sense, requires infinitely many orders of magnitude of power-law scaling, and a consequent interpretation of experimental results as indicating fractality requires many orders of magnitude." From our point of view, that is a key point for considering a set as fractal or prefractal. Although mathematically a fractal object needs to span infinite orders of magnitude of scale, a prefractal object has finite number of iterations and thus is limited in range.

4. Capillary Pressure Curve

One of the important properties in a porous medium is the relationship between capillary pressure P_c and fluid phase saturation, the fractional volume of pores occupied by the corresponding fluid. Consider an invading fluid (e.g., mercury or gas) that slowly invades a porous medium initially saturated with a defending fluid (e.g., gas or water). Note that both invader and defender are immiscible. In two-phase flow, depending on the wetting characteristics of the fluids, there are two different types of displacement. In the drainage process a non-wetting invading fluid displaces a wetting fluid (Aker 1996) by advancing through the porous medium and invading the least resistant accessible pore throats. As capillary pressure increases more pore throats are invaded and thus the saturation of the defending fluid decreases. Both experiments and simulations showed that the drainage process in porous media could be well modeled via invasion percolation (see e.g., Wilkinson and Willemsen 1983, Lenormand 1990, Zhou and Stenby 1993, Bakke and Øren 1997, Glass et al. 1998). In imbibition, however, a wetting invading fluid displaces a non-wetting fluid. At low injection rate the invading fluid would invade the narrowest accessible pore throats (Lenormand et al. 1983).

In the literature, various models were proposed to describe the capillary pressure curve in porous media, such as rocks and soils. In soil physics and hydrology, the most widely used model is that empirically proposed by van Genuchten (1980), which has a continuous sigmoid-shape form. In petroleum engineering, however, the discontinuous model of Brooks and Corey (1964) – that provides a better understanding of air invasion than the van Genuchten model – has been extensively applied. Although the Brooks and Corey (1964) model was proposed empirically, based on the power-law behavior of the capillary pressure curve, later de Gennes (1985), Tyler and Wheatcraft (1990), and Bird et al. (2000) found that the power-law behavior of the capillary pressure curve can be formulated within fractal geometry, as we illustrate in what follows.

In addition to empirical models, probability density functions, such as lognormal and exponential, were used to develop mathematical frameworks and relate the capillary pressure to the volume of pores and the fluid saturation in the medium. For example, Kosugi (1994) used a truncated lognormal distribution including three parameters i.e., mean μ, standard deviation σ and an upper cutoff (maximum pore radius $r_{\max}$) and proposed the following capillary pressure curve

$$\frac{S_w - S_{wr}}{1 - S_{wr}} = \frac{1}{2}\operatorname{erfc}\left(\frac{\ln\left[(P_a - P_c)/(P_a - P_0)\right] - \sigma^2}{\sqrt{2}\sigma}\right), \; |P_c| \geq |P_a| \tag{12}$$

where S_w is the water saturation, S_{wr} is the residual water saturation, erfc is the complementary error function, P_0 is the capillary pressure corresponding to the most probable pore size on the lognormal distribution and P_a is the entry pressure corresponding to the maximum accessible pore radius.

In addition to lognormal and exponential probability density functions, concepts from fractal geometry and power-law distribution were also used to derive mathematical models to characterize the capillary pressure curve. In Table 1, we summarize some of the unimodal models developed in the literature. Although they are similar in form, they were derived using different terminologies. For bimodal capillary pressure curve model, see Millán and González-Posada (2005), Russell (2010) and Hunt et al. (2013).

de Gennes (1985) is probably the first to apply fractal characteristics of the pore-solid interface to model the capillary pressure curve. His model is

$$S_w = \frac{\theta}{\varphi} = \left(\frac{P_c}{P_a}\right)^{D_s - 3}, \; |P_c| \geq |P_a| \tag{13}$$

in which D_s is the surface fractal dimension characterizing the pore-solid interface, θ is the water content and φ is the total porosity. Using 172 soil samples from several databases, Ghanbarian-Alavijeh and Millán (2009) showed that the surface fractal dimension – determined from fitting Eq.

(13) to capillary pressure data – was strongly correlated to the water content retained at $P_c = 1500$ kPa and the soil clay content by logarithmic functions with $R^2 = 0.97$ and 0.88, respectively.

Table 1. Fractal capillary pressure curve models existing in the literature.

Reference	*Model*	*Model parameters*	*Fractal dimension*
de Gennes (1985)	$\theta = \varphi(P_c/P_a)^{D_s-3}$	P_a, D_s	Surface
Tyler and Wheatcraft (1990)	$\theta = \varphi(P_c/P_a)^{D_m-3}$	P_a, D_m	Mass
Rieu and Sposito (1991)	$\theta = \varphi - 1 + (P_c/P_a)^{D_m-3}$	P_a, D_m	Mass
Perrier et al. (1996)	$\theta = \varphi - C_P[1 - (P_c/P_a)^{D_p-3}]$	P_a, C_P, D_p	Pore space
Perfect (1999)	$\theta = \varphi\ [(P_c^{D_m-3} - P_{c\max}^{D_m-3})/(P_a^{D_m-3} - P_{c\max}^{D_m-3})]$	P_a, P_{cmax}, D_m	Mass
Bird et al. (2000)	$\theta = \varphi - C_B[1 - (P_c/P_a)^{D_s-3}]$	P_a, C_B, D_s	Surface
Cihan et al. (2007)	$\theta = \varphi - C_C[1 - (P_c/P_a)^{D_d-3}]$	P_a, C_C, D_d	Mass
Deinert et al. (2008)	$\theta = \varphi - C_D[1 - (P_c/P_a)^{(D_p-3)/(3-D_s)}]$	P_a, C_D, D_s, D_p	Surface and pore space
Ghanbarian-Alavijeh et al. (2012)	$\theta = \varphi - C_G[1 - (P_c/P_a)^{D_p-3}]$	P_a, C_G, D_p	Pore space

P_c: capillary pressure, P_{cmax}: maximum capillary pressure corresponding to the minimum accessible pore size, P_a: entry pressure corresponding to the maximum accessible pore size in the medium, θ: water content, φ: porosity, D_m: mass fractal dimension, D_p: pore space fractal dimension, D_s: surface fractal dimension, D_d: fractal dimension of drained pore space, C_P, C_B, C_C, C_D and C_G are constant coefficients in different models

Equation (13) that is similar in form to the model of Brooks and Corey (1964) sheds light on the empirical parameter λ (= $3 - D_s$) referred to by Brooks and Corey as the pore-size distribution index, being small for media with a wide range of pore sizes and large for a pore space with relatively uniform pore-size distribution.

Later in 1990, Tyler and Wheatcraft proposed Eq. (13) using a different methodology based on properties of Sierpinski carpet and Menger sponge. Within the Tyler and Wheatcraft (1990) terminology, the fractal dimension in the exponent in Eq. (13) represents the mass fractal dimension characterizing both the solid matrix and pore space (see Table 1). Therefore, using the Tyler and Wheatcraft (1990) approach one has $\lambda = 3 - D_m$ and thus one may induce that $D_m = D_s$. Katz and Thompson (1985) are probably first to claim that pore space and surface fractal dimensions were the same. They stated that, "We further argue that the pore volume [space] is a fractal with the same fractal dimension as the pore-rock interface. This conclusion is supported by correctly predicting the porosity from the fractal parameters and by directly showing that the fractal dimension measured by autocorrelation of pores on thin sections agree with that measured on fracture surfaces." However, as Hansen and Skjeltorp (1988) pointed out, the results of Katz and Thompson (1985) were later disputed both in connection with SANS experiments (see Wong et al. 1986) and theory (see Roberts 1986). Recently, Dathe and Thullner (2005) determined the fractal dimension of solid matrix, pore space and the interface between the two using the box counting method and thin section images of various types of soils. They found that the smallest fractal dimension belonged to the pore-solid interface and fractal dimension of the solid phase was greater than that of the pore phase, in agreement with the results of Crawford and Matsui (1996) and Anderson et al. (2000). Dathe and Thullner (2005) concluded that, "… only the pore-matrix interface is a fractal, whereas the pore and the matrix phases are composed by a fractal component defined by the dimension of the interface and an Euclidean component. The porosity of a medium was found to be the weighting factor for these components. Although not being fractal, pore and matrix phase can be well approximated by a power law with the "fractal" dimension being a function of the dimension of the interface, the porosity, the size of the initiator, and the spatial resolution of the measurement."

As can be observed in Table 1, the Perrier et al. (1996), Bird et al. (2000), Cihan et al. (2007) and Ghanbarian-Alavijeh et al. (2012) models have essentially the same power-law form. However, the interpretation of the model parameters is different. For example, in the Perrier et al. (1996) model the pore space is fractal and the constant C_P is equal to V_0/V_t in which V_t is the total sample volume and V_0 is an upper bound on pore volume as the minimum pore radius tends to zero ($\varphi \leq C_P \leq 1$). In the general drainage model of Cihan et al. (2007), however, the fractal dimension for drained pore space D_d characterizes the capillary pressure curve and the constant coefficient C_C is a complex function of the ratio of the drained pore space to the total pore space ($0 \leq p \leq 1$), the scaling factor b in the Menger sponge, D_d and the mass fractal dimension D_m. In the Ghanbarian-Alavijeh et al. (2012) model, pore space is fractal similar to the Perrier et al. (1996) model and $C_G = \varphi r_{\max}^{3-D_p} / \left(r_{\max}^{3-D_p} - r_{\min}^{3-D_p}\right)$.

Crawford et al. (1995) argued that the capillary pressure curve "is a complicated function of both the pore-size distribution and the connectivity, and does not depend in a simple way on the spatial correlation of structure." They further concluded that the interpretation of the capillary pressure curve is ambiguous because a power-law relationship between capillary pressure and water content could be a consequence of a fractal pore space, a fractal solid matrix, a fractal pore-solid interface, or a non-fractal self-similar pore-solid interface. They measured fractal dimensions of both the pore space and the solid matrix from soil thin sections, and found that the fractal dimension of the solid matrix was a better predictor of the exponent in the Tyler and Wheatcraft (1990) model than the fractal dimension of the pore space. However, they did not measure the surface fractal dimension from 2D images, and it is not certain that the Tyler and Wheatcraft model is the best basis for comparison. Building on the Crawford et al. (1995) results, Deinert et al. (2008) found that the exponent of the fractal capillary pressure curve model i.e., $(D_p - 3)/(3 - D_s)$ is produced not only when either the pore space or the pore-solid interface is fractal, but also when both are fractal (see Table 1). Thus the question whether the surface fractal dimension determined from soil images could provide an accurate characterization of soil water retention curve remains unanswered.

4.1. Fractal Capillary Pressure Curve Model

In what follows, we derive a fractal capillary pressure curve model for which the probability density function of the pore sizes, $f(r)$, follows the power-law given by Eq. (8) in which $r_{\min}$ and $r_{\max}$ are the smallest and largest accessible pore radii, representing the lower and upper bounds of the power law, respectively. Recall that the larger the fractal dimension, the broader the pore-size distribution is.

Here, we ignore the effect of thin films of water on the internal surface of the pores and represent the pores by cylindrical tubes to each of which a radius r is attributed. Following Katz and Thompson (1986, 1987) and Hunt (2001), the geometrical characteristics of pores of self-similar isotropic porous media scale with one scaling factor in all directions, implying that larger pores necessarily have larger lengths. Thus, one may assume pore length l_p proportional to pore radius r ($l_p \propto r$). Accordingly, the volume of a single pore $V_p \propto r^3$, and the volume fraction of all the pores between the lower and upper cutoffs of the fractal scaling (the porosity φ of the pore space) is computed by integrating $f(r)$ given in Eq. (8):

$$\varphi = \int_{r_{\min}}^{r_{\max}} sr^3 f(r)dr = sD_p r_{\min}^{D_p}\left(r_{\max}^{3-D_p} - r_{\min}^{3-D_p}\right) \tag{12}$$

in which s is a shape factor.

The volume fraction of the pores filled by water is, however, a function of some upper limit $r \le r_{\max}$

$$\theta = \int_{r_{\min}}^{r} sr^3 f(r)dr = sD_p r_{\min}^{D_p}\left(r^{3-D_p} - r_{\min}^{3-D_p}\right) \tag{13}$$

where θ is the volumetric water content. Note that for the sake of simplicity, we assumed that each pore of the porous medium is occupied by only water or air. As is well-known, when water is the wetting phase, it occupies the smallest pores and air, as the non-wetting phase, fills the largest ones in the medium. This pore occupancy hypothesis obviously only approximates a more realistic situation in which air occupies the central part of the pore and water fills the corners, but we ignore such complexities.

Combining Eqs (12) and (13) results in

$$S_w = \frac{\theta}{\varphi} = 1 - \frac{\beta}{\varphi}\left(1 - \left(\frac{r}{r_{\max}}\right)^{D_p - 3}\right) \tag{14}$$

where S_w is the water saturation and β is $\varphi r_{\max}^{3-D_p} / \left(r_{\max}^{3-D_p} - r_{\mathrm{mix}}^{3-D_p}\right)$. The larger the fractal dimension, the broader is the pore-size distribution. Note that in practice $r_{\min}$ and $r_{\max}$ are the radii of the smallest and largest *accessible* pores in the medium, with the accessibility defined in the percolation sense (that is, the existence of paths from the external surface to the pores).

It was recently proposed (Cheng et al. 2004) that capillary pressure should be related to fluid interfacial area rather than pore radius r under equilibrium conditions. Deinert et al. (2005) also indicated that the variation in fluid interfacial area with fluid volume is a significant factor that determines equilibrium capillary pressure. Following Mandelbrot's (1983) relationship between surface area A_s and volume V of a fractal object, $A_s \propto V^{\frac{D_s}{3}}$, Deinert et al. (2008) found that the capillary pressure P_c at equilibrium is proportional to a power of the pore volume (V_p)

$$P_c \propto V_p^{(D_s-3)/3} \propto r^{D_s-3} \tag{15}$$

In the limiting case $D_s = 2$, Eq. (15) has the capillary pressure inversely proportional to $V_p^{1/3}$ and consequently r, analogous to the Young-Laplace equation.

Combining Eqs (14) and (15) yields the capillary pressure curve model for fractal porous media as follows:

$$S_w = \frac{\theta}{\varphi} = 1 - \frac{\beta}{\varphi}\left(1 - \left(\frac{P_c}{P_a}\right)^{\frac{D_p-3}{3-D_s}}\right), \quad |P_a| \le |P_c| \le |P_{c\max}| \tag{16}$$

where P_a is the entry capillary pressure, and $P_{c\max}$ is the maximum capillary pressure (the upper limit of capillary pressure). Equation (16) is similar to the Perrier et al. (1996), Bird et al. (2000), Cihan et al. (2007) and Deinert et al. (2008) capillary pressure curve models. However, as we pointed out earlier the interpretation of model parameters is different. Note that all parameters in Eq. (16) have physical meaning. Equation (16) has the same form as the Brooks and Corey (1964) and Rieu and Sposito (1991) models when β is set equal to φ and 1, respectively. As can be seen in Eq. (16), precise estimation of the capillary pressure curve requires sufficient information of the fractal nature of the porous medium including on both pore space and pore-solid interface (i.e., D_p and D_s). We also address accessibility effects on the capillary pressure curve estimation in Section 5.

We should caution that the fractal model of capillary pressure, Eq. (16), is based upon geometrical properties and a power-law distribution of the pore throat sizes. Therefore, it does not address features such as wettability alteration, the history and rate of change of saturation, and the effect of the pore accessibility and connectivity.

5. Comparison with Pore-network Models

Here we compare Eq. (16) with pore-network simulations under drainage conditions. For this purpose, a complex network constructed of pore throats of various sizes following a power-law probability density function i.e., Eq. (8) with $D_p = 0.5$ was used. The minimum and maximum pore radii in the network are respectively 0.1 and 1 μm that indicate a relatively narrow pore throat-size distribution spanning one order of magnitude. Since pore throats in the network are constructed of cylindrical tubes with perfectly smooth wall the pore-solid interface fractal dimension $D_s = 2$. The drainage curve simulations were perfumed using PoreFlow (Raoof et al. 2013) – a complex pore network modeling of flow and transport developed by Amir Raoof and his coworkers (Raoof and Hassanizadeh 2010, 2012, Li et al. 2014). Figure 7 shows the pore-network model at $S_w =$ 0.2 and 0.5.

Given that $r_{min} = 0.1$, $r_{max} = 1$ μm and $D_p = 0.5$ in the pore-network model, one would have $\beta/\varphi = r_{max}^{3-D_p} \Big/ \left(r_{max}^{3-D_p} - r_{min}^{3-D_p}\right) = 1.003$. P_a in Eq. (16) is the entry pressure corresponding to the maximum accessible pore throat in the pore network. Since its value is unknown, we set $D_p = 0.5$, $D_s = 2$ and $\beta/\varphi = 1.003$ and fit Eq. (16) to the simulated capillary pressure curve to determine the value of P_a. As can be seen in Fig. 8, Eq. (16) fit well the simulated data with $P_a = 300$ kPa corresponding to a pore throat of radius of 0.48 μm, which is near the half of the maximum pore throat size ($r_{max} = 1$ μm) available in the network.

Larson and Morrow (1981) showed that the capillary pressure curve is not only a function of geometrical and wetting properties of individual pores but also a function of the pores' connections to the surface of the rock/soil sample. Since pores' connections (accessibility) depend on the distance from the sample surface, the capillary pressure curve should be sensitive to the sample dimensions. In fact, pores that are close to the surface of a sample are more often accessible than interior pores. In natural porous media some large pore bodies are connected to others via

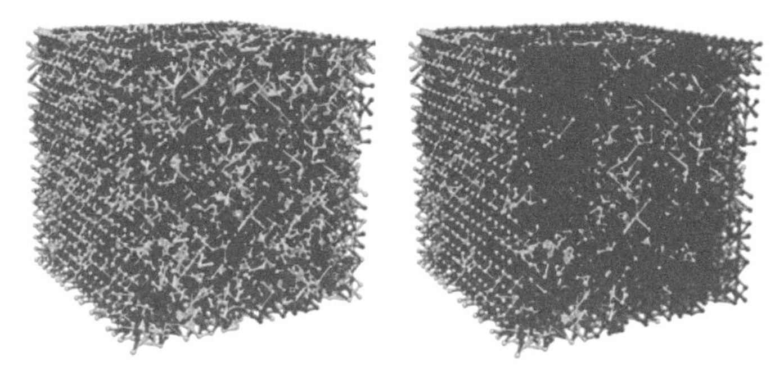

Figure 7: Pore-network model used in this study to simulate the capillary pressure curve at two different water saturations: S_w = 0.2 (left) and S_w = 0.5 (right). The light grey represents the non-wetting, being injected from the left side, and the dark grey denotes the wetting phase (Courtesy of Amir Raoof).

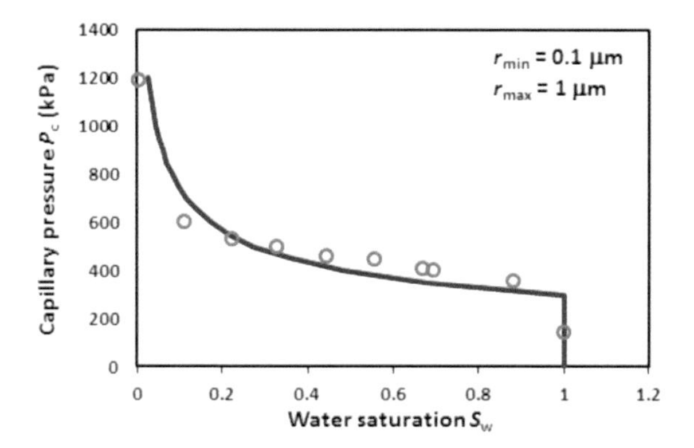

Figure 8: The capillary pressure P_c as a function of water saturation S_w using pore-network model denoted by unfilled circles and Eq. (16) with D_p = 0.5, D_s = 2, P_a = 300 kPa and β/φ = 1.003 shown by solid curve. The only parameter optimized through the fitting process is P_a whose value is a function of accessibility in the pore network (see context for further discussion).

small pore throats. These pores are not drained until a sufficiently large pressure appropriate to the largest pore throats is executed. Therefore, their volumes are assigned to the smaller pore (body) part of the pore-size distribution (at larger pressures) incorrectly (Larson and Morrow 1981).

6. Comparison with Sphere Packs

We also compare Eq. (16) with numerical simulations in sphere packs reported by Mawer et al. (2015) who carried out computation of the capillary pressure curve for 15 sphere packs. The value of porosity of each sphere pack is given in Table 2. In order to generate a partially saturated

Table 2: The fractal capillary pressure curve model parameters for 15 packs from Mawer et al. (2015).

Pack	*Porosity*	*Condition*	β	D_p	$P_a\,(m)$	R^2
Finney	0.36	Drainage	0.36	-0.16	0.04	0.998
		Imbibition	0.37	1.12	0.03	0.998
1	0.33	Drainage	0.36	-0.09	1.38	0.998
		Imbibition	0.37	1.68	0.22	0.993
2	0.35	Drainage	0.351	-0.69	1.37	0.997
		Imbibition	0.37	1.44	0.09	0.996
3	0.23	Drainage	0.24	0.49	1.87	0.999
		Imbibition	0.26	1.74	0.29	0.997
4	0.25	Drainage	0.26	0.62	1.86	0.996
		Imbibition	0.28	1.67	0.29	0.996
5	0.26	Drainage	0.26	-0.58	1.84	0.999
		Imbibition	0.28	1.60	0.29	0.997
6	0.28	Drainage	0.28	-0.82	1.80	0.998
		Imbibition	0.30	1.49	0.28	0.998
7	0.30	Drainage	0.30	-1.41	1.83	0.997
		Imbibition	0.32	1.38	0.28	0.998
8	0.32	Drainage	0.32	-1.29	1.75	0.996
		Imbibition	0.35	1.65	0.25	0.993
9	0.34	Drainage	0.34	0.21	1.51	0.992
		Imbibition	0.36	1.53	0.25	0.993
10	0.36	Drainage	0.36	-0.53	1.49	0.999
		Imbibition	0.39	1.65	0.23	0.991
11	0.38	Drainage	0.38	-0.63	1.45	0.998
		Imbibition	0.39	1.27	0.24	0.995
12	0.42	Drainage	0.42	-0.36	1.25	0.993
		Imbibition	0.44	1.30	0.22	0.995
13	0.44	Drainage	0.44	-0.61	1.25	0.999
		Imbibition	0.46	1.17	0.22	0.995
14	0.46	Drainage	0.46	-1.64	1.24	0.996
		Imbibition	0.47	0.99	0.21	0.995
Average	0.34	Drainage	0.34	-0.50	1.46	0.997
		Imbibition	0.36	1.45	0.23	0.995

sphere pack, Mawer et al. (2015) saturated the pore space and carried out drainage and imbibition simulations to capture hysteresis. Further details are given by Mawer et al. (2015). The parameters of the power-law model of the capillary pressure curve, namely, D_p, β and P_a, were estimated by directly fitting Eq. (16) to the numerically simulated data. We should point out that since spheres are smooth we thus set $D_s = 2$ reducing Eq. (16) to the Ghanbarian-Alavijeh et al. (2012) model (see Table 1).

The calculated fractal capillary pressure curve model parameters are presented in Table 2 for each sphere pack under drainage and imbibition processes. As can be observed, the pore space fractal dimension D_p for imbibition is greater than that for drainage indicating that the pore throat-size distribution under imbibition conditions is broader than that under drainage conditions. This can be confirmed via the average D_p values reported for imbibition and drainage (1.45 vs. –0.50) in Table 2. One should, however, bear in mind that the drainage pore throat-size distribution in other types of porous materials might be broader than that of imbibition. Regarding the negative fractal dimension values reported in Table 2, we should note that using the relationship among porosity, r_{min}/r_{max}, and pore space fractal dimension, Ghanbarian-Alavijeh and Hunt (2012a) showed that $D_p < 0$ is theoretically permissible (see their Eq. (7) and Table 1; see also Mandelbrot 1990). The values of the correlation coefficient $R^2 > 0.99$ presented in Table 2 also demonstrate that Eq. (16) fit well the simulated capillary pressure data of sphere packs.

Regarding the estimation of the capillary pressure curve in natural porous media, Ghanbarian-Alavijeh and Hunt (2012b) recently evaluated various methods to estimate the capillary pressure curve from the particle-size distribution. More specifically, the fractal dimension characterizing the capillary pressure curve (pore space, mass, or surface fractal dimension) was estimated from that derived from the particle-size distribution. Their experimental results showed that the former was greater than the latter, in agreement with the results of Hunt and Gee (2002). Ghanbarian-Alavijeh and Hunt (2012b) also showed that capillary pressure curve might be estimated reasonably well from particle-size distribution, if the estimated capillary pressure curve is modified for non-equilibrium conditions.

In Sections 5 and 6, we compared Eq. (16) with pore-network results and sphere pack models and showed that the fractal capillary pressure curve fits the data well. Practical estimation of the capillary pressure curve in natural porous media, however, requires accurate characterization of both pore space and pore-solid interface as well as knowledge of pore accessibility and connectivity. In spite of advances in direct measurements of pore structural properties using X-ray computed tomography and image segmentation, estimation of pore space and pore-solid interface fractal dimensions from two- and three-dimensional images are still ambiguous, depending on image resolution, thresholding and segmentation, etc. (see e.g., Baveye et al. 1998, Ogawa et al. 1999, Dathe and Baveye 2003, Tarquis et al. 2008). Further investigations are indeed required to more rigorously determine fractal properties of porous media from non-destructive microtomography techniques. Whether pore space and pore-solid interface fractal dimensions determined from high-resolution images could estimate the capillary pressure curve accurately remains unanswered and further investigations are required.

Acknowledgements

The authors are grateful to Amir Raoof, Utrecht University, for sharing his pore-network model results, to Chloe Mawer, Silicone Valley Data Science, for providing the results of numerical simulations of drainage and imbibition, and to Timothy A. Cousins, University of Texas at Austin, for sharing Figs 5 and 6.

REFERENCES

Aker, E. 1996. A simulation model for two-phase flow in porous media. PhD dissertation, Department of Physics, University of Oslo.

Anderson, A.N., A.B. McBratney and E.A. FitzPatrick. 1996. Soil mass, surface, and spectral fractal dimensions estimated from thin section photographs. *Soil Science Society of America Journal* 60(4): 962–969.

Anderson, A.N., A.B. McBratney and J.W. Crawford. 1997. Applications of fractals to soil studies. *Advances in Agronomy* 63: 1–76.

Anderson, A.N., J.W. Crawford and A.B. McBratney. 2000. On diffusion in fractal soil structures. *Soil Science Society of America Journal* 64(1): 19–24.

Avnir, D., O. Biham, D. Lidar and O. Malcai. 1998. Is the geometry of nature fractal? *Science* 279: 39–40.

Bakke, S. and P.E. Øren. 1997. 3-D pore-scale modelling of sandstones and flow simulations in the pore networks. *SPE Journal* 2(02): 136–149.

Barton, C.C., R. Paul and L. Pointe. 1995. (Eds.) Fractals in the earth sciences. New York: Plenum Press.

Barton, C.C. and P.R. La Pointe. 2012. (Eds.) Fractals in petroleum geology and earth processes. Springer Science & Business Media.

Baveye, P., C.W. Boast, S. Ogawa, J.Y. Parlange and T. Steenhuis. 1998. Influence of image resolution and thresholding on the apparent mass fractal characteristics of preferential flow patterns in field soils. *Water Resources Research* 34(11): 2783–2796.

Baveye, P., C.M. Boast, S. Gaspard, A.M. Tarquis and H. Millán. 2008. Introduction to fractal geometry, fragmentation processes and multifractal measures: Theory and operational aspects of their applications to natural systems. In: Senesi, N., Wilkinson, K. J. (eds.). Biophysical Chemistry of Fractal Structures and Processes in Environmental Systems. John Wiley and Sons, Chichester, UK.

Bird, N., M.C. Díaz, A. Saa and A.M. Tarquis. 2006. Fractal and multifractal analysis of pore-scale images of soil. *Journal of Hydrology* 322(1): 211–219.

Bird, N.R.A., E. Perrier, and M. Rieu. 2000. The water retention function for a model of soil structure with pore and solid fractal distributions. *European Journal of Soil Science* 51: 55–63.

Bittelli, M., G.S. Campbell and M. Flury. 1999. Characterization of particle-size distribution in soils with a fragmentation model. *Soil Science Society of America Journal* 63(4): 782–788.

Brock, J.R. 1971. On size distribution of atmospheric aerosols. *Atmospheric Environment* 5: 833–841.

Brooks, R.H. and A.T. Corey. 1964. Hydraulic properties of porous media. Hydrol. Pap. 3, Dep. of Civ. Eng., Colo. State Univ., Fort Collins.

Bunde, A. and S. Havlin. 2012. (Eds.) Fractals and disordered systems. Springer Science & Business Media.

Bunde, A. and S. Havlin. 2013. (Eds.) Fractals in science. Springer.

Cheng, J.T., L.J. Pyrak-Nolte, D.D. Nolte and N.J. Giordano. 2004. Linking pressure and saturation through interfacial areas in porous media. *Geophysical Research Letters* 31(8): L08502.

Cihan, A., E. Perfect and J.S. Tyner. 2007. Water retention models for scale-variant and scale-invariant drainage of mass prefractal porous media. *Vadose Zone Journal* 6(4): 786–792.

Crawford, J.W. and N. Matsui. 1996. Heterogeneity of the pore and solid volume of soil: Distinguishing a fractal space from its non-fractal complement. *Geoderma* 73(3): 183–195.

Crawford, J.W., B.D. Sleeman and I.M. Young. 1993. On the relation between number-size distributions and the fractal dimension of aggregates. *Journal of Soil Science* 44(4): 555–565.

Crawford, J.W., N. Matsui and I.M. Young. 1995. The relation between the moisture-release curve and the structure of soil. *European Journal of Soil Science* 46(3): 369–375.

Crovelli, R.A. and C.C. Barton. 1995. Fractals and the Pareto distribution applied to petroleum accumulation-size distributions. In: Fractals in Petroleum Geology and Earth Processes. Barton, C.C. and La Pointe, P.R. (eds). Plenum Press, New York, pp. 59–72.

Dathe, A. and P. Baveye. 2003. Dependence of the surface fractal dimension of soil pores on image resolution and magnification. *European Journal of Soil Science* 54(3): 453–466.

Dathe, A. and M. Thullner. 2005. The relationship between fractal properties of solid matrix and pore space in porous media. *Geoderma* 129(3): 279–290.

de Gennes, P.G. 1985. Partial filling of a fractal structure by a wetting fluid. In: Adler, P.M., Fritzsche, H., Ovshinsky, S.R. (Eds.), Physics of Disordered Materials. Plenum Press, New York, pp. 227–241.

Deinert, M.R., J.Y. Parlange and K.B. Cady. 2005. Simplified thermodynamic model for equilibrium capillary pressure in a fractal porous medium. *Physical Review* E72(4): 041203.

Deinert, M.R., A. Dathe, J.Y. Parlange and K.B. Cady. 2008. Capillary pressure in a porous medium with distinct pore surface and pore volume fractal dimensions. *Physical Review* E77(2): 021203.

Drazer, G. and J. Koplik. 2000. Permeability of self-affine rough fractures. *Physical Review* E62(6): 8076–8085.

Durner, W. 1994. Hydraulic conductivity estimation for soils with heterogeneous pore structure. *Water Resour. Res.* 30: 211–223.

Falconer, K.J. 1990. Fractal Geometry, Mathematical Foundations and Applications. John Wiley and Sons, Chichester, UK.

Feder, J. 2013. Fractals. Springer Science & Business Media.

Filgueira, R.R., Y.A. Pachepsky, L.L. Fournier, G.O. Sarli and A. Aragon. 1999. Comparison of fractal dimensions estimated from aggregate mass-size distribution and water retention scaling. *Soil Science* 164(4): 217–223.

Filgueira, R.R., Y.A. Pachepsky and L.L. Fournier. 2003. Time-mass scaling in soil texture analysis. *Soil Science Society of America Journal* 67(6): 1703–1706.

Fredlund, D.G., and A. Xing. 1994. Equations for the soil-water characteristic curve. *Canadian Geotechnical Journal* 31(4): 521–532.

Ghanbarian-Alavijeh, B. and H. Millán. 2009. The relationship between surface fractal dimension and soil water content at permanent wilting point. *Geoderma* 151(3): 224–232.

Ghanbarian-Alavijeh, B. and A.G. Hunt. 2012a. Comments on "More general capillary pressure and relative permeability models from fractal geometry" by Kewen Li. *Journal of Contaminant Hydrology* 140: 21–23.

Ghanbarian-Alavijeh, B. and A.G. Hunt. 2012b. Estimation of soil-water retention from particle-size distribution: Fractal approaches. *Soil Science* 177(5): 321–326.

Ghanbarian, B. and H. Daigle. 2015. Fractal dimension of soil fragment mass-size distribution: A critical analysis. *Geoderma* 245: 98–103.

Ghanbarian-Alavijeh, B., H. Millán and G. Huang. 2011. A review of fractal, prefractal and pore-solid-fractal models for parameterizing the soil water retention curve. *Canadian Journal of Soil Science* 91(1): 1–14.

Ghanbarian-Alavijeh, B., T.E. Skinner and A.G. Hunt. 2012. Saturation dependence of dispersion in porous media. *Physical Review* E86(6): 066316.

Ghanbarian, B., A.G. Hunt, T.E. Skinner and R.P. Ewing. 2015. Saturation dependence of transport in porous media predicted by percolation and effective medium theories. *Fractals* 23: 1540004.

Giménez, D., E. Perfect, W.J. Rawls and Y. Pachepsky. 1997. Fractal models for predicting soil hydraulic properties: A review. *Engineering Geology* 48(3): 161–183.

Giménez, D., R.R. Allmaras, D.R. Huggins and E.A. Nater. 1998. Mass, surface, and fragmentation fractal dimensions of soil fragments produced by tillage. *Geoderma* 86(3): 261–278.

Giménez, D., J.L. Karmon, A. Posadas and R.K. Shaw. 2002. Fractal dimensions of mass estimated from intact and eroded soil aggregates. *Soil and Tillage Research* 64(1): 165–172.

Giri, A., S. Tarafdar, P. Gouze and T. Dutta. 2015. Multifractal analysis of the pore space of real and simulated sedimentary rocks. *Geophysical Journal International* 200(2): 1106–1115.

Glass, R.J., M.J. Nicholl and L. Yarrington. 1998. A modified invasion percolation model for low-capillary number immiscible displacements in horizontal rough-walled fractures: Influence of local in-plane curvature. *Water Resources Research* 34(12): 3215–3234.

Guber, A.K., Y.A. Pachepsky and E.V. Levkovsky. 2005. Fractal mass–size scaling of wetting soil aggregates. *Ecological Modelling* 182(3): 317–322.

Hansen, J.P. and A.T. Skjeltorp. 1988. Fractal pore space and rock permeability implications. *Physical Review* B38(4): 2635–2638.

Heiba, A.A., M. Sahimi, L.E. Scriven and H.T. Davis. 1992. Percolation theory of two-phase relative permeability. *SPE Journal* 7(01): 123–132.

Hirmas, D.R., D. Giménez, V. Subroy and B.F. Platt. 2013. Fractal distribution of mass from the millimeter- to decimeter-scale in two soils under native and restored tall grass prairie. *Geoderma* 207: 121–130.

Hunt, A.G. 2001. Applications of percolation theory to porous media with distributed local conductances. *Advances in Water Resources* 24(3): 279–307.

Hunt, A.G. and G.W. Gee. 2002. Water-retention of fractal soil models using continuum percolation theory. *Vadose Zone Journal* 1(2): 252–260.

Hunt, A.G., B. Ghanbarian and K.C. Saville. 2013. Unsaturated hydraulic conductivity modeling for porous media with two fractal regimes. *Geoderma* 207: 268–278.

Hunt, A., R. Ewing and B. Ghanbarian. 2014. Percolation Theory for Flow in Porous Media, 3rd ed. Vol. 880, Springer, Berlin.

Kantelhardt, J.W. 2012. Fractal and multifractal time series. In: Mathematics of complexity and dynamical systems (pp. 463-487). Springer New York.

Katz, A. and A.H. Thompson. 1985. Fractal sandstone pores: implications for conductivity and pore formation. *Physical Review Letters* 54(12): 1325–1328.

Katz, A.J. and A.H. Thompson. 1986. Quantitative prediction of permeability in porous rock. *Physical Review* B 34(11): 8179–8181.

Katz, A.J. and A.H. Thompson. 1987. Prediction of rock electrical conductivity from mercury injection measurements. *Journal of Geophysical Research* 92(B1): 599–607.

Kosugi, K.I. 1994. Three-parameter lognormal distribution model for soil water retention. *Water Resources Research* 30(4): 891–901.

Kozak, E., Z. Sokołowska, W. Stepniewski, Y.A. Pachepsky and S. Sokołowski. 1996. A modified number-based method for estimating fragmentation fractal dimensions of soils. *Soil Science Society of America Journal* 60(5): 1291–1297.

Krohn, C.E. 1988a. Fractal measurements of sandstones, shales, and carbonates. *Journal of Geophysical Research: Solid Earth* 93(B4): 3297–3305.

Krohn, C.E. 1988b. Sandstone fractal and Euclidean pore volume distributions. *Journal of Geophysical Research: Solid Earth* 93(B4): 3286–3296.

Larson, R.G. and N.R. Morrow. 1981. Effects of sample size on capillary pressures in porous media. *Powder Technology* 30(2): 123–138.

Lenormand, R. 1990. Liquids in porous media. *Journal of Physics: Condensed Matter* 2(S): SA79–A88.

Lenormand, R., C. Zarcone and A. Sarr. 1983. Mechanisms of the displacement of one fluid by another in a network of capillary ducts. *Journal of Fluid Mechanics* 135: 337–353.

Li, S., A. Raoof and R. Schotting. 2014. Solute dispersion under electric and pressure driven flows; pore scale processes. *Journal of Hydrology* 517: 1107–1113.

Lindquist, W.B., A. Venkatarangan, J. Dunsmuir and T.F. Wong. 2000. Pore and throat size distributions measured from synchrotron X-ray tomographic images of Fontainebleau sandstones. *Journal of Geophysical Research: Solid Earth* 105(B9): 21509–21527.

Madadi, M. and M. Sahimi. 2003. Lattice Boltzmann simulation of fluid flow in fracture networks with rough, self-affine surfaces. *Physical Review* E67(2): 026309.

Malamud, B.D. and D.L. Turcotte. 1999. Self-affine time series: I. Generation and analyses. *Advances in Geophysics* 40: 1–90.

Mandelbrot, B.B. 1983. The fractal geometry of nature. W.H. Freeman, San Francisco.

Mandelbrot, B.B. 1985. Self-affine fractals and fractal dimension. *Physica Scripta* 32(4): 257–260.

Mandelbrot, B.B. 1990. Negative fractal dimensions and multifractals. *Physica A: Statistical Mechanics and its Applications* 163(1): 306–315.

Mawer, C., R. Knight and P.K. Kitanidis. 2015. Relating relative hydraulic and electrical conductivity in the unsaturated zone. *Water Resour. Res.* 51: 599–618.

Millán, H. and R. Orellana. 2001. Mass fractal dimensions of soil aggregates from different depths of a compacted Vertisol. *Geoderma* 101(3): 65–76.

Millán, H. and M. González-Posada. 2005. Modelling soil water retention scaling: Comparison of a classical fractal model with a piecewise approach. *Geoderma* 125(1): 25–38.

Millán, H., M. González-Posada, M. Aguilar, J. Domınguez and L. Cespedes. 2003. On the fractal scaling of soil data: Particle-size distributions. *Geoderma* 117(1): 117–128.

Mitzenmacher, M. 2004. A brief history of generative models for power law and lognormal distributions. *Internet Mathematics* 1(2): 226–251.

Ogawa, S., P. Baveye, C.W. Boast, J.Y. Parlange and T. Steenhuis. 1999. Surface fractal characteristics of preferential flow patterns in field soils: Evaluation and effect of image processing. *Geoderma* 88(3): 109–136.

Omuto, C.T. 2009. Biexponential model for water retention characteristics. *Geoderma* 149(3): 235–242.

Pachepsky, Y., J.W. Crawford and W.J. Rawls. 2000. Fractals in soil science (Vol. 27). Elsevier.

Pelletier, J.D. and D.L. Turcotte. 1999. Self-affine time series: II. Applications and models. *Advances in Geophysics* 40: 91–166.

Perfect, E., M. Dıaz-Zorita and J.H. Grove. 2002. A prefractal model for predicting soil fragment mass-size distributions. *Soil and Tillage Research* 64(1): 79–90.

Perrier, E., M. Rieu, G. Sposito, and G. de Marsily. 1996. Models of the water retention curve for soils with a fractal pore size distribution. *Water Resources Research* 32(10): 3025–3031.

Perrier, E.M.A. and N.R.A. Bird. 2002. Modelling soil fragmentation: The pore solid fractal approach. *Soil and Tillage Research* 64(1): 91–99.

Poon, C.Y., R.S. Sayles and T.A. Jones. 1992. Surface measurement and fractal characterization of naturally fractured rocks. *Journal of Physics D: Applied Physics* 25(8): 1269–1275.

Raoof, A. and S.M. Hassanizadeh. 2010. A new method for generating pore-network models of porous media. *Transport in Porous Media* 81(3): 391–407.

Raoof, A. and S.M. Hassanizadeh. 2012. A new formulation for pore-network modeling of two-phase flow. *Water Resources Research* 48(1): W01514.

Raoof, A., H.M. Nick, S.M. Hassanizadeh and C.J. Spiers. 2013. PoreFlow: A complex pore-network model for simulation of reactive transport in variably saturated porous media. *Computers & Geosciences* 61: 160–174.

Rieu, M. and G. Sposito. 1991. Fractal fragmentation, soil porosity, and soil water properties: I. Theory, II. Applications. *Soil Science Society of America Journal* 55: 1231–1244.

Rieu, M. and E. Perrier. 1997. Fractal models of fragmented and aggregated soils. In: P. Baveye et al. (eds). Fractals in soil science. CRC Press, Boca Raton, FL. Pp. 169-201.

Roberts, J.N. 1986. Comment about fractal sandstone pores. *Physical Review Letters* 56(19): 2111.

Russell, A.R. 2010. Water retention characteristics of soils with double porosity. *European Journal of Soil Science* 61(3): 412–424.

Sahimi, M. 2003. Heterogeneous materials II. Nonlinear and breakdown properties and atomistic modeling. Springer. 637 pp.

Sahimi, M. 2011. Flow and Transport in Porous Media and Fractured Rock: From Classical Methods to Modern Approaches. Wiley-VCH Publisher. Pp. 709.

Schmittbuhl, J., S. Rouxand and Y. Berthaud. 1994. Development of roughness in crack propagation. *Europhysics Letters* 28(8): 585–590.

Stanley, H.E. 1984. Application of fractal concepts to polymer statistics and to anomalous transport in randomly porous media. *Journal of Statistical Physics* 36(5–6): 843–860.

Tarquis, A.M., R.J. Heck, J.B. Grau, J. Fabregat, M.E. Sanchezand and J.M. Antón. 2008. Influence of thresholding in mass and entropy dimension of 3-D soil images. *Nonlinear Processes in Geophysics* 15(6): 881–891.

Tarquis, A.M., M.E. Sanchez, J.M. Antón, J. Jimenez, A. Saa-Requejo, D. Andina and J.W. Crawford. 2012. Variation in spectral and mass dimension on three-dimensional soil image processing. *Soil Science* 177(2): 88–97.

Tuli, A., K. Kosugi and J.W. Hopmans. 2001. Simultaneous scaling of soil water retention and unsaturated hydraulic conductivity functions assuming lognormal pore-size distribution. *Advances in Water Resources* 24(6): 677–688.

Turcotte, D.L. 1986. Fractals and fragmentation. *Journal of Geophysical Research: Solid Earth* 91(B2): 1921–1926.

Turcotte, D.L. 1992. Fractals and Chaos in Geology and Geophysics. Cambridge University Press, Cambridge, UK.

Tyler, S.W. and S.W. Wheatcraft. 1989. Application of fractal mathematics to soil water retention estimation. *Soil Science Society of America Journal* 53(4): 987–996.

Tyler, S.W. and S.W. Wheatcraft. 1990. Fractal processes in soil water retention. *Water Resources Research* 26(5): 1047–1054.

Tyler, S.W. and S.W. Wheatcraft. 1992. Fractal scaling of soil particle-size distributions: Analysis and limitations. *Soil Science Society of America Journal* 56(2): 362–369.

van Genuchten, M. 1980. A closed-form equation for predicting the hydraulic conductivity of unsaturated soils. *Soil Science Society of America Journal* 44(5): 892–898.

Wilkinson, D. and J.F. Willemsen. 1983. Invasion percolation: A new form of percolation theory. *Journal of Physics A: Mathematical and General* 16(14): 3365–3376.

Wong, P.Z., J. Howard and J.S. Lin. 1986. Surface roughening and the fractal nature of rocks. *Physical Review Letters* 57(5): 637–640.

Wu, Q., M. Borkovec and H. Sticher. 1993. On particle-size distributions in soils. *Soil Science Society of America Journal* 57(4): 883–890.

Yu, B. 2008. Analysis of flow in fractal porous media. *Applied Mechanics Reviews* 61(5): 050801.

Zhou, D. and E.H. Stenby. 1993. Interpretation of capillary pressure curves using invasion percolation theory. *Transport in Porous Media* 11(1): 17–31.

CHAPTER
3

Two- and Three-phase Fractal Models of Porous Media: A Multiscale Approach

Edith Perrier
Director of Research Emeritus at IRD, UMI UMMISCO,
32 avenue Henri Varagnat, 93143 Bondy Cedex, France
edith.perrier@ird.fr

1. Introduction

A porous medium, that is a material including voids, is by definition a binary system, made of two components, the solid phase and the pore (void) phase. Many natural substances such as rocks, soils, and some biological tissues can be modeled as porous media, even if most of these real media are highly complex, because most of their properties can only be rationalized by considering them to be binary porous media. Within such a categorization, many man made materials such as concrete, ceramics and geotextiles may also be considered as porous media.

Porous media are also a typical example for percolation theory, which deals in a very general and abstract way with two-phase binary media, where one phase is a conductor and the other phase is an insulator as regards the transmission of any quantity of fluid, electricity, information, etc. Percolation theory provides tools to analyze the connectivity of porous media handled as networks of black/white, conducting/non-conducting sites or bonds, with applications in various other domains. So, as mentioned by Wikipedia, the "concept of porous medium" is used in many areas of applied science and engineering, to mean a spatialized set of one solid and one pore phase.

So why are we talking here about two- or three-phase models of porous media?

At a macroscopic scale where the details of the pore space are not considered, porous media can be modeled as continuous systems, and defined by macroscopic variables, such as their porosity, permeability, etc. Therefore, in this case, we could say that they are modeled as a single phase even if one can use a probabilistic approach with 1 and 0 values for random functions to mathematically define a porous medium (Matheron 1967, Delhomme and deMarsily, 2005). At a microscopic scale where one can "see" some pore and solid "objects", 2-D or 3-D images are often binarized to distinguish the pore and solid spaces, whichever way they are experimentally obtained initially, and whatever the thresholding algorithms (Baveye et al. 2010). We could have used the latter word "space" instead of "phase" in the title of this chapter, because some fractal models consider the geometry of the porous medium and its spatial organization, but other fractal models consider only the pore or solid size distributions without reference to space, so we have kept the more general word "phase".

The (fractal or nonfractal) geometry of a porous medium determines its functional properties, namely with regards to fluid flow through the pores, the transport of solute, organic or non-organic matter, heat diffusion, etc. This is why the topic has been addressed for years by many researchers. We are interested in fluid flows, which are driven by an intrinsic property of the porous medium, known as permeability, and also in multiphase fluid flow, when various fluid phases coexist within the pore space, for example, three air-oil-water phases in rocks, or air-water phases in unsaturated soils. However, we consider that the fluids moving or stocked inside a porous medium are not part of it, they can be removed or added, they are external to the porous medium itself. This is why we caution the reader to be aware that we do not use here the expression 2-phase or 3-phase fractal model of porous media (i.e. pore fractal, solid fractal, pore-solid fractal models) in the same way as it is sometimes addressed in the literature where the expression multi-phase porous media can refer at the same time to the solid and pore phase of the porous medium, or to the fluid phases inside the porous medium (Yu et al. 2009, Mei et al. 2010, Zhou et al. 2010). When fluids will be considered, the expression fluid-phase will be explicitly used.

So which is the possible third phase of a model of porous media that we introduce in this chapter?

The third phase here is a mere abstraction, just a matter of scale. It has been introduced in fractal models to represent an "apparent third phase" at a given resolution of an image, when large pores and solid particles can be detected whereas smaller pores or solids could be revealed only by increasing the image resolution (or with a series of images obtained at different scales with different technical means). Fractal models of porous media have been introduced to take into account such multiscale media, knowing that most of natural media are distributed over a large range of length scale. In rocks or soils, except when they can be considered as monoscale (for example with packings of grains of the same size generating some void space), the existence of this third phase is well known: it is called the matrix. The matrix can be represented as a continuous medium in so-called dual porosity models where only two scales are considered (e.g. Gerke and Van Genuchten 1993). The fractal approach is intrinsically based on a multiscale approach: As regards the modeling of the geometry of porous media at a microscopic scale, it relies first on the self-similarity principle where it is assumed that zooming in the matrix will reveal the porous medium at least statistically identical to the whole. Secondly a fractal model represents, in a recurrent manner, a large if not infinite range of scale within a single simplified model of the complex real medium.

Throughout the present chapter we will use the PSF (Pore Solid Fractal) model (Perrier et al. 1999), which is explicitly a three-phase model of porous media ($Phase_P$, $Phase_S$, $Phase_F$), and we will discuss its contribution to the concepts of key scaling properties in fractal porous media.

The first section will present relevant properties of the PSF model, which can be mathematically derived. Then in the second section, we will consider computer simulations of two fluid-phases inside such a theoretical porous medium variably saturated by two immiscible fluids, a wetting fluid-phase and non-wetting fluid-phase. We will simulate the capillary pressure curve and the permeabilities relative to both fluids, in the unsaturated case as well as in the saturated case which is a particular case of the latter. Finally, in the third section, we will address the issue of applying the model of real data, namely from images of natural porous media.

2. Mathematical Properties of the Pore Solid Fractal (PSF) Model

The three-parameter pore-solid-fractal model was imagined about twenty years ago by Perrier (1994) to overcome some theoretical inconsistencies noticed in classical fractal modeling in soil science, namely when the pore phase or the solid phase was vanishing with infinite iterations of the model. The goal was to actually build a geometrical model of a two-phase porous medium, and to avoid the mere use of power-law statistical relationships as proof of a fractal behavior (Baveye and Boast 1998). This model was later found to be equivalent to the multiscale percolation system developed independently by Neimark (1989) in a purely theoretical context, then Perrier et al. (1999) used Neimark's simpler and better formalism to establish a first list of generic properties for fractal models of soil structure. A series of papers were then published in the following years on the PSF model applications to soils (Bird et al. 2000, Perrier and Bird 2002, Bird and Perrier 2003, Perrier and Bird 2003). Several years later, the PSF model still remains an appropriate approach, at least as a didactical tool, to learn about fractal structure of porous media, since it includes most of previously developed fractal models as particular cases.

2.1 Definitions

In the classical Solid Mass Fractal model MF (see Fig. 1a) and in the Pore Solid Fractal PSF Model (see example in Fig. 1b), a squared area is divided into N parts, a proportion F of the N are subdivided in an iterative way, copying the pattern (or the generator) designed at the first iteration. The difference between the two models is that the $(1 - F)N$ subparts which are kept undivided in the Solid MF model represent pores, whereas in the PSF model we distribute the $(1 - F)N$ subparts into $(P)N$ pores and $S(N)$ solids. Obviously one has:

$$P + S + F = 1 \tag{1}$$

The copy can be done in a statistical way as illustrated in Fig. 1b, in which a random PSF has been obtained by assigning probabilities P, S and F to each subpart (probabilities to be respectively a white pore, a black solid, or gray matrix to be decomposed at next iteration).

The fractal dimension is in both cases easy to calculate, as for all simple self-similar fractal sets found in fractals textbooks. It is the log-log ratio of the number of smaller copies of the whole (NF) versus the similarity ratio $N^{1/d}$ which divides the linear size of the whole to give the linear size of each copy, as:

$$D = d\frac{\log NF}{\log N} \tag{2}$$

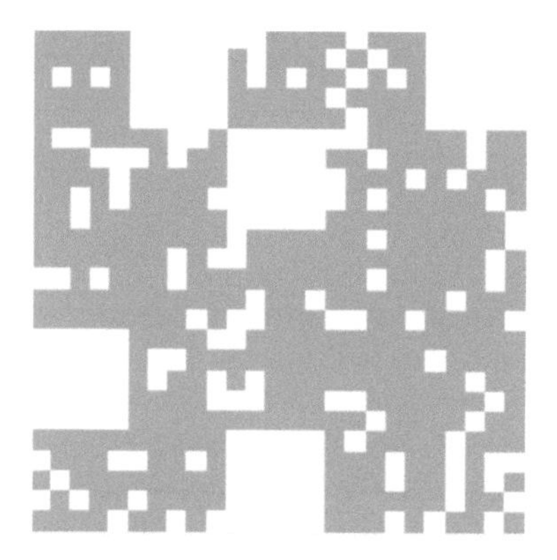

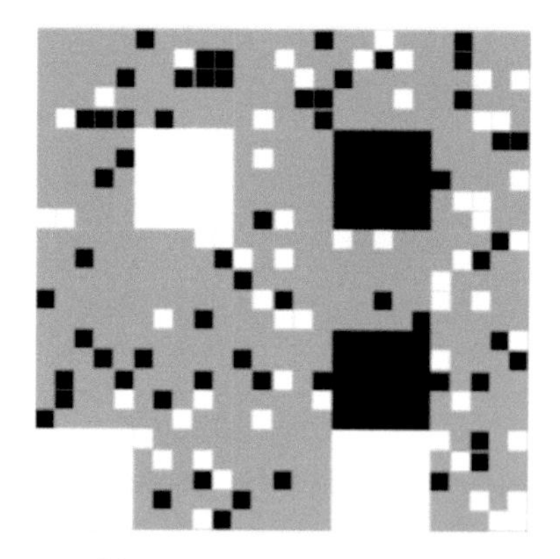

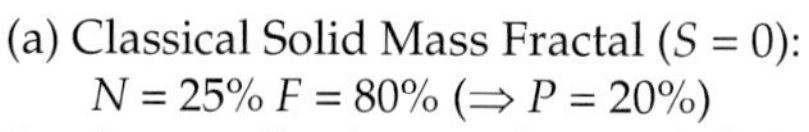

(a) Classical Solid Mass Fractal ($S = 0$): $N = 25\%$ $F = 80\%$ ($\Rightarrow P = 20\%$)

(b) Pore Solid Fractal $N = 25$ $F = 80\%$ $S = 10\%$ ($\Rightarrow P = 10\%$)

Random realizations with a probability $F = 0.80$ for each of the 25 subparts to be decomposed at each iteration. Here *iter* = 2 iterations.

Figure 1: Examples of fractal models of porous medium $\left(D = 2\frac{\log 20}{\log 25} \approx 1.86\right)$.

As a special case, when $S = 0$, the PSF model reduces to the classical solid mass fractal model (Solid MF), as shown in Fig. 1a and in Fig. 2b, a model which is widely used in soil science (e.g. Rieu and Sposito 1991, Crawford et al. 1995).

Another special case is when $P = 0$, causing the PSF model to reduce to the pore mass fractal model (Pore MF), which is just the dual of the Solid MF, which means that the solid and pore phases have been inverted. The Pore MF (see example Fig. 2c) is often used in geology to characterize porous rocks (e.g. Katz and Thompson 1985, Jin et al. 2015).

2.2 Number of Phases

In the context of the present chapter, we would like to emphasize the number of phases shown in Fig. 1, that is: Two white/gray phases in the Solid MF model (Fig. 1a) and three black/white/gray phases in the PSF model (Fig. 1b). The gray phase corresponds in both cases to the fractal part, which will vanish if we keep zooming in the image with successive iterations of the decomposition process. Similarly in Fig. 2, where the results of four iterations have been plotted, one has three black/white/gray phases in the general PSF model (Fig. 2a), and two phases in the special cases where $S = 0$ or $P = 0$, that is: two white/gray phases in the Solid MF model (Fig. 2b), and two black/gray phases in the Pore MF model (Fig. 2c).

Let us imagine now more iterations for the models plotted in Figs 1 and 2. It is obvious that, if $S = 0$ or $P = 0$, and if the fractal decomposition is iterated infinite times, the model becomes mono-phase. Each scale reveals

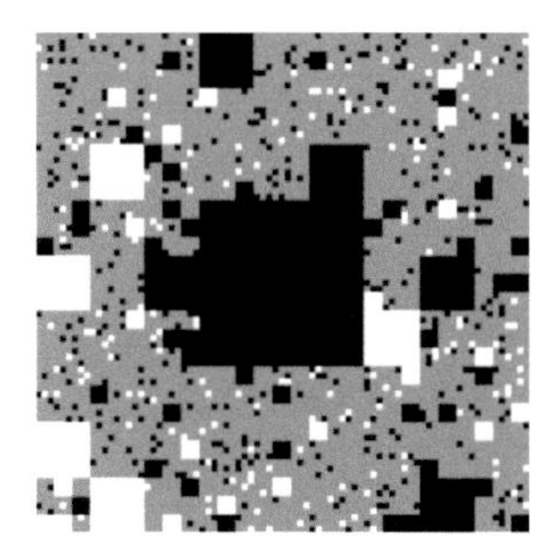

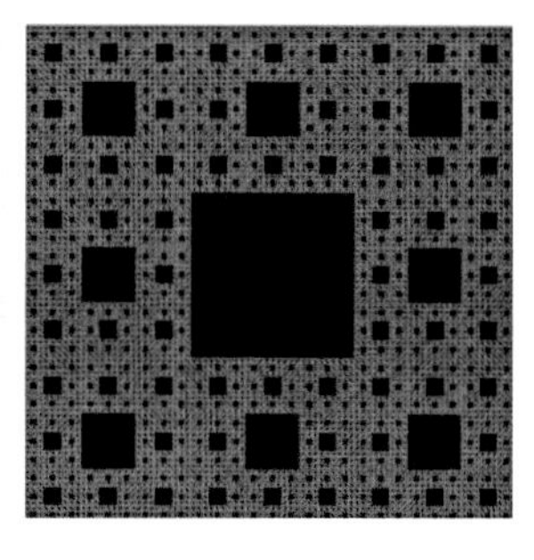

(a) $F = 88\%$ $S = 8\%$ ($P = 4\%$) $D = 2 \log (0.88*9)/\log 9 = 1.88\ldots$

(b) $F = 8/9$ $S = 0$ ($P = 1/9$) $D = 2 \log 8/\log 9 = 1.89\ldots$ N = 9, *iter* = 4 iterations

(c) $F = 8/9$ $S = 1/9$ ($P = 0$) $D = 2 \log 8/\log 9 = 1.89\ldots$

If the number *iter* of iterations tends towards infinity, the fractal part (colored gray) disappears: The three-phase PSF model (a) tends towards a two-phase model and the two-phase solid (b) and pore (c) mass fractal models tend towards a mono-phase model.

Figure 2: Number of phases in fractal models of porous media.

smaller objects of one phase and the mass of the second phase vanishes. The calculated porosity (respectively the density) would even reach 100% if the fractal model Solid MF (respectively the Pore MF) is used as pure mathematical objects. This is the first reason to consider that the strictly speaking PSF model with non-zero P and non-zero S might be a more realistic model of fractal porous medium, since it can be iterated infinite times to represent arbitrary small pore and solid sizes while matching a given experimental porosity.

When the model represents a real porous medium, one introduces cut-offs of scale to account for the larger and smallest observed pore sizes, and the truncated model is often called a prefractal (e.g. Kim et al. 2011).

Regarding the Solid MF and Pore MF truncated models, the introduction of a lower cut-off of scale enables the representation of a two-phase porous medium. The fractal part becomes the solid phase in the solid MF or the pore phase in the pore MF. The gray color is usually no more used to plot this fractal part because white and black colors are typically used to denote pores and solids.

Regarding the strictly speaking PSF model (that is $P \neq 0$ and $S \neq 0$), when it is truncated, three phases have to be considered, and the third, gray color, is still necessary to plot the fractal part. As mentioned in the Introduction section, this third phase becomes an abstraction of the matrix, which is considered as a continuous medium, not explicitely decomposed into pores and solids.

However, several modeling assumptions can be done as regards the porous material inside the matrix, which can be described according to other observed types of porous structures, using fractal or non-fractal models, continuous or discrete models. One modeling assumption out of many possible ones is that the matrix may be a scaled copy of the whole, which means that the conceptual model is an infinitely iterated PSF as before, even if one works on a finite number of model iterations for practical reasons.

Let us note here that if we work on computer simulations of PSF structures (see section 3) or if we work on real images of porous media (see section 4), it is impossible to deal with an infinite range of scales as with the mathematical approaches presented in the present section, so handling the third matrix phase will be unavoidable.

To conclude about the number of phases in the models,

If infinite iterations are carried on, the gray fractal part disappears and one gets the following results:

1. The Solid MF ($S = 0$) and Pore MF ($P = 0$) models become mono-phase models. Consequently, for $P = 0$, the Pore MF represents only a power law distribution of grain sizes while for $S = 0$, the Solid MF represents only a power law distribution of pore sizes.

2. The strict PSF model ($P \neq 0$ and $S \neq 0$) represents two phases: a solid one and a pore one.

If a finite number of iterations are carried on, the gray fractal part does not disappear, so it has to be interpreted, which is done as follows:

1. If $S = 0$, the truncated or prefractal Solid MF (respectively Pore MF if $P = 0$) becomes a two-phase model. The gray fractal part is interpreted as representing the solids (resp. the pores) and it is classically re-colored black (resp. white).
2. If $P \neq 0$ and $S \neq 0$ the truncated or prefractal PSF model is a three-phase model. The gray fractal part is considered as representing the matrix phase, which is added to the solid phase and the pore phase resulting from the definition of the model. Different modeling assumptions will be made later as regards the properties of such a matrix phase and its porosity at a lower scale.

2.3 Structural Properties

Several calculations can be done directly on the 3-phase PSF model, without the need to binarize it as for a classical porous medium model. The PSF is a pure theoretical porous medium, a simple mathematical object which can be used to do right calculations and to make proofs. A list of properties that have been proved by simple mathematical summations, recurrence, and logical arguments are what follows.

If the porous medium is well represented by an infinitely iterated PSF (or, since it is a strong assumption, if the porous medium was a PSF mathematical object) then:

1. The porosity Φ is constant at all scales: $\Phi = \dfrac{P}{P+S}$ and the density ρ is also scale invariant $\rho = \dfrac{S}{P+S}$ (Perrier et al. 1999)
2. The pore phase and the solid phase have a finite volume, which is, respectively ΦL^d and $(1 - \Phi)L^d$ where L^d is the total volume of the porous medium sample of linear size L in a space of Euclidean dimension d (Perrier et al. 1999). Note that neither the pore phase nor the solid phase are mass fractals.
3. The particle size distribution and the pore size distribution are power-laws with the same exponent $-D$ i.e. $N(= r) \propto r^{-D}$ and the cumulative number-size distributions are approximately the same power law $N(< r) \propto r^{-D}$ (Perrier et al. 1999, Bird et al. 2000).
4. If $D > d - 1$, the length ($d = 2$) or the area ($d = 3$) of the pore–solid interface approaches infinity as a power-law function of the resolution scale. This interface is fractal of dimension D (Perrier et al. 1999).

5. Considering a scale invariant fragmentation process of dimension D_f, one gets a power-law form for the cumulative distribution of fragment mass versus size, with exponent $d - D_f : M(<r) \propto r^{d-D_f}$ (Perrier and Bird 2002).

Finally, if the PSF is truncated, that is not iterated ad infinitum, and if the (third gray) matrix phase is assigned a constant porosity and density different from those given in (1), the total density and porosity of the porous medium is no more scale invariant (Bird and Perrier 2003).

2.4 Pore Size Distribution and Capillary Pressure Curve

Promising results were obtained under partially saturated conditions in fractal porous media for the expression of the capillary pressure curve (also known as water retention curve, particularly in soil science). If we assume a simple model based on a bundle of parallel capillary tubes[1] for the repartition of the two fluid-phases (see e.g. Jury and Horton 2004), and if the porous medium is a PSF, the analytical capillary pressure curve model is given by (Perrier et al. 1996, Bird et al. 2000):

$$\theta(h) = \Phi - \frac{P}{P+S}\left\{1 - \left(\frac{h}{h_{min}}\right)^{D-d}\right\} \text{ for } h_{min} \le h \le h_{max} \tag{3}$$

where θ is the wetting fluid content (cm^3/cm^3) within the capillary pressure range $[h_{min}, h_{max}]$ inversely associated with the maximum and minimum pore size values r_{max} and r_{min} respectively.

It has been shown (Perrier 1994) that Eq. (3) is valid even if pores of size smaller than r_{min} are added in the matrix phase of the truncated PSF model.

In the special case where $S = 0$, Eq. (3) reduces to the expression derived by Rieu and Sposito (1991): $\theta(h) = \Phi - \left\{1 - \left(\frac{h}{h_{min}}\right)^{D-d}\right\}$ whereas if the strict PSF is iterated ad infinitum Eq. (3) becomes the simple power law observed in many experimental studies $\theta(h) = \Phi\left(\frac{h}{h_{min}}\right)^{D-d}$.

Let us note that the previous analytical expressions have been calculated according to a simple model of parallel capillary tubes,

---

1 Using a simple capillary model (based on the physical Laplace law in parallel cylindrical tubes), when two non-miscible fluids are present in porous media (one of the two fluids is always more wetting than the other one), the wetting fluid (for example water if the second fluid-phase is air) is located in the pores smaller than a critical size inversely proportional to the pressure at the two fluids interface.

which means that these expressions are just another way to describe cumulative pore size distributions. More generally, mathematical models for fractal capillary curves are mere equivalents to fractal models of pore size distributions. They may be no more valid if the pores are no more considered as independent but interconnected in networks where hysteretical effects appear (see section 3).

Let us also note that capillary pressure curve fractal models are presented in Chapter 2 of the present book by Ghanbarian and Millan.

2.5 Permeability

Another important property of a porous medium is its permeability. It is a key property regarding fluid flow and transport phenomena.

Let us recall here that the value of the intrinsic permeability is independent of the fluids properties; it is a characteristic of the porous medium. It might be theoretically derived by integrating the spatial distribution of the pore phase but it is practically impossible to solve mathematically the equations of fluid mechanics in pore complex geometries, so the issue is mainly handled numerically. We are not aware of any mathematical calculation of the intrinsic permeability for fractal porous media and we refer to section 3 of the present chapter to consider numerical models.

Nevertheless, assuming that the intrinsic permeability has been determined experimentally, theoretical work has been done to model the scaling law of fluid relative permeabilities as a function of the intrinsic value for the porous medium and of the volumetric content for a given fluid. Rieu and Sposito (1991) introduced tortuosity parameters in the MF model to be fitted to experimental data. Hunt, Ghanbarian and collaborators (For example Ghanbarian and Hunt 2012, Hunt et al. 2013) employed critical path analysis from percolation theory to model the water relative permeability of partially saturated media (also known as unsaturated hydraulic conductivity). We present here only the results obtained for a special case when a PSF model iterated ad infinitum and porosity $\Phi = \dfrac{S}{P+S}$ volume fraction-based percolation threshold θ_t for simple unimodal and mono-fractal model (Ghanbarian et al. 2015):

$$K = K_S \left(\frac{\theta - \theta_t}{\Phi - \theta_t} \right)^{2+\frac{D}{3-D}} \quad (4)$$

The prediction of both the intrinsic and relative permeabilities will be also addressed in section II by means of computer simulation in three-phase networks.

2.6 Connectivity

The permeability of a porous medium is not easy to calculate mathematically because it depends not only on the pore size distribution but also on the way the pore space is spatially organized, or if we consider a discrete set of individualized pores, on the way the pores are connected. When the pores are disconnected in a given sample, there is no path for any fluid to traverse from one side of the sample to another side, meaning that the macroscopic permeability obviously equals zero. The connectivity can be defined as a lower order property, which equals 0 or 1 respectively when the medium is permeable or impermeable. If the connectivity equals 1, the exact non-null value of the permeability may remain unknown. The connectivity issue can be addressed more easily than the permeability issue: if one works on binarized images as shown on Fig. 3a, it is easy to scan the white and black pixels or voxels to check if a continuous path exists from one side to another one, using an invasion percolation algorithm to simulate fluid progression in the pore space. There exists a percolation threshold which depends on factors such as the Euclidean dimension, pore space geometry and topology. For example, for site percolation theory and a square lattice constructed of randomly distributed conducting sites, the system percolates as soon as the occupation probability p becomes greater than $p_c \approx 60\%$. More generally, using percolation theory, it has been shown that in random (infinite) porous media, there exists a critical threshold for percolation below which the probability to percolate equals 0 and above which the probability to percolate equals 1. Above but near the threshold, however, the sample-spanning cluster forms a connected pathway, and the porous medium properties (e.g., permeability) increase as power-laws with either universal or non-universal exponents.

The exact values of the percolation threshold can be calculated mathematically in very few and very simple cases. Another approach used by physicists consists in using renormalization theory on a network. It has been shown that the threshold is at least approximately the invariant point of a renormalization function. For example, for the squared bidimensional grid, the renormalization function is calculated by an upscaling method described in Fig. 3c on a square lattice constructed of randomly distributed sites, where cells are regrouped 4 by 4 in a recurrent way. Knowing p the probability for each cell to be locally conducting, the probability $f(p)$ for a set of four cells to conduct from left to right is calculated using all possible configurations (see Fig. 3b). In this example, the function $f(p)$ is $2p^2$-p^4 (see, e.g., Turcotte 1992) and the invariant point in this case is 0.618, only 3% greater than the actual value. The actual value is the one obtained for infinite networks, which are impossible to handle by computer simulation. Anyway, when mathematical derivations are not available, as in the present example of a simple bidimensional square

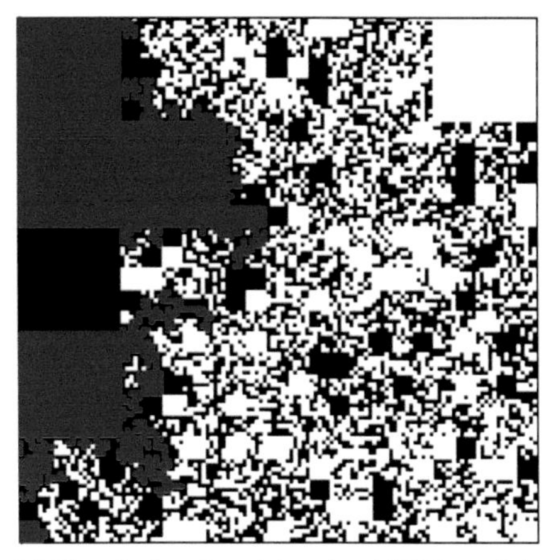

(a) Simulation in a binarized PSF model. Invasion of a (blue) fluid from the left

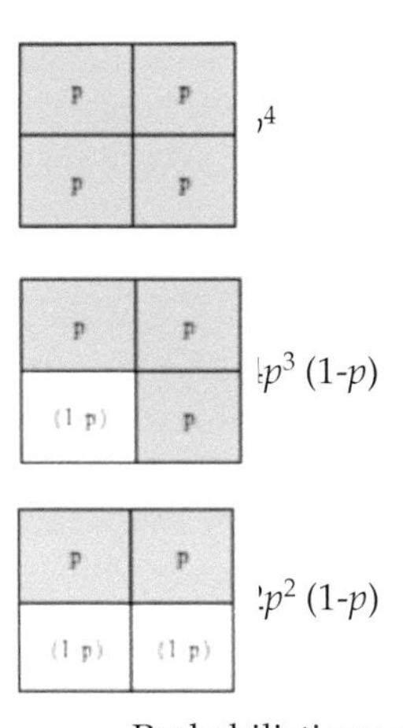

Probabilistic upscaling
$P = p^{4+} + 4p^3(1-p) + 2p^2(1-p) = p^2(2-p^2) = f(p)$

(b) Upscaling function on a squared grid

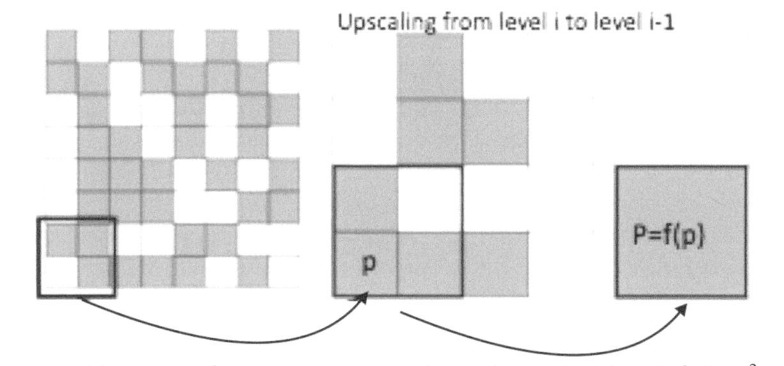

(c) Renormalization on a squared two-dimensional grid: $f(p) = p^2(2-p^2)$

Figure 3: A renormalization function $f(p)$ to estimate the percolation threshold.

lattice of conducting and non-conducting sites, the best estimation of the threshold is given by computer scientists! By increasing gradually the size of the lattice, more and more precise approximations of the threshold are reached (p_c = 0.59275...), and the number of reliable digits goes on increasing with the use of extensive computer simulations.

Some results have been obtained to model the connectivity of fractal models using percolation theory and adapted renormalization functions. Details can be found in related papers (Bird and Perrier 2010, Perrier et al. 2010). The main achievement is that the results are different when

2 This might give ideas to build impermeable artificial porous media with low level of solid material (especially if the material is expensive)...

the porous medium is multiscale. Namely in a mass fractal model, the renormalization function becomes $f'(p) = (1 - F) + F.f(p)$ where F is the fractal parameter given in Eq. (1), and a rather strange behavior has been demonstrated: Fractal media can have an arbitrary large porosity while being impermeable[2]. The renormalization function for the PSF model would be $f'(p) = P + F.f(p)$ and again the probability for such a porous medium to be impermeable or connected can be estimated a priori as a function of the parameters of the model.

Inspired by the renormalization approach, the first attempt to use upscaling methods in numerical simulations of the permeability of the PSF model will be presented in next section.

Fractals applications to percolation theory are presented in Chapter 5 of the present book by Hunt and Yu.

3. Computer Simulation of 1 or 2 Fluid Phases in 1 or 2 or 3-phase Fractal Models of Porous Media

3.1 Fluids in One-phase Models of Porous Media

Fluids flow through the voids of the porous medium, and numerous studies have been done considering only the pore phase, which can be modeled as a discrete set of pore objects. It has already been mentioned in sections 2.4 and 2.5 that an equivalent pore size distribution can be inferred from a capillary pressure curve and that conversely fluid properties can be estimated from pore size distributions. With computer simulations, it is possible to take into account the connectivity of the pore phase by constructing pore networks. For example, at the geological scale, a large amount of modeling has been done on fracture networks, in which fractures are represented by linear segments with different orientations, lengths and apertures to estimate the percolation threshold above which fluids flow is controlled by the fracture size (Berkowitz 2002). The number size distribution of fractures has often been found to follow a scaling power-law distribution, so they may be called fractals (see Bonnet et al. 2001 for a review). Let us note that fracture networks are discussed in details by Sahimi (2011). In soil modeling, many models are also based on pore networks. Some of them design pores with two types of void objects, using large pore bodies linked by thin throats or channels (see, e.g, Johnson et al. 2003, Matthews et al. 2006, Sholokhova et al. 2009, Raooh and Hassanizadeh, 2010), but such models have been applied in a monoscale context. They do not consider multiscale or fractal porous media. Pore network modeling was used by Perrier (1995) to simulate the hydraulic properties of Rieu and Sposito's fractal model of soils (1991), but the geometry of the model was not realistic enough, and as soon as the larger pores were empty, all the porous soil aggregates were disconnected and the hydraulic conductivity dropped to a null value.

Pore models classically avoid the representation of the solid phase, which appears not useful since the solid phase is just the dual of the pore space. More realistic pore networks can be extracted from binarized images where the solid phase has been identified (Delerue and Perrier 2002, Sholokhova et al. 2009, Raooh and Hassanizadeh, 2010) but once the pore network has been extracted, the solid phase is forgotten in flow modeling on the pore networks. A main problem with pore networks extracted from image analysis is that one can account only for the pores visible at a given resolution. For example in Delerue and Perrier's work (2002), as soon as the visible macropores were empty, the relative permeability or conductivity dropped to a null value and there was no way to account for the matrix conductivity. Another concern lies in the extreme simplification used to represent pore shapes, in order to use integrated analytical laws for the local conductance of spheres, cylinders, or parallepipeds. This is an intrinsic limitation of all pore network models, even if some numerical results are now available for some more complex pore shapes (Sholokhova et al. 2009).

3.2 Fluids in Two-phase Models of Porous Media

In order to account for the complex geometry of the actual continuous pore phase in real porous media, one has to carry out fluid flows simulations at the microscopic scale. Fluids flows in the pore phase are theoretically ruled by the Navier-Stokes equations, and various numerical solvers of these PDEs can be used. Whatever the method, one has to work on binary images of the porous medium, where the pore-solid interface gives the limiting conditions for fluid flow. Among the various methods used in the last decades (see Sahimi 2011 for a review) the Lattice Boltzmann models are increasing popular due to the increasing power of computer calculations and they are now used as a reference to test other approaches. Sholokhova et al. (2009) worked on monoscale images of sandstones and they concluded that the Lattice Boltzmann based methods "produce bulk absolute permeability values that fit published data more accurately" than pore body–channel network models. Such methods were also used in fractal porous media, the images of which are presented on Fig. 4 (Jin et al. 2015, Kim et al. 2011). We refer to the related papers to find valuable information on the numerical scheme.

Regarding the properties of the PSF model presented here in section 2, Kim et al. (2011) found that, for a similar given porosity, the intrinsic permeability of the strict PSF was higher than for the MF model. For both cases, the permeability was positively correlated with the porosity. Due to the time requirement for simulations of flow using the Lattice Boltzmann method, only some sets of parameters were investigated, on truncated 3-D

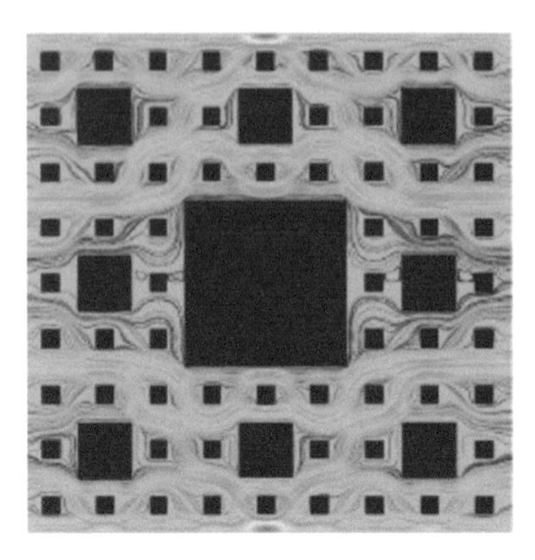

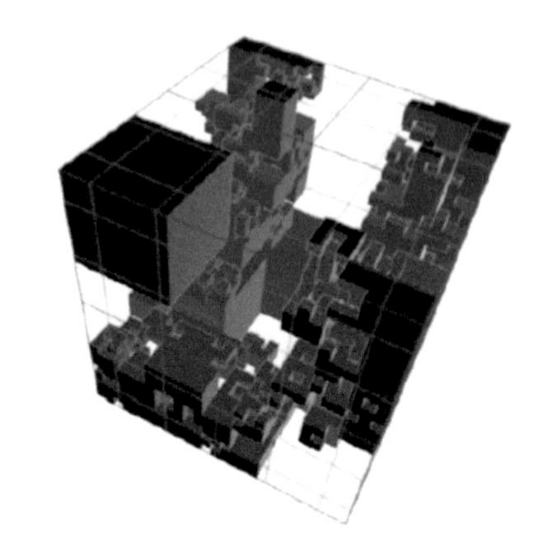

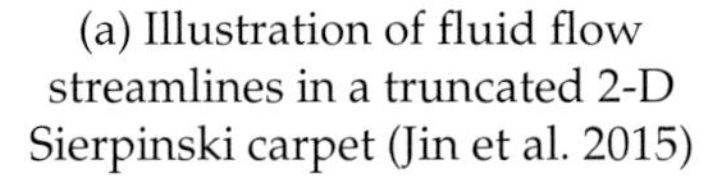

(a) Illustration of fluid flow streamlines in a truncated 2-D Sierpinski carpet (Jin et al. 2015)

(b) A 3-D truncated PSF model of porous media used for flow simulations by Kim et al. (2011)

Figure 4: Simulation of fluid flow in binarized fractal porous media using Lattice Boltzmann methods.

prefractal models constructed with three iterations and the hydrodynamic characteristics were estimated using 10 realizations for each set.

Computer time and memory requirements are obviously the main limitations of these studies conducted at the microscopic level, as far as heterogeneous and multiscale porous media are considered. For real data, Beckingham et al. (2013) have shown that the image resolution has a strong effect on the results since "large discrepancies in predicted permeabilities resulted from small variations in image resolution", and they had to tune the resolution of their CT 3-D images in a way that may be data-specific. For simulated data, Kim et al. (2011) had to consider fractal models where a few iterations were enough to reach the percolation threshold. When more iterations of the fractal model are needed, we suggest to take into account three phases of the porous medium as explained in following section 3.

3.3 Fluids in Three-phase Models of Porous Media

3.3.1 Construction of a Network Introducing the Third Matrix Phase

Going back to the network approach depicted in section 3.1, a three-phase network can be built, as illustrated here using the PSF model, nodes are created to handle three types of objects (Pore, Solid, Matrix). An example of such a network is given in Fig. 5a, in which, as shown, the nodes are points located at the center of each square. Pore and solid nodes represent pore and solid square objects colored white and black respectively. In the PSF, although there is a broad distribution of sizes, only three sizes are represented in Fig. 5a. Matrix nodes are located at the center of matrix squares, whose size is determined by the last level of the PSF model and which are colored gray. Once the network is created, it becomes an abstract object with only nodes and links (see Fig. 5b for a tripartite network with three types of nodes colored black, dark gray and light gray built from a random model).

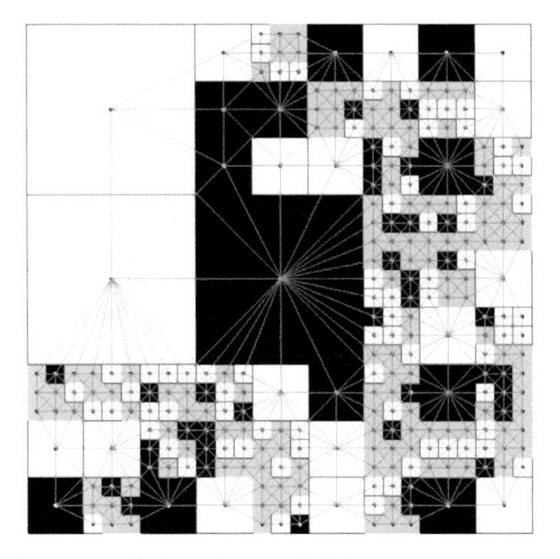

(a) A random realization of the PSF model (parameters: $N = 9$, $P = 30$, $S = 15$) and the 3-phase spatial network associated to three iterations of the model.

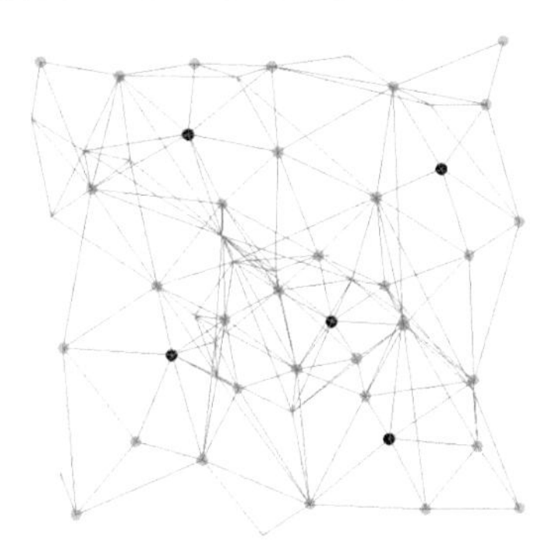

(b) A network abstraction (black solid, light gray pore and dark gray matrix nodes) handled in computer simulations

Figure 5: Illustrations of 3-phase networks.

3.3.2 Estimation of the Intrinsic Permeability with One Fluid

The permeability k is calculated in a classical way by simulation on the network by electrical analogy (see for example Perrier et al. 1995, Sholokhova et al. 2012). Similar to experimental measurements, the medium is first saturated by a given fluid in order to calculate the local fluxes q and macroscopic flux Q at an imposed pressure gradient ΔP

between two sides of a sample of length L and cross-sectional area A. Then the fluid property i.e., its viscosity μ is removed to determine the fluid-independent permeability (also known as intrinsic permeability) characterizing only the geometry of the porous medium and expressed in square meters (or Darcy). One gets the following equation:

$$k = \frac{Q\mu}{\Delta P}\frac{L}{A} \tag{5}$$

Using a classical network modeling approach, each node i is assigned a local conductance g_i. The local flux between two connected nodes i and j is calculated as a function of the pressure p_i and p_j at the nodes, of the distance l_{ij} between the nodes and of a mean conductance g_{ij} as follows:

$$q_{ij} = -\frac{g_{ij}}{\mu}\frac{p_i - p_j}{l_{ij}} \tag{6}$$

The mean conductance g_{ij} between two nodes is calculated here as the mere harmonic mean of the nodes conductances, to account for the fact that the fluxes are mainly driven by the lowest conductances encountered in fluid paths.

$$g_{ij} = l_{ij}\left(\frac{l_i}{g_i} + \frac{l_j}{g_j}\right) \tag{7}$$

To calculate the macroscopic flux Q and consequently the permeability k, one has to solve a linear set of n equations at n nodes filled with the fluid, where p_i and p_j are n unknown variables, and one equation is written for each conducting node i by imposing that the sum of positive input and negative output fluxes from neighbors j equals zero: $\sum_i q_{ij} = 0$.

In the present network approach, a local intrinsic conductance g_i $[L^4]$ is given to each node i. The originality is to assign conductances not only to the pores but also to all the components of the porous medium. For the solids, the local conductance is obviously $g_S = 0$. For the pore nodes, the conductance is approximated on the basis of the Poiseuille law, which has been established rigorously for cylinders or some other simple geometries: their local conductance g_P was selected as a power 4 of their size r: $g_P \propto r^4$. As regards the matrix nodes obtained after a finite number *iter* of iterations of the fractal model, they represent the porous medium at a lower scale, that is a mixture of pores and solids of smaller sizes, and their conductance g_F cannot be anymore approximated using the Poiseuille law. We consider three options:

Option 1. $g_F = g_P$. This means that we assign to the matrix objects of a given size the conductance of pores of the same size. In this case they will conduct more than expected and the result of the simulations will be higher than expected, that is we will get an upper bound k_{max} of the actual permeability k.

Option 2. $g_F = g_S$. This means that we assign to the matrix objects a null conductance. In this case they will obviously conduct less than expected and the result of the simulations will be smaller than expected, that is we will get a lower bound k_{min} of the actual permeability k.

These first two options are easy to handle, but they give only lower and upper bounds for k:

$$k_{min} \leq k \leq k_{max} \tag{8}$$

Anyway, the larger the number *iter* of iterations of the fractal model, the better the approximation of k as shown in Fig. 6.

In all simulated cases, when *iter* increases, k_{min} increases and k_{max} decreases, which is intuitively correct. However for many sets of parameters, it is impossible to reach an asymptotic value for k, due to the limitations of computer power.

Option 3. $g_F = g(k_F)$. This means that we will assign a local conductance g_F calculated as a function of the permeability k_F of the matrix. If k_F is

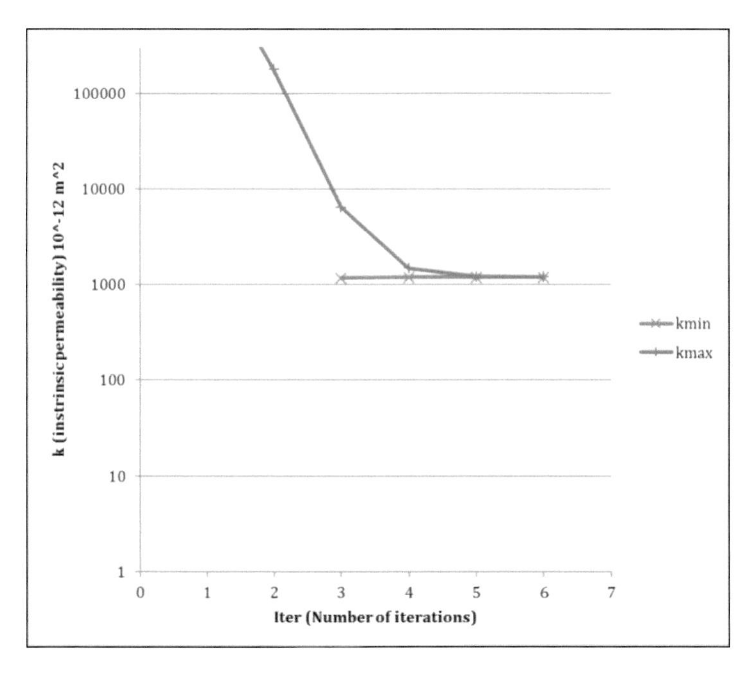

Plot of k_{min} and k_{max} as functions of the iteration level *iter* for the PSF example shown in Fig. 5 (*iter* = 3) and Fig. 6 (*iter* = 4)

Figure 6: Simulation of lower and upper bounds for the intrinsic permeability.

known, g_F is calculated using network simulations in an inverse way, that is g_F is the value given to each matrix node in order to obtain a macroscopic permeability equal to k_F for a network composed only of matrix nodes.

This third option is more interesting but more difficult to implement. We will just present here some formal arguments as follows:

(a) The value of the total permeability k is obviously dependent on the value given to the conductance g_F assigned to the matrix nodes, which is calculated numerically from the permeability of the matrix phase k_F. Thus the total intrinsic permeability k can be considered a function f_{sim} of k_F determined empirically by simulation, since it is obtained by a numerical integration of the local conductances on the network: $k = f_{sim}(k_F) = \int_{network} g_P, g_S, g_F(k_F)$

(b) The matrix objects in a fractal model are smaller copies of the whole. So their permeability should be obtained from the permeability k of the whole using some scaling function f_{scale}. Considering that the size of each matrix object obtained at a given iteration *iter* of the fractal model equals the size of the whole divided by $(N^{1/d})^{iter}$, and that the permeability k of the PSF model scales as power 2 of its linear size, our assumption is: $k_F = f_{scale}(k) = k/(N^{1/d})^{2*iter}$

These formal considerations lead to the following set of three equations (Eq. (10)) where Eq. (10c) is derived from Eq. (10a) and Eq. (10b):

$$k = f_{sim}(k_F) = \int_{network} g_P, g_S, g_F(k_F) \tag{10a}$$

$$k_F = f_{scale}(k) = k/(N^{1/d})^{2*iter} \tag{10b}$$

$$k = f_{sim}(kf) = f_{sim}(f_{scale}(k)) = (f_{sim} \text{ o } f_{scale})(k) \tag{10c}$$

Thus it might be possible to calculate k by iterative simulations, as the invariant point of the function f_{sim} o f_{scale}. So far such a function is impossible to calculate analytically, and there is no proof that an invariant point exists for all sets of parameters. In the example shown in Fig. 6, and at the third iteration where simulation runs are achieved within a few minutes, $k_{min} = 1172\ 10^{-12}\ m^2$ and $k_{max} = 6487\ 10^{-12}\ m^2$ (see Fig. 6). Iterative simulations give an invariant point $k = 1175\ 10^{-12}\ m^2$. This might be considered as an acceptable approximation of the best estimate $1203\ 10^{-12}\ m^2$ obtained with six iterations after runs of several hours. And this best estimate is also an invariant point of the numerical implementation of f_{sim} o f_{scale} at the sixth iteration of the model. But the numerical results depend strongly on the selected parameters and the connectivity of random realizations of the model. In some cases, the total permeability is driven by the matrix permeability, which remains undetermined, and in some other cases where the large pores are connected through a percolation path, the value of the conductance of the matrix has almost no effect on the total permeability. Further research has still to be done systematically to address this third theoretical option to account for the matrix permeability in three-phase networks.

3.3.3 Relative Permeabilities and Capillary Pressure Curves with Two Fluids

Let us now address the case of unsaturated porous media filled by two immiscible fluids.

The mathematical derivation of the associated capillary pressure curve based upon the power-law pore size distribution was addressed in section 2.4. In the simulation shown in Fig. 7, considering the effect of connectivity, water invasion into an oil-saturated porous medium is simulated according to the same capillary model, based on pore sizes. However, as can be seen in the illustration, some pores of small sizes are not accessible by the wetting fluid injected from the top of the simulated sample.

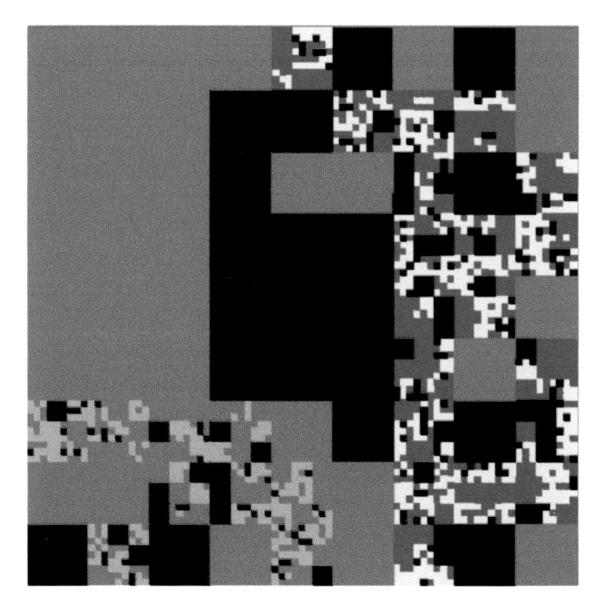

Water is colored light blue in the matrix and dark blue in the pores
The medium was initially oil-saturated (brown in the pore phase, light brown in the matrix phase)

Figure 7: Simulation of water invasion on the PSF network model shown in Fig. 5a with 3 iterations and here with 4 iterations.

In the network simulation, nodes have been given size attributes, but also neighbors. Under any given modification of the equilibrium state corresponding to a new capillary pressure at the two fluids interface, the nodes states (filled by the wetting fluid or by the non-wetting one) are exchanged only if they are connected to the invasion front. Hence, our simulations yield the hysteretic behavior of the capillary pressure curve shown in Fig. 8a, where the curve associated to the wetting fluid invasion (imbibition) is well below the curve simulated for the non-wetting fluid invasion (drainage). In the same way, during a drainage process, large pores are not accessible at low pressure values, and the wetting fluid content (water content in Fig. 8a) is much greater than that expected from a simplified capillary model ignoring the connectivity effect. Let us note first that the simulated capillary pressure curves of the PSF structure iterated only four times shows four distinct steps corresponding to four pore sizes, and that it also accounts for the pores hidden in the matrix porous objects. The porosity of the matrix objects has been assigned to be equal to $\Phi = \frac{P}{P+S}$ as explained in section 2.3, and the pore size in a matrix object of size l_i was approximated in an arbitrary way by $\sqrt{\frac{P}{P+S}}l_i$. In the way as for the pore objects explicitly represented in the network, they are reachable or not during an invasion process depending on the connectivity of the network. The theoretical curve obtained by means of continuous mathematical calculations (Eq. 3) is obviously smoothed[3]. Nevertheless the simulated data fits rather well on the pore size range corresponding to the fractal model above the truncation at the fourth iteration.

As regards the permeabilities relative to each fluid phase, they are calculated in the same way as in section 3.3.2, but on the sub-network of fluid-filled nodes, which is extracted from the whole network at each equilibrium state. One example of results is shown in Fig. 8b for the permeabilities relative to oil and water normalized by the total (intrinsic) permeability in the drainage process[4], using the best estimate of the matrix permeability and the associated matrix conductance as shown previously.

The results presented in this section remain mainly qualitative.

Further research should handle tridimensional networks, since this is the more realistic way to address the connectivity issue. Identifying

3 Let us note that these four steps are associated to four pore sizes in the square PSF model. Statistical extensions of the model could be done using decompositions into polygons instead of squares as done by Perrier et al. (1996) who built statistical extensions of the Rieu and Sposito (1991) cubic fractal model. This would add a variability of pore sizes at each iteration of the model, leading to smoothed simulated capillary pressure curves more comparable to the theoretical ones.

4 The superimposition of the curves obtained in the imbibition process would exhibit only a slight hysteresis.

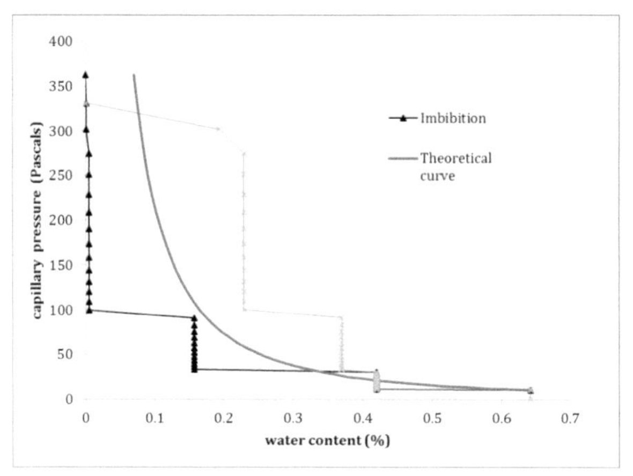

(a) Capillary pressure simulated during drainage and imbibition, with four steps each corresponding to the four iterations of the truncated PSF model. The theoretical curve has been calculated using Eq. (3).

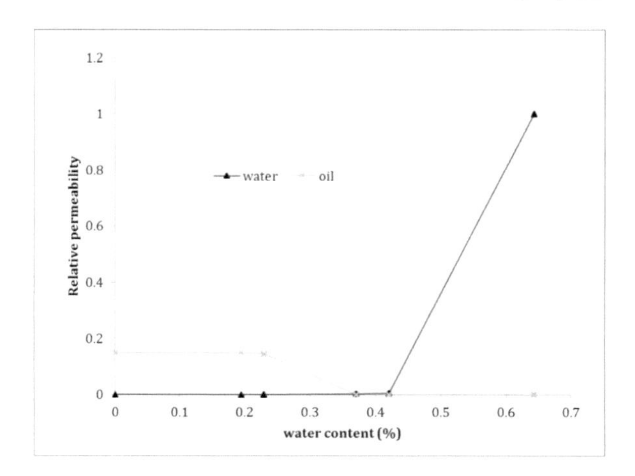

(b) Simulated relative permeabilities during a drainage process.

Figure 8: Simulated fluid properties of the PSF model shown in Fig. 7.

neighbors in a 3-D geometrical space is a more complicated task than in 2-D. Anyway, once the neighboring network is built in the computer memory, all algorithms would work in the same way regarding the simulation of permeabilities and fluid properties, since each node has local properties (type, shape, size, fluid state if any, etc.) and neighbors, independently of the actual space dimension. This would also allow the comparison with other simulation methods, namely those based on Lattice Boltzmann approaches (e.g. Kim et al. 2011).

Further quantitative research should also investigate the variability of the results in the 3-parameters space, and the variability for random realizations associated to each set of three parameters.

The first simulations described in this section indicate that accounting for a third phase in pore network models could assist to solve the fundamental upscaling issue, which is encountered in porous media modeling. As a matter of fact, limited-resolution images of porous media cannot cover all the scales encountered in reality, and if they could, the computer power will be always finite, as well as regards data storing or algorithms processing.

4. Applications

4.1 Multiscale Networks Created from 2-D Image Data

The same methodology can apply to real images provided that pore, solid and matrix parts can be distinguished. An arbitrary example is given in Fig. 9a, where only pores are visible, since there are no solids of comparable size. The objects could be extracted using dedicated contour algorithms used in image analysis; however in the present example the largest pores have been detected "by hand", just to describe the methodological approach. Their shapes are irregular and are modeled using polygons instead of squares as in the PSF model. Several algorithms have been developed to detect pore objects, their size and shape from a binary image. Those based on the skeleton of the pore space (see e.g., Lindquist et al. 2000, Delerue and Perrier 2002, Monga et al. 2007) also enable the extraction of the pore network including the connections between neighboring individualized pore objects. The methodology illustrated here on 2-D images would consist in also taking into account the dual of the pore space, that is the matrix space in the example shown in Fig. 9a. We also used a skeleton, but the skeleton of the matrix phase, in order to partition it into conducting matrix objects. They are built around the pore objects (and the solid objects if any) by decomposing the third continuous matrix phase into a set of discrete matrix objects. This is obtained automatically using a Voronoi tesselation algorithm with seeds located on the boundaries of the matrix phase, to extract the skeleton then create the matrix objects. The higher the number of seeds, the higher the number of created polygonal matrix objects, which can be slightly distinguished in Fig. 9b.

The algorithms presented in section 3.3 work on any network. Again porosity and permeability values have to be assigned to the matrix objects. Since the visible pores are disconnected vugs on the example shown in Fig. 9, the simulations on the network saturated with one fluid just confirm that the total intrinsic permeability is almost equal to the matrix permeability. We assume in this example that the matrix properties can be known by

(a) A photograph of a stone with macro and microporosity as pure methodological example (30×30 cm, 2448×2448 pixels)

There are two visible phases (matrix + pores) at the photograph resolution.

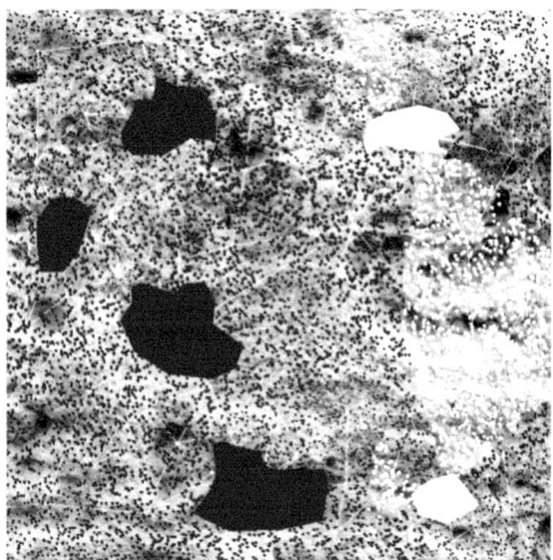

(b) Simulation of air invasion (colored white) in an initially water-saturated medium on a 2-phase network (pore and matrix).

The porous matrix areas were built using a Voronoï tessellation around the largest pores detected on the image: they are plotted using thin line segments colored white, visible behind the water colored blue.

(c) Simulated (hysteretic) capillary

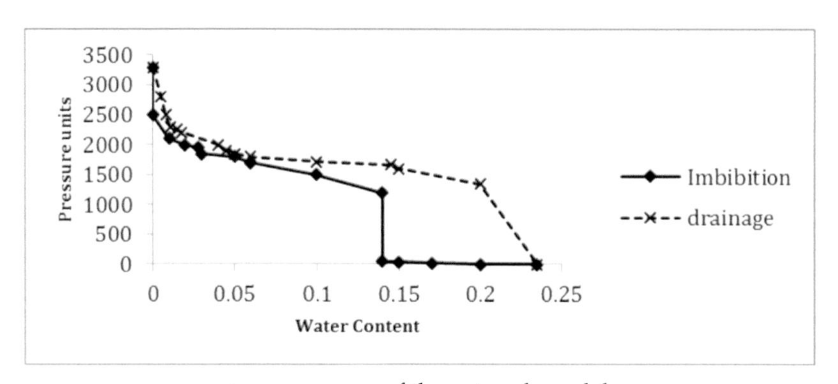

pressure curve of the network model.

Figure 9: A multi-phase network built from a 2D image of real porous medium.

experiments conducted at a lower scale. Regarding the simulation of fluid invasion on the network of pore and matrix objects (air invasion in an initially water-saturated medium in Fig. 9b), it is again done according to pore sizes and network connectivity.

Pore objects function as fluid reservoirs, which are reachable from the top side of the simulated sample only when the smallest pores inside the matrix have been filled with air, so a large hysteresis can be seen on the simulated capillary pressure curves shown in Fig. 9c.

The latter example considers a dual-porosity model through a network modeling approach. If fractal properties have been measured at the scale of a given image of porous medium, a fractal approach might consist in making copies of the whole into each matrix object, and to work in a recursive way as explained for the PSF model in section 3.3. More generally, such a multiscale approach using multiphase networks can rely on other data obtained at a lower scale to infer the properties of the matrix.

4.2 Pore and Solid Size Distributions

Despite the attempt presented in the previous section, fluid flow modeling in multiscale media derived from a single image obtained at a given scale may not be accurate. This is why models, including fractal models, have been introduced. The next question is: How to calibrate models from real data?

For this purpose, key parameters should be found in the power-law expressions of the cumulative pore and solid size distributions, which exhibit the same fractal dimension in the PSF model. Solid size distributions can be obtained using sieves of different scale or by laser-based techniques on undisturbed cores. According to the PSF theory, one could, theoretically, predict the capillary pressure curves from such solid granulometries (Bird et al. 2000). This would be a true and useful prediction; unfortunately no significant results have been obtained so far. In some soils, both the solid size distribution and the water capillary pressure curve can be well fitted by the power-laws analytical models associated with the PSF geometrical model, but the returned fractal dimensions are significantly different. One main reason may be that the data do not fit the PSF model, that is pore and solid do not follow the same scaling laws in such real soils: It is obvious that the reality is more complex than the idealized model, which should be further improved. Another possible reason is that reliable data are very difficult to obtain. For example, pore size distributions can be obtained using specifically designed porosimeters. But it would be tricky to say that such a porosimetry predicts the water capillary pressure curve, since this is an indirect measure, based on the measurement of another capillary pressure curve, where the invading fluid is mercury replacing air in the pore phase.

In the same way air replaces water in a drainage process, and the difference just lies in the ratio of interfacial tensions values. More generally, the derivation of water capillary pressure curves from pore size distributions is made using the simple model of parallel capillary tubes already mentioned, and conversely the pore size distributions calculated from any fluid pressure curve are only approximated. Pore size distributions can also theoretically be obtained directly by image analysis, but the results vary according to the algorithms designed to binarize the images, and according to the definition of pore sizes in complex geometries in 2-D or 3-D images. Moreover handling the whole range of scales in pore sizes remains a difficult issue. Improvements are thus also needed regarding data acquisition, since reliable pore size distributions could be used to better parameterize a model of porous medium and to try to predict its permeability or other properties (Matthews et al. 2006).

Regarding the modeling of the capillary pressure curve, by fitting various fractal-based models to measured capillary pressure curve, Ghanbarian et al. (2011) found that the PSF model described measurements over the entire range of saturation well, particularly at lower water contents.

Regarding the prediction of relative permeability, Hunt et al. (2013) concluded that "further investigation is required to study its prediction. In such an effort it should be possible to eliminate the uncertainty in interpretation by using more accurately measured pore-size distributions, e.g., from 3-D soil images".

4.3 Measure of Fractal Dimensions on Images

Many fractal dimensions have been calculated on images of porous media. Similarly many multifractal analyses have been done on porous media using associated Renyi dimensions (e.g. Dathe et al. 2006, Perrier et al. 2006). We will not present in this chapter the numerous results and applications of such analysis, but we will mainly discuss a methodological issue by focusing on the following question: Which is the object phase whose fractal dimension is calculated?

In some cases, a gray level image of porous medium is analyzed as a whole piece of informational data and the fractal dimension is calculated from the whole image. For example, Oleschko et al. (2008) and Torres-Arguelles et al. (2010) converted a bi-dimensional gray-level image into a one-dimensional array providing a time series (called firmagram) whose successive values are the gray intensity of successive pixels. The fractal dimension is then calculated from the Hurst exponent of such time series.

In some other cases, fractal dimensions are calculated on binarized images, measuring the area of the pore or solid phase at different resolutions through a box-counting method (Dathe and Thullner 2005,

Zamora-Castro et al. 2008, Perfect and Donnelly 2015). The pore and the solid phases exhibit different fractal dimensions which both contribute to the quantitative characterization of the porous medium. Measuring the area (or the volume in 3-D) of the pore or solid phase of the image implicitly rely on the mass fractal model presented in section 2.1. So the measure of the solid mass fractal dimension should help to predict the capillary pressure curve using the Rieu and Sposito model presented in section 2.4.

In other cases, the object whose fractal dimension is calculated is the pore-solid interface (e.g., Dathe et al. 2001, Dathe and Thullner 2005). Such a measure on binarized images should help to predict the capillary pressure curve (Eq. 3) using the generalized PSF-based model presented in section 2.4

Finally Oleschko's team from Mexico developed a fractal toolbox including a series of algorithms designed to obtain as automatically as possible most of the available fractal, multifractal, or other quantitative parameters descriptors which can be calculated from images of porous media (e.g. Torres-Arguelles et al. 2010).

Extensive studies have been based on fractal dimensions of the pore phase, solid phase, and pore-solid interface objects measured at different scales, some fractal dimensions are derived from pure pore and solid size power-laws distributions (even if the word fractal may not always be admitted in this latter case). Let us conclude this chapter based on the existence of a third conceptual phase in porous media by the following case-study exercise on the PSF model.

Here we address the issue of determining the fractal dimension from an image of the PSF model obtained with a finite number of iterations, as if it was an image of a real porous medium, that is an image obviously truncated for technical reasons, since the resolution cannot be infinite. An example is given as an illustration in Fig. 10. It deals with the three-phase PSF image already shown in Fig. 2a, obtained after four iterations, whose theoretical fractal dimension, given by construction, equals 1.88. To use a classical box counting method, one needs to decide which object one measures by covering it with boxes of different sizes, that is one needs to binarize the image to select the measured object and its complement. Three attempts are illustrated.

- In Fig. 10a, the object is the "strict" solid phase (which was colored black in Fig. 2). The log-log fit obtained using the box counting method is remarkably good ($R^2 = 0.9931$) and the estimated mass solid fractal dimension would be 1.73, which is considerably less than the actual fractal dimension theoretically determined for the same PSF model presented in Fig. 2a.

- In Fig. 10b, the PSF model was completed at the last iteration in the classical way by replacing the porous gray matrix phase with pores and solids in order to respect the same total theoretical porosity. To rephrase it, the matrix phase (which was colored gray in Fig. 2a) was binarized and replaced by the same proportion of pores (*P*/*P*+*S*) and solids (*S*/*P*+*S*) as the whole, in order to respect the theoretical porosity of the fully iterated PSF. The object measured in this case is the sum of the larger solids created after four iterations of the model plus smaller solids inside the matrix. The log-log fit obtained using the box counting method is also very good ($R^2 = 0.9996$). However, the calculated mass solid fractal dimension $D = 1.92$ is greater than the actual fractal dimension $D = 1.88$.

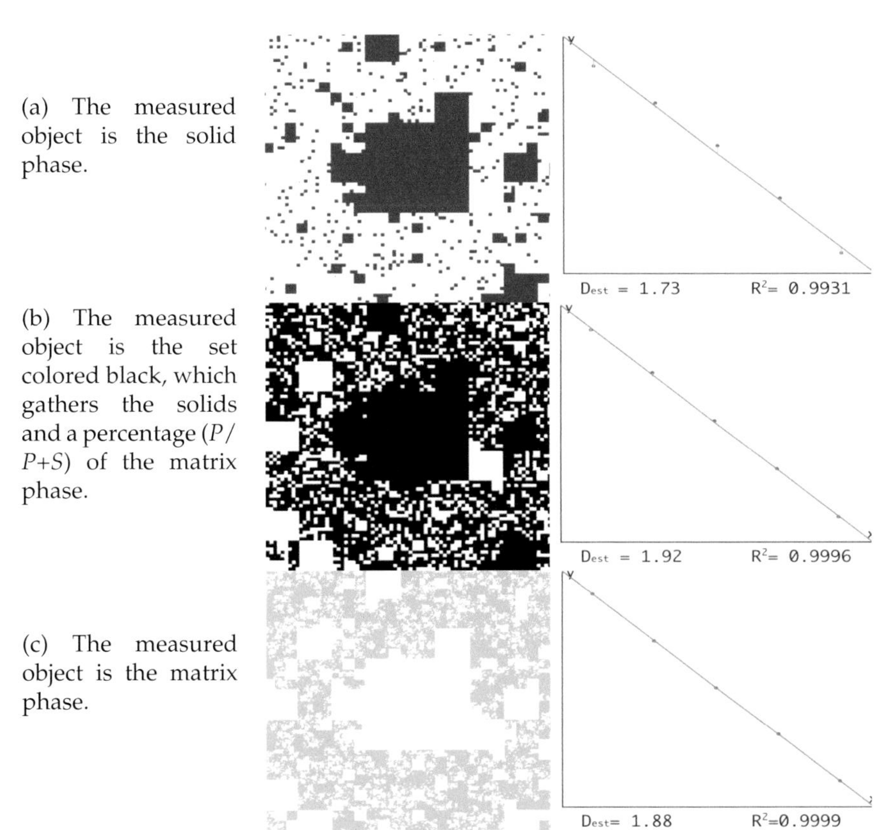

A fractal dimension is calculated using either two approximations of the solid phase (a) and (b) or the matrix phase (c) of the PSF model realization given in Fig. 2a.

Figure 10: Estimating fractal dimensions on different objects using the same box-counting method.

- In Fig. 10c, the measured object is the mere matrix phase. The log-log fit obtained using the same box counting method is again excellent (R^2 = 0.9999) and the calculated fractal dimension is 1.88.

Although all three fits are statistically excellent[5], only the fractal dimension determined from the gray matrix phase is very close to the expected theoretical value 1.88 given in Fig. 2a! These results appear logical, since the fractal part in the PSF model is actually the (vanishing) third phase, but such an object has never been used so far to estimate the fractal dimension. Such a new type of fractal analysis on real images would enable the parametrization of the PSF model from real data. The novelty would be to trinarize images[6] instead of binarizing them (as done in other fields e.g., by Higo et al. 2014) to avoid pure solid phase or pure pore phase.

Beyond the theory presented here, the main issue remaining open is: if we assume that some porous media, at least at some scales, could be well represented by a three-phase model, how to identify in a reliable way pore, solid and matrix phases in a real image? This requires further investigation.

To conclude, we have, in this chapter, mainly presented work done on the Pore Solid Fractal (PSF) model, and methodological discussions derived from this type of modeling point of view of a porous medium.

The explicit introduction of the third phase of the PSF model called matrix enabled the clarification on several theoretical issues as regards the modeling of the capillary pressure curve in fractal porous media. Rieu, whose legacy was acknowledged by Cheverry et al. (2003) said "the introduction of the gray part solved our problems: The real world is not always black and white, it can also be gray" (oral communication, 1999). And let us add here that the gray part is "the fractal part" in the PSF model, by definition.

This might have practical applications, namely regarding the determination of fractal dimensions, not only by measuring pore and size distributions, or pore-solid interfaces, but also by looking for new ways to analyze gray-level images. If the PSF is an appropriate model for a porous medium, its fractal dimension should be estimated on the image object obtained by removing the visible solids and pores and keeping only the matrix part. Such a fractal dimension would be useful to better calibrate the model from real data.

---

5 Many similar log-log fits are found statistically excellent in the literature about fractal models, maybe because measures varying as power-laws of scale are expected even for Euclidean objects.

6 It is expected that using two thresholds instead of one would lead to even more conceptual problems.

As regards fluids in porous media, exact results were obtained to predict the capillary pressure curve (when its hysteresis is neglected) as a function of the fractal dimension. The first attempt to calculate the intrinsic permeability of the PSF model by computer simulation on a network has also been presented in this chapter.

The on-going work presented here on three-phase networks open research paths to deal with multiscale porous media. Beyond the applicability of fractal models to improve our knowledge about the quantitative properties of porous media, such models might help to address the very difficult issue of upscaling methodology, in order to integrate data measured on a very large range of scales in complex porous media.

REFERENCES

Baveye, P., M. Laba, W. Otten, L. Bouckaert, P. Dello Sterpaio, R.R. Goswami, D. Grinev, A. Houston, Y. Hu, J. Liu, S. Mooney, R. Pajor, S. Sleutel, A. Tarquis, W. Wang, Q. Wei and M. Sezgin. 2010. Observer-dependent variability of the thresholding step in the quantitative analysis of soil images and X-ray microtomography data. *Geoderma* 157(1-2): 51–63.

Baveye, P. and P. Boast. 1998. Concepts of fractals in soil science: Demixing apples and oranges. *Soil Science Society of America Journal* 62(5): 1469–1470.

Beckingham, L.E., C.A. Peters, W. Umb, K.W. Jones and W.B. Lindquist. 2013. 2-D and 3-D imaging resolution trade-offs in quantifying pore throats for prediction of permeability. *Advances in Water Resources* 62: 1–12.

Berkowitz, B. 2002. Characterizing flow and transport in *fractured* geological media: A review. *Adv. Water Resour.* 25: 861–884.

Bird, N. and E. Perrier. 2003. The PSF model and soil density scaling. *European Journal of Soil Science* 54(3): 467–476.

Bird, N. and E. Perrier. 2010. Multiscale percolation properties of a fractal pore network. *Geoderma* 160(1): 105–110.

Bird, N., E. Perrier and M. Rieu. 2000. The water retention curve for a model of soil structure with Pore and Solid Fractal distributions. *European Journal of Soil Science EJSS* 55(1): 55–65.

Bonnet, E., O. Bour, N.E. Odling, P. Davy, I. Main, P. Cowie and B. Berkowitz. 2001. Scaling of fracture systems in geological media. *Review of Geophysics* 39: 347–383.

Cheverry, C., E. Perrier, P. Boivin, G. Vachaud and C. Valentin. 2003. Michel Rieu (1943-1999), his vision and his legacy. *European Journal of Soil Science*, 54(3): 439–442.

Crawford, J.W., N. Matsui and I.M. Young. 1995. The relation between the moisture release curve and the structure of soil. *Eur. J. Soil Sci.* 46: 369–375.

Dathe, A., E. Perrier and A. Tarquis. 2006. Multifractal analysis of the pore- and solid-phases in binary two-dimensional images of natural porous structures. *Geoderma* 134(3-4): 318–326.

Dathe, A., S. Eins, J. Niemeyer and G. Gerold. 2001. The surface fractal dimension of the soil-pore interface as measured by image analysis. *Geoderma* 103(1-2): 203–229.

Dathe, A. and M. Thullner. 2005. The relationship between fractal properties of solid matrix and pore space in porous media. *Geoderma* 129: 279–290.

Delerue, J.F. and E. Perrier. 2002. DXSoil, a library for image analysis in soil science. *Computers&Geosciences* 28(9): 1041–1050.

Delhomme, J.P. and G. de Marsily. 2005. Flow in porous media: An attempt to outline Georges Matheron's contributions. In: Space, Structure and Randomness, Bilodeau, Meyer, Schmitt (Eds). *Lecture Notes in Statistics*, Springer Ed. 183: 69–97.

Gerke, H.H. and M.T. Van Genuchten. 1993. A dual-porosity model for simulating the preferential movement of water and solutes in structured porous media. *Water Resources Research* 29(2): 305–319.

Ghanbarian-Alavijeh, B., H. Millan and G. Huang. 2011. A review of fractal, prefractal and pore-solid-fractal models for parameterizing the soil water retention curve. *Canadian Journal of Soil Science* 91(1): 1–14.

Ghanbarian-Alavijeh, B. and A. Hunt. 2012. Unsaturated hydraulic conductivity in porous media: Percolation theory. *Geoderma* 187–188: 77–84.

Ghanbarian, B., A. Hunt, T.E. Skinner, R.P. Ewing et al. 2015. Saturated dependence of transport in porous media predicted by percolation and defective medium theories. *Fractals* 23: 1540004.

Higo, Y., F. Oka, R. Morishita, Y. Matsushima and T. Yoshida. 2014, Trinarization of lX-ray CT images of partially saturated sand at different water-retention states using a region growing method. *Nuclear Instruments and Methods in Physics Research* B324: 63–69.

Hunt, A., B. Ghanbarian and K. Saville. 2013. Unsaturated hydraulic conductivity modeling for porous media with two fractal regimes. *Geoderma* 207–208: 268–278.

Jin, Y., Y.B. Shu, X. Li, J.L. Zheng and J.B. Dong. 2015. Scaling Invariant Effects on the Permeability of Fractal Porous Media. *Transport in Porous Media* 109: 433–453.

Johnson, A., I.M. Roy, G.P. Matthews and D. Patel. 2003. An improved simulation of void structure, water retention and hydraulic conductivity in soil with the Pore-Cor three-dimensional network. *European Journal of Soil Science* 54: 477–48.

Jury, W. and R. Horton. 2004. Soil Physics, 6th edition. John Wiley and Sons.

Katz, A.J. and A.H. Thompson. 1985. Fractal sandstones pores: Implication for conductivity and pore formation. *Phys. Rev. Lett.* 54 _12: 1325–1328.

Kim, J.W., M.C. Sukop, E. Perfect, Y.A. Pachepsky and H. Choi. 2011. Geometric and Hydrodynamic Characteristics of Three-dimensional Saturated Prefractal Porous Media Determined with Lattice Boltzmann Modeling. *Transport in Porous Media* 90: 831–846.

Lindquist, W.B., A. Venkatarangan, J. Dunsmuir and T.F. Wong. 2000. Pore and throat size distributions measured from synchrotron X-ray tomographic images of Fontainebleau sandstones. *Journal of Geophysical Research: Solid Earth* (1978–2012), 105(B9): 21509–21527.

Matheron, G. 1967. Elements pour une théorie des milieux poreux, Ed. Masson, Paris.

Matthews, G.P., C.F. Canonville and A.K. Moss. 2006. Use of a void network model to correlate porosity, mercury porosimetry, thin section, absolute permeability, and NMR relaxation time data for sandstone rocks. *Physical Review* E 73(3).

Mei, M., B. Yu, J. Cai and L. Luo. 2010. A hierarchical model for multi-phase fractal media. *Fractal-complex Geometry Patterns and Scaling in Nature and Societies* 18(1): 53–64.

Monga, O., F. Ndeye and J.-F. Delerue. 2007. Representing geometric structures in 3-D tomography soil images: Application to pore-space modeling. *Computer&Geosciences* 33(9): 1140–1161.

Neimark, A.V. 1989. Multiscale percolation systems. *Sov. Phys.-JETP* 69: 786–791.

Oleschko, K., G. Korvin, A. Munoz, J. Velazquez, M. Miranda, D. Carreon, L. Flores, M. Martınez, M. Velasquez-Valle, F. Brambila, J.-F. Parrot and G. Ronquillo. 2008. Mapping soil fractal dimension in agricultural fields with GPR. *Nonlin. Processes Geophys.* 15: 711–725.

Perfect, E. and B. Donnelly. 2015. Bi-phase box counting: An improved method for fractal analysis of Binary images. *Fractal-complex Geometry Patterns and Scaling in Nature and Societies*, 23(1).

Perrier, E. 1994. Structure géométrique et comportement hydrique des sols. Simulations exploratoires. Thèse Université Paris VI 1994, 250 pages (see also Ed. Orstom Collection Etudes et Theses 1995).

Perrier, E., C. Mullon, M. Rieu and G. de Marsily. 1995. Computer construction of fractal soil structures: Simulation of their hydraulic and shrinkage properties. *Water Resources Research* 31(12): 2927–2943.

Perrier, E., M. Rieu, G. Sposito and G. de Marsily. 1996. Models of the Water Retention Curve for soils with a fractal pore-size distribution. *Water Resources Research* 32(10): 3025–3031.

Perrier, E., N. Bird and M. Rieu. 1999. Generalizing the fractal model of soil structure: The PSF approach. *Geoderma* 88: 137–164.

Perrier, E. and N. Bird. 2002. Modelling soil fragmentation: The PSF approach. *Soil and Tillage Research* 64: 91–99.

Perrier, E. and N. Bird. 2003. The PSF model of soil structure: A multiscale approach. In: Scaling methods in soil physics. Radcliffe, Selim, Pachepshy (Eds). CRC Press, pp. 1–18.

Perrier, E., A. Tarquis and A. Dathe. 2006. A Program for Fractal and Multifractal Analysis of Two-Dimensional Binary Images: Computer Algorithms versus Mathematical Theory. *Geoderma* 134(3–4): 284–294.

Perrier, E., N. Bird and T. Rieutord. 2010. Percolation properties of 3-D multiscale pore networks: How connectivity controls soil filtration processes. *Biogeosciences*, 7(10): 3177–3186.

Raooh, A. and S.M. Hassanizadeh. 2010. A new method for generating Pore-Network Models of Porous Media. *Trans Porous Med* 81: 391–407.

Rieu, M. and G. Sposito. 1991. Fractal fragmentation, soil porosity, and soil water properties. I. Theory & II. Applications. *Soil Sci. Soc. Am. J.* 55: 1231–1238 and 1239–1244.

Sahimi, M. 2011. Flow and Transport in Porous Media and Fractured Rock. From Classical Methods to Modern Approaches, 2nd Edition 2011, Wiley, 733 pp.

Sholokhova, Y., D. Kim and W.B. Lindquist. 2009. Network flow modeling via lattice-Boltzmann based channel conductance. *Advances in Water Resources* 32: 205–212.

Torres-Arguelles, V., K. Oleschko, A. Tarquis, G. Korvin, C. Gaona, J.-F. Parrot and E. Ventura-Ramos. 2010. Fractal Metrology for biogeosystems analysis. *Biogeosciences* 7: 3799–3815.

Turcotte, D.L. 1992. Fractals and Chaos in Geology and Geophysics. Cambridge.

Yu, B., J. Cai and M. Zou. 2009. On the physical properties of apparent two-phase fractal porous media. *Vadose Journal Zone* 8(1): 177–186.

Zamora-Castro, S.A., K. Oleschko, L. Flores, E. Ventura and J.-F. Parrot. 2008. Fractal Mapping of Pore and Solid attributes. *Multiscale Mapping, Vadoze Zone Journal* 7(2): 473–492.

Zhou, H., E. Perfect, L. Baoguo and L. Yizhong. 2010. Comment on "On the physical properties of apparent two-phase fractal porous media". *Vadose Journal Zone* 9(1): 192–193.

CHAPTER

4

The Mathematics of Complexity in the Study of Soil and Granular-porous Media

Miguel Ángel Martín
Departamento de Matemática Aplicada, Escuela Técnica Superior de Ingenieros Agrónomos, Universidad Politécnitca de Madrid, 28040 Madrid, Spain
miguelangel.martin@upm.es

1. Introduction: Granular Media as Complex Systems

Soil, sediments, rocks and powders produced by grinding process are some examples of media constructed of solid particles of various sizes and shapes. Due to their broad applications in different fields and research areas e.g., geology, porous catalysts, soil and material sciences, and industry, investigating mathematical heterogeneity and complexity of these types of granular porous media is of great interest. However, heterogeneity and complexity, both understood in an ambiguous sense, are traditionally claimed to be ubiquitous at all scales in these kinds of systems and their modelling.

The complex systems approach is commonly applied to a system with a large amount of components which, by means of repeated interactions among its constituents, build up a state with novel macroscopic properties. According to the theory of no equilibrium thermodynamics (see Prigogine 1945), the balance of entropy production in dissipative systems may lead to stationary states or intermediate emergent structures characterized by an entropic or organization level: the maximum level of disorder that the constrains allow the system (maximum entropy principle). As a result of optimality and randomness those are characterized by scaling forms (or events) over a wide range of length (or time) scales: they are fractal patterns.

Iteration is a key feature in the evolution of systems, and fragmentation falls within iterative processes. The structures, which form under iterative processes, may be far from disordered forms. Randomness, which is commonly present in fragmentation processes in nature, is, however, the other key component for the complex evolution. Iterative actions, randomness, and optimality primary principles are main ingredients in complex systems. Fragmentation energetic laws have an important role in fragmentation processes. Theoretical results in this respect have been found by Bertoin and Martínez (2005) who demonstrated a power law dependence with respect to the size of the particle and the energy needed to fragment it. The total energy available is a natural constrain. In fact entropy maximization methods have been used to explain the nature of size distributions by sudden breakage (Englman et al. 1988).

All these facts are consistent with an explanation of fragmentation processes under the point of view of the complex system approach. But, what about the resulting emergent structures? The simple visual perception indeed suggests a special geometrical arrangement. This configuration was termed *granulography* in Andreasen and Andersen (1930) where it is seen as a characteristic geometric feature of grain distribution or "granulometry" along different ranges of grain sizes. Using modern terms, this means "characteristic heterogeneity" which, as we discuss later, is directly related to entropic or organization level. Also, behind that geometrical configuration one may find the essential meaning of the, now called, "geometric self-similarity" appearing as an "emergent property" in complex approach terms.

In what follows, we apply different theoretical elements and tools of the Mathematics of Complexity in order to better understand what kind of heterogeneity is expected in soils and granular media, how to measure it, and how heterogeneity may be characterized and simulated.

2. Patterns of Size Distributions

As particulate systems, investigating the particle size distribution PSD (also known as grain-size distribution) – the distribution that defines the relative amounts of particles present and sorted according to size – is essential. As a matter of fact, soil hydraulic properties, and physical and chemical interactions at the pore-solid interface in granular materials and ceramics, are closely related to the PSD (see e.g., Bittelli et al. 1999, Segal et al. 2009).

Here our focus is on real granular media (natural or artificial ones) whose PSD is broad, so that it may be supposed that all grain sizes under a certain size are possible. Obviously, a single pattern (or a monomodal distribution) does not describe the entire particle size distribution in natural materials (see e.g., Wu et al. 1993, Bittelli et al. 1999, Miller and Schaetzl 2012). In the literature, several probability distributions, such as power-law or pareto (e.g., Tyler and Wheatcraft 1992) and log-normal (e.g., Hwang and Choi 2006) have been used to fit to PSD depending on the nature of the particle system. Thus, one may develop ideal models supported more or less by a stronger rational basis, which are used to study the fitness of sample data to those models. Its reliability should rest in some kind of compromise between the simplicity of the model hypothesis and the properties of the PSD, which is able to fairly (efficiently) predict.

Some of the models were postulated on the basis of geometric features observed in such materials. In the pioneering work (Andreasen and Andersen 1930), the differential equation

$$\frac{dQ}{d(\log d)} = \alpha Q \tag{1}$$

was proposed as a semi empirical model for the cumulative mass-size distribution function $Q(d)$ of certain granular products with grain size below a given limit. The differential equation is formulated for those products whose grain-size distribution conformed in such a way that by adding a portion of greater grains, the resulting product (grain distribution) is geometrically similar to the previous one, so that, a photography of both products seem equal (they have the same *granulography*, the term those authors used). It was empirically shown that this ideal property, mathematically formulated by the equation (1), is closely followed by granular materials produced by a grinding process and, further, the parameter α serves to predict the void fraction of the packing structure.

The *hyperbolic distribution*, formalised mathematically by Barndoff-Nielsen (1977), was originally developed by Bagnold (1941) where it was found that the logarithm of the histogram of the experimental PSD of sand deposits resembles a hyperbola. Bagnold and Barndorff-Nielsen (1980) later proposed such a model for the size distribution of sediments. There also dynamical explanations for the occurrence of hyperbolic distribution are given. The influence of factors, as transport of particles by wind and water, on the final pattern of PSD is explained on the basis of theoretical results previously established by Barndoff-Nielsen (1977, 1978). More recently, the skew log-Laplace model was proposed (see e.g., Fieller et al. 1984) as an alternative to the hyperbolic distribution for soil sediments PSD.

On the other hand, in many different natural scenarios, the fragment size distribution has been shown to follow fractal or power-law behaviour (Mandelbrot 1982, Turcotte 1986). In soils, the number and mass-size distributions of particles and/or aggregates scale as $N(x > X) \approx X^{-\alpha}$ and $M(x < X) \approx X^{\beta}$ in which $N(x > X)$ and $M(x < X)$ are the cumulative number and mass-size distributions of soil aggregates and/or particles of size x

greater or less, respectively, than a characteristic size X (see Anderson et al. 1998).

Several models have been proposed to rationalize such empiric evidence. Thus, multiplicative models, together with large deviation theory, have been used to understand these observations suggesting that the observed power-law distributions of fragment sizes should correspond to the superposition of probability density functions that are log-normal in the "centers" and take the power-law form in the "tail". It would be the result of a natural mixing of simple multiplicative processes that take place along the fragmentation of different particles (Frish and Sornette 1997, Sornette 2000). However, as indicated in the above references, there is no accepted theoretical description.

The model of Andreasen and Andersen (1930) for granular media also leads to a fractal (power-law) scaling, which is shown to be closely followed by grain size distribution of granular materials produced by a grinding process. Interestingly, it is the first recognized fractal behaviour reported 50 years before the term *fractal* was coined by Mandelbrot (Mandelbrot 1983).

The fractal nature of PSD may lead to similar hierarchical arrangement of the intergranular pore space. This fact has strong influence on transport processes as well on the distribution of other constituents that might be incorporated or dispersed. In the case of soil, for instance, the spatial variability of organic matter, chemicals, nutrients and pollutants have been shown to follow high complexity patterns (Kravchenko et al. 1999, Lehmann et al. 2008).

3. Measuring Heterogeneity: The Entropic Level

The first important issue is addressing the heterogeneity measurement with adequate concepts from the Theory of Information. This theory deals with the quantification of information in any system and rests on the concept of information entropy introduced in the pioneering work of C.E. Shannon (1948). We shall use it for giving a precise meaning to the term heterogeneity commonly used in an ambiguous way.

For simplicity let us assume that the unit square S in two dimensions is the support of a distribution μ with highly heterogeneous features. The entropy analysis for PSD will be addressed later.

In order to investigate the heterogeneity of the unit square S, we consider a collection (mesh) of $2^k \times 2^k$ ε-boxes, $P_\varepsilon = \{R_i: i = 1, 2,\ldots, 2^{2k}\}$, of side length $\varepsilon = 2^{-k}$, representing a partition of S for each value k, $k = 1, 2, 3, \ldots$ (see Fig. 1).

When the mass $\mu(R_i)$ inside any box R_i is known, the *Shannon entropy* (Shannon 1948) of μ with respect to a fixed partition P_ε is given by

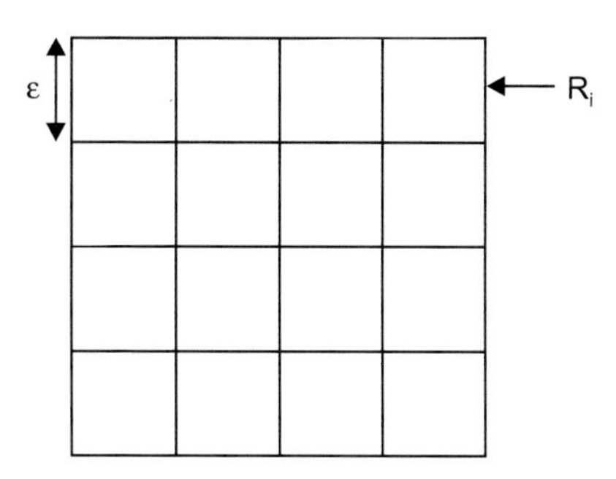

Figure 1: Partition of the support S by squares of side length ε.

$$H_\mu(P_\varepsilon) = \sum_{i=1}^{2^{2k}} \mu(R_i) \log \mu(R_i) \tag{2}$$

provided $\mu(R_i) \log \mu(R_i) = 0$ if $\mu(R_i) = 0$.

The number $H_\mu(P_\varepsilon)$ is expressed in information units (bits) whose boundary values are log 2^{2k}, which corresponds to the most even (homogeneous) case – where all the squares have the same cumulative mass – and 0, which corresponds to the most uneven (heterogeneous) case – where the whole mass is concentrated in a single square. The Shannon entropy $H_\mu(P_\varepsilon)$ is widely accepted measure of evenness or heterogeneity in the mass distribution μ at the scale level given by each partition P_ε. In fact, it can be shown that any measure of heterogeneity with the natural properties for such goal must be a multiple of $H_\mu(P_\varepsilon)$ (Khinchin 1957).

By increasing the values of k (or decreasing the values of ε) one can obtain an increasing amount of information about the distribution as $H_\mu(P_\varepsilon)$ tends to infinity. If such increase is not erratic, but rather conforms to a scaling or asymptotic behaviour of $H_\mu(P_\varepsilon)$ when $\varepsilon \downarrow 0$, then the *entropy or information dimension* of μ is defined (Rényi 1957) as

$$D \approx \frac{-H_\mu(P_\varepsilon)}{\log \varepsilon} \tag{3}$$

where "≈" means that – $H_\mu(P_\varepsilon)$ would be linearly proportional to log ε.

The entropy dimension gives account of the evolution of heterogeneity along scales. It may be seen as the entropic or organization level of the complex structure: the greater the entropy dimension is, the higher is the degree of uncertainty in the system.

4. Why Should Emergent Heterogeneity Structures be Expected?

Predictive algorithms, empiric facts, and models are needed in order to better understand why heterogeneity is produced and what kind of heterogeneity should be expected. In the following we discuss these issues by giving some examples corresponding to different aspects of soil complexity. As we demonstrate, they shed light on different reasons yielding emergent heterogeneity structures.

Fragmentation algorithms may help in understanding the origin of emergent structures in the PSD. Kolmogorov (1941) mathematically showed that the asymptotic PSD conforms to the log-normal distribution, if random rules for the fragmentation process are independent of the ratio between the size of the particle and the size of the particle obtained from it. However, questions raised in Kolmogorov´s paper about what PSD could be expected when such ratio has a power-law dependence on the size of the particle, remains still unanswered. Nonetheless, fragmentation algorithms inspired in Kolmogorov´s question have recently been used (Martín et al. 2009, 2015a).

To quantify heterogeneity, Martin et al. (2009) considered the following algorithm: let N be a non-negative natural number greater than one and $\alpha \geq 0$. Each particle of size r is divided in k smaller particles of size $\frac{r^{1+\alpha}}{k}$, given that k is a number chosen randomly between 1 and N, with equal probability for all the possible choices. The ratio between the size r of a particle and the size of the particles obtained from it is proportional to r^{α}, certain power of r, as suggested by Kolmogorov (1941). For $\alpha = 0$ the resulting sample distribution was log-normal. However, for $\alpha = 0$ results showed distributions far from log-normality indicating, on the contrary, great structured complexity for simulated PSD that may be characterized by entropy like parameters.

Recalling the results in Bertoin and Martínez (2005), a curious and interesting coincidence is noted between the power-law dependence with respect to the size of a particle of energy needed for fragmentation and the question posed by Kolmogorov mentioned above. In the opinion of the author, it deserves a great amount of research linking both aspects in order to give a wider answer to the resulting size distribution.

The transference of PSD complexity to the pore space is naturally expected since obviously pore size distribution should be highly influenced by the PSD. In order to characterize pore space heterogeneity different methods have been used.

Valuable and detailed information about the pore space geometry can be provided by 2-D image analysis of thin soil sections. Scaling analysis can be used to characterize the heterogeneous 2-D spatial arrangement of solid and void phases (Muller and McCauley 1992, Posadas et al. 2003) and the complexity of pore size distribution (Caniego et al. 2001, 2003).

Recent advances in visualization techniques such as X-ray computed tomography (CT) (see e.g., Peyton et al. 1994) has led to multiple cross-sectional images of soil samples. Cylindrical soil columns may be scanned and a large number of images of the sections perpendicular to the axis of the cylinder obtained which led to a 3-D structure of soil pore space.

The analysis of series representing the 2-D sectional porosity *vs.* depth (Fig. 2) shows a complex behaviour. In particular these series present structured randomness with memory (San José et al. 2011).

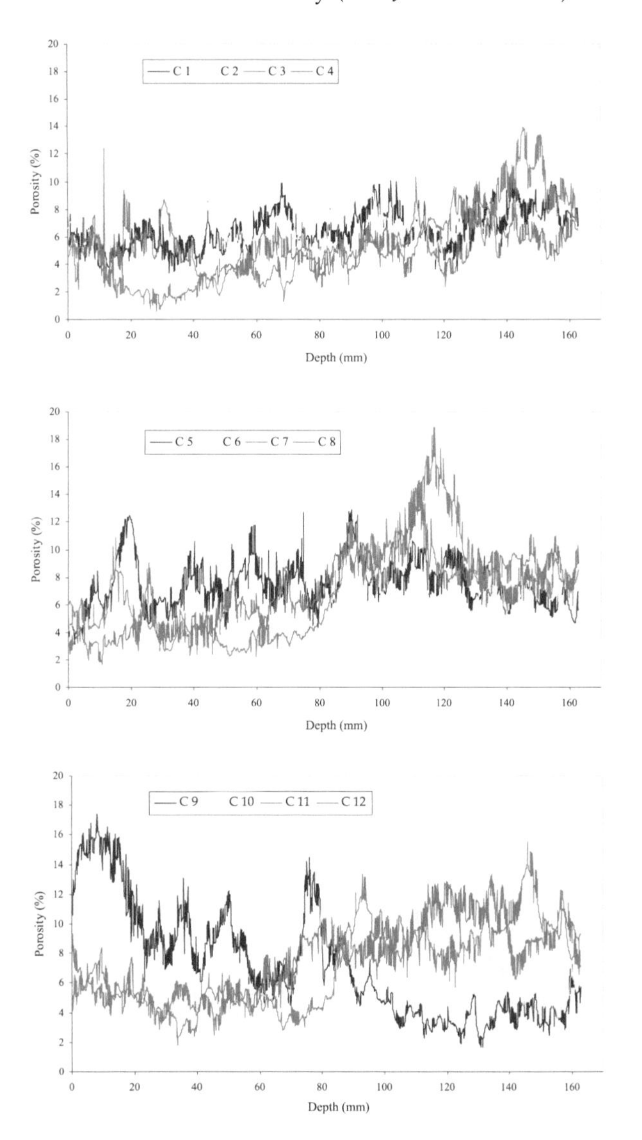

Figure 2: Series representing the 2-D sectional porosity *vs.* depth (San José et al. 2011).

We shall illustrate using a simple model e.g., "the intermittent pluri-sink model" (Martín et al. 2015b), how simple linear actions mixed with a random component lead to the emergence of complex heterogeneity patterns. It turns out that these patterns have a well-defined "entropic level" that takes account of the final balance of the "tense fight" between the deterministic and random components.

Let us suppose that S is a soil area square shaped. Further assume that at any of the four corners there is a sink i (i = 1, 2, 3, 4) randomly acting in an intermittent manner. Each of the sinks acts with p_i the relative frequency of the appearance of such action. A pollutant deposit ("pollutant seed") is supposedly located at an arbitrary point of the square. When a given sink i acts, its suction action is able to attract the pollutant matter to another point reducing the distance to the sink by a factor $r_i < 1$, where the pollutant rests until a new (or the same) sink acts. This factor reflects the mean value of the suction power of the respective sink. However, the "flying" pollutant matter leaves a unit of pollutant at any point where the pollutant "rests" along his travelling.

Although a much more sophisticated model might be constructed for a more realistic performance under the same essential idea, we prefer to emphasize how complexity may appear under quite simple and natural actions evolving in time.

First it can be observed that for realizations with only a few number of iterations the random component dominates the result and two different realizations may give quite different results (Fig. 3): randomness imposes its rules and dominates the results.

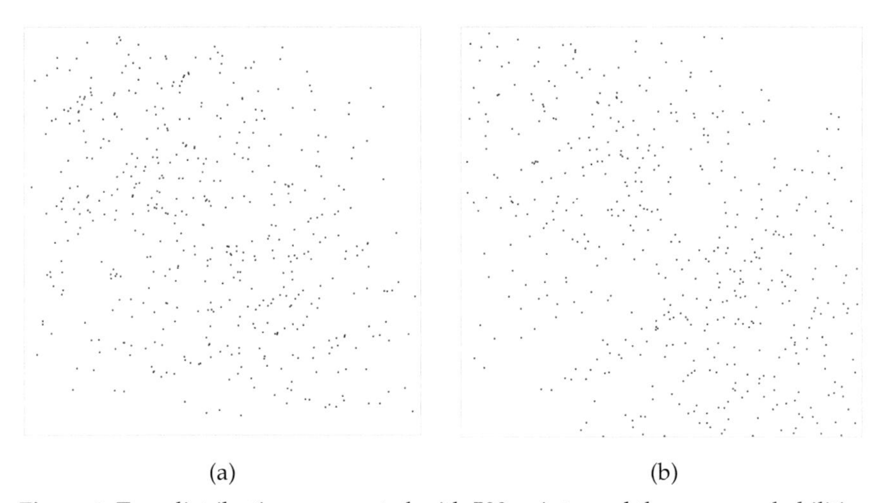

(a) (b)

Figure 3: Two distributions generated with 500 points, and the same probabilities and factors ($p_1 = 0.29$, $p_2 = 0.21$, $p_3 = 0.29$, $p_4 = 0.21$, $r_1 = 0.7$, $r_2 = 0.5$, $r_3 = 0.7$, $r_4 = 0.5$).

However, when the number of iterations increases, similar distributions for different realizations seem to appear: this is the emergence of a fixed heterogeneity pattern.

Figure 4 illustrates some implementations.

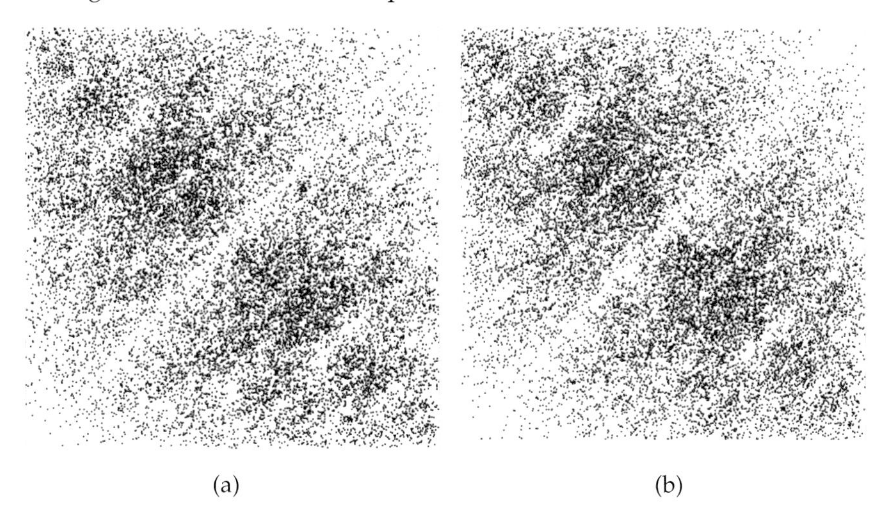

(a) (b)

Figure 4: Two distributions generated with 20,000 points, and the same probabilities and factors ($p_1 = 0.29$, $p_2 = 0.21$, $p_3 = 0.29$, $p_4 = 0.21$, $r_1 = 0.7$, $r_2 = 0.5$, $r_3 = 0.7$, $r_4 = 0.5$).

The last statement, however, needs to be supported by adequate mathematical analysis.

In any of the selected choices (different values of p_i and $r_i < 1$), the scaling analysis of entropy $-H_\mu(P_\varepsilon)$ against log ε shows linear behaviour, which gives the entropy dimension D value. Table 1 summarizes the obtained results.

Table 1: The entropy dimension for some values of p_i and r_i

p_i	r_i	N	D	R^2
0.29-0.21-0.29-0.21	0.7-0.5-0.7-0.5	500	1.412	0.9645
			1.447	0.9578
		20000	1.937	0.9999
			1.936	0.9999

As can be observed, when the number of points N increases, the goodness of the linear fit (denoted by R^2) increases as well and the entropic level tends to a fixed value. The results of the scaling analysis clearly show the emergence of a mass distribution with a well-defined structured heterogeneity. In fact the robustness of the obtained results are based on theoretical results (see Falconer 1997).

This paradigmatic example has important consequences: (1) there are many reasons (some simpler than others) causing complex heterogeneity structures, (2) a suitable method is needed to analyze complex data to catch the implicit uncertainty degree they hide, and (3) above knowledge may facilitate the use of new theoretically founded predictive methods by means of computer simulations.

5. Dealing with Soil PSD Heterogeneity

Mathematically, the PSD of granular media may be considered as a continuous mass particle-size distribution μ supported in the interval of grain sizes. Limited information on PSD is usually determined over a list of size ranges that covers all the sizes present in the sample. Grains sorted according to the size thus appear distributed in size classes $J_1, J_2, ..., J_k$ defined by those size ranges. Different methods of analysis may provide the mass fractions $p_1, p_2, ..., p_k$, respectively. Similarly as we pointed out in section 2 the Shannon entropy (Shannon 1948) of the partition is defined by

$$H = -\sum_{i=1}^{k} p_i \log p_i \tag{4}$$

Shannon's entropy is an information–theoretical parameter that may be suitably interpreted as a measure of the complexity of a distribution. In fact, entropy has already been proposed in the life sciences as a plausible quantity of biodiversity (Margalef 1958) in the sense of evenness or heterogeneity of the diversity of species in an ecosystem. The same use of entropy to measure pedodiversity has been recently discussed (Ibañez et al. 1998, Martín and Rey 2000).

5.1 Balance Entropy Index of PSD

Notice that the Shannon entropy of heterogeneity does not take into account the length of the size intervals which is, in fact, an important issue. Shannon's entropy serves as a good measure of evenness, if all class intervals are equal. Otherwise, the entropy may become a distorted evenness measure. The typical representation of soil texture in terms of clay, silt, and sand contents employs size ranges <0.002 mm, 0.002 to 0.05 mm and 0.05 mm to 2 mm, respectively, according to the USDA soil texture classification. This results in extremely unequal class sizes of 0.002 mm for clay, 0.048 mm for silt, and 1.95 mm for sand.

Martín et al. (2005a) proposed to normalize the Shannon entropy value with a multiplier that explicitly takes into account the differences in class sizes:

$$\beta = -\frac{H}{\sum p_i \log \varepsilon_i} = \frac{\sum p_i \log p_i}{\sum p_i \log \varepsilon_i} \tag{5}$$

in which β is the Balance Entropy Index (BEI), and ε_i is the proportion of the i-th class size interval in the total range of sizes. Theoretically, β ranges between 0 and 1 for any distribution: the closer to 1 the value of β the more even the distribution. $\beta = 0$ corresponds to the most uneven distributions, where all the mass is concentrated in a single size interval, while $\beta = 1$ corresponds to the most even distribution, in which the mass inside each interval is proportional to each size interval.

For the USDA soil texture classification with three clay, silt and sand categories ($i = 1, 2, 3$), $\varepsilon_1 = \frac{0.002}{2} = 0.001$, $\varepsilon_2 = \frac{0.048}{2} = 0.024$, and $\varepsilon_3 = \frac{1.950}{2}$ = 0.975. Using the proportions of clay, silt, and sand denoted by p_i, one can compute the balanced entropy from q. (5) for any texture defined by clay, silt and sand contents. Results of such computations are shown in the USDA textural triangle in Fig. 5. Note that the values of β approach unity at the rightmost corner of the textural triangle. Inside the textural triangle, the balanced entropy tends to increase as sand content increases. The balanced entropy provides a continuous parameterization within the textual triangle.

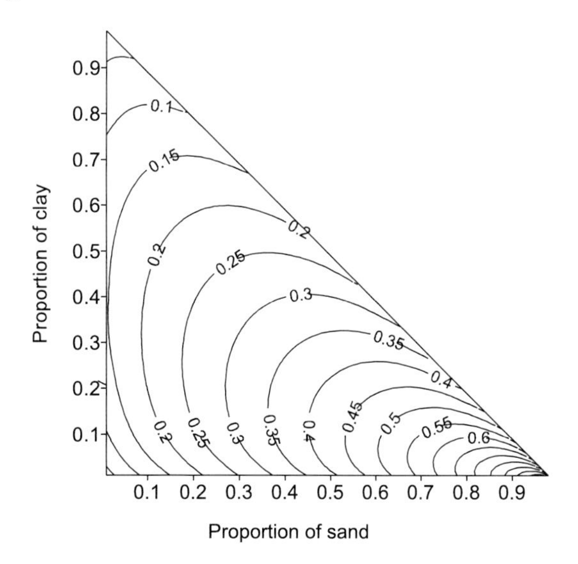

Figure 5: The results of BEI computations in the USDA textural triangle.

Besides the natural use of the BEI for the characterization of heterogeneity, Martin et al. (2005b) showed that the balanced entropy was the best single and the most important predictor of volumetric water contents at –33 kPa which are notoriously difficult to estimate. Their results suggested that the BEI might be a natural index for the packing arrangement of soil particles. Thus, the balanced entropy may be a promising approach to improve the accuracy of estimated soil hydraulic properties.

5.2 Renyi Dimensions

Nowadays, new technologies are of invaluable help providing highly calibrated field or laboratory measurements, and one can get a huge amount of PSD data e.g., from laser diffraction analysis. As a consequence, mathematical tools are required to analyse and interpret those data as well as to construct prediction models. If accurate data are available, one may further characterize PSDs in more detail.

Scaling analysis of particle-size distributions over an interval of size I is commonly made through successive partitions of the interval in dyadic scaling down (Evertsz and Mandelbrot 1992). If L is the diameter of interval I, dyadic partitions in k stages ($k = 1, 2, 3,.$) generate a number of cells $N(\varepsilon) = 2^k$ with diameter $\varepsilon = L \cdot 2^{-k}$ that cover the initial interval I. Given a certain measure μ (e.g., the mass of soil particles) distributed over the interval of size I, the measure (e.g., local mass) of each cell $\mu_i(\varepsilon)$ is supposed to be supplied by available data. In soil particle size distributions the measure in each region or subinterval of sizes would be the mass of soil particles of characteristic size in such subinterval.

Renyi dimensions, $D(q)$, may be used, among other multifractal parameters, to characterize measures. Renyi dimensions, also called generalized dimensions, may be computed through parameter q by

$$D(q) = \frac{1}{q-1} \lim_{\varepsilon \to 0} \frac{\log \sum_{i=1}^{N(\varepsilon)} \mu_i(\varepsilon)^q}{\log \varepsilon} \tag{6}$$

where q could be any real integer.

Parameter q (commonly $-10 \le q \le 10$) acts as a scanning tool analyzing the denser and rarer regions of the measure μ (Chhabra and Jensen 1989). For $q \gg 1$, regions with a high degree of concentration are amplified, while regions with a small degree of concentration are magnified for $q \ll 1$.

Most often used Renyi dimensions are D_0 and D_1. D_0 is called box-counting dimension which quantifies the scaling properties of those cells (subintervals from dyadic divisions) that contain size particles. It represents the dimension of the set of sizes with non-zero relative volume. The value of D_1 is the entropy dimension of the measure and gauges the scaling in the concentration of the measure by taking into account the amount of measure in each cell (Evertsz and Mandelbrot 1992). The higher the entropy dimension D_1, the more even the distribution. Both dimensions range between 0 and 1 for measures supported on the line.

Renyi dimensions spectra are, in general, non-increasing functions with a characteristic sigmoidal form (see Fig. 6). When studied distributions are close to monofractal measures (i.e., a uniform distribution of mass on a fractal set), Renyi dimensions spectra are closer to horizontal lines, so that $D(q) \approx D_0$.

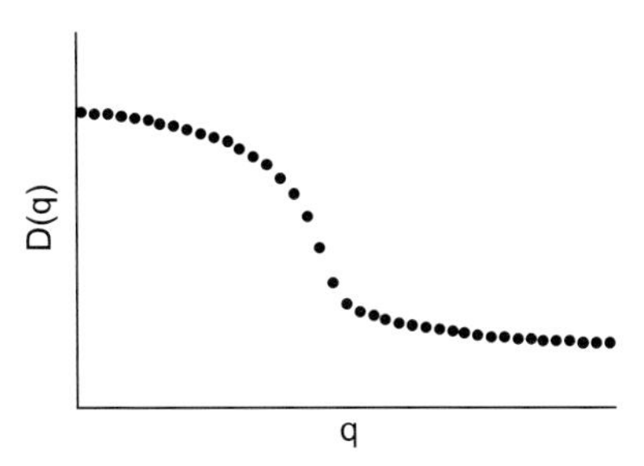

Figure 6: The Renyi dimensions spectrum.

Although entropy dimension D_1 may be used as an index to characterize measures and their heterogeneity (Martín and Rey 2000, Martín et al. 2001), the narrow intervals of values achieved by D_1 in these constructed measures and the possibility of calculating a spectra of values suggest that this parameter should not be used as a single parameter to discriminate among measures. Instead, laser diffraction analysis and multifractal techniques interaction offer a great potential to characterize distributions through a spectrum of multifractal parameters such as Renyi dimensions.

It has been recently shown that the Renyi dimensions can be properly defined from real data for grain size soil distributions (Martín and Montero 2001). In fact, Renyi dimension analysis is one of the different versions of multifractal analysis that have been used in soil sciences (see for instance Caniego et al. 2001, Caniego et al. 2003, Martin and Montero 2001, and references therein).

6. Self-similar Modelling of Particle-size Distribution (PSD)

6.1 Self-similarity of PSD as an Emergent Structure

Self-similarity is a feature that causes a part of an object (or structure) to be exactly or statistically similar to the whole and thus the object looks the same on any scale, which is called scale-invariant. In geometrical self-similar objects (for instance, in a fractal tree) scale invariance property is physically perceived when, after cutting the greater branches, the basic tree structure remains unaltered. Likewise, self-similarity in granular media might be interpreted by assuming that after an arbitrary sieving, the PSD of the fraction of particles below a certain size, is "similar" to the PSD of the whole sample (see Fig. 7).

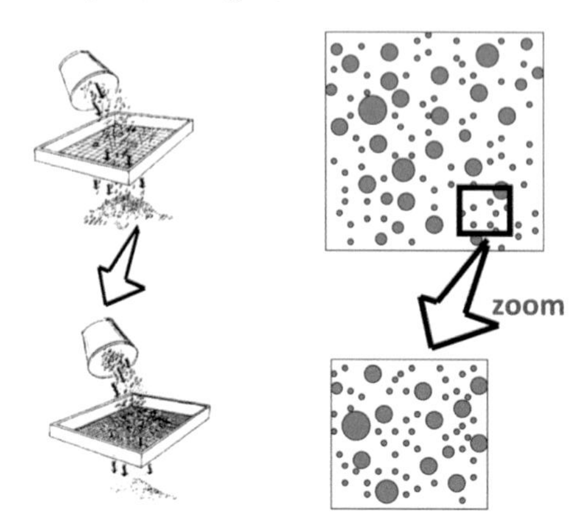

Figure 7: Illustration of the scale invariance property.

The power-law behaviour of grain sizes was commonly identified in the literature as *fractal* or *selfsimilar*, both terms coming from Mandelbrot (1982). However, Mandelbrot himself realized that the precise meaning of self-similarity of fractal objects needed to be established in a more intellectual and mathematical way, after which the power-law behaviour would appear as a consequence of such a property. Namely, Mandelbrot suggested Hutchinson to figure out a mathematical formulation, which was finally published by Hutchinson (1981). Such a mathematical formulation allows summarizing, under a simple equation, the self-similar nature of the object: E is selfsimilar if $E = \bigcup_{i=1}^{N} w_i(E)$ being w_i linear functions. This formulation allows one to derive geometrical and analytical power-law scaling of the measurable properties showing fractal behaviour.

In the case of granular media most fractal approaches directly deal with different power-laws related to the mass-size distribution of grains i.e., the cumulative mass of particles of a certain size and less versus the size. However, the most common perception of regularity along scales is based on the fact that the visual (geometrical) appearance seems to be the same for the smaller as for the greater fractions. This is an intuitive way of understanding self-similarity conceptually previous to the power-law behaviour. We shall refer to this property as *mass-geometrical self-similarity*.

The pioneering work by Andreasen and Andersen (1930) surprisingly demonstrates the essential meaning of the concept of self-similarity applied to granular media. Two granular products are considered similar if photographs of both products seem equal when the photography is presented in a scale where the unit length is the size of the greatest grains (they have the same *granulography*, a term there coined).

6.2 Self-similar Modeling, Characterization and Simulation of Soil PSD

If we observe a soil sample and we consider the particles grouped in different classes according to their sizes, the heterogeneity structure of the distribution at this scale is revealed by the mass of soil in each of these classes. The self-similarity hypothesis in which the PSD model given in Martín and Taguas (1998) rests is that such heterogeneity structure is statistically repeated along smaller scales. Then the model is a self-similar mass-size distribution in the interval $I = [0, 2]$ (mm). These sorts of distributions, intensively studied in fractal geometry, satisfy that the mass distribution on the basic textural intervals is reproduced within each one of them (suitably rescaled) and it is again reproduced within each one of the rescaled basic intervals within them, and so on. In this way, a fractal self-similar distribution is obtained within the interval [0, 2] (mm), which matches the textural data, assigning the right mass to each one of the basic intervals, and it also replicates this mass distribution structure within smaller size intervals.

Formally the model is constructed as follows. Let us suppose that from textural data for a soil we have selected a set of N relative proportions of mass corresponding to N consecutive size classes. In order to simplify, let us further suppose that $N = 3$. First, we shall present how to apply mathematically the above idea to real PSD data. Later, we shall discuss how the results depend on the data selected and how to manage these ideas in order to get better practical results. Let us denote by $I_1 = [0, a]$, $I_2 = [a, b]$ and $I_3 = [b, c]$ the subintervals of sizes corresponding to the three size classes and p_1, p_2 and p_3 the relative proportions or probabilities ($p_1 + p_2 + p_3 = 1$) of mass for the intervals I_1, I_2 and I_3, respectively. Associated with these definitions, one may consider the following functions

$$\varphi_1(x) = r_1 x \qquad \varphi_2(x) = r_2 x + a \qquad \varphi_3(x) = r_3 x + b \tag{7}$$

where $r_1 = a/c$, $r_2 = (b - a)/c$ and $r_3 = (c - b)/c$ and x is any point (or value) of the interval $[0, c]$. That is, φ_1, φ_2 and φ_3 are the linear functions (similarities) which transform the points of the interval $[0, c]$ in the points of the subintervals I_1, I_2 and I_3, respectively. The set $\{\varphi_1, \varphi_2, \varphi_3; p_1, p_2, p_3\}$ is called an iterated function system (IFS).

By means of the similarities φ_i and the probabilities p_i, an IFS determines how a fractal distribution reproduces its structure at different scales. As is shown in Martín and Taguas (1998), the set of textural data together with the self-similarity assumption determines unequivocally a self-similar fractal distribution, which may be considered a model for the corresponding PSD.

The model building is very flexible, in the sense that the distribution can be constructed from any number of size classes (three in the USDA system), each of any length. The lengths of the classes may and do vary depending on national classifications, available data, etc.

The mass proportion of any interval $J = [e, f]$ may be computed using the associated IFS as follows: (a) take any starting value x_0 of $[0, c]$; (b) choose, at random, an integer number i of the set 1, 2, 3, with probability p_i, that is, the outcome may be 1 with probability p_1, may be 2 with probability p_2 and 3 with probability p_3. We denote by x_1 the value $\varphi_i(x_0)$; (c) repeat the random experiment of (b), and suppose the new outcome is j and compute $\varphi_j(x_1)$, which we denote by x_2. We obtain in this way a sequence $x_0, x_1, \ldots, x_n$. Then, if m_n is the number of x_i's which belong to any interval J, the ratio m_n/n approaches the mass of the interval J as the number of iterations n goes to infinity.

In practice, the estimation of mass of the interval J is achieved quickly. In fact, a computation of mass is practically invariable after $n = 3000$, and it gives the same value if we repeat this apparently random computation starting with a different point x_0. The details on the PSD model and the supporting theory can be seen in Martín and Taguas (1998). The performance of the fractal self-similar model compared to real data is shown in Fig. 8.

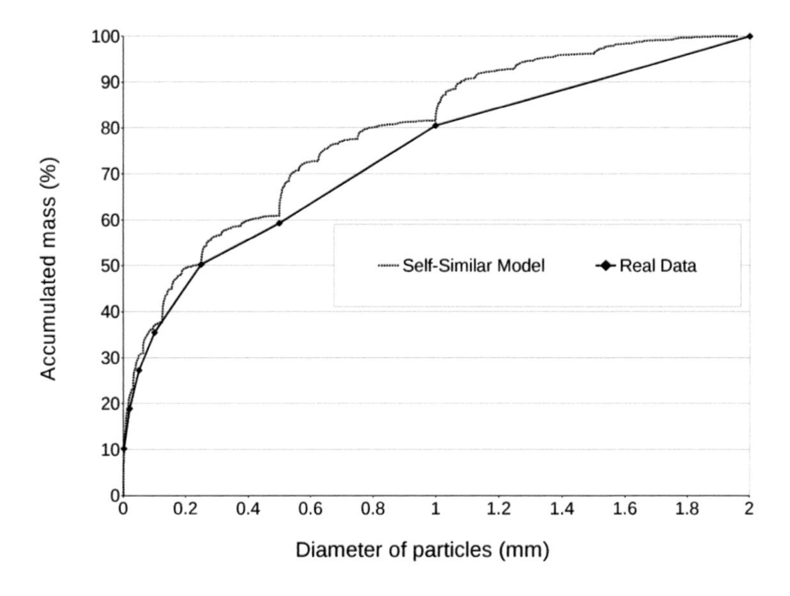

Figure 8: The mass-size distribution simulated by self-similar model.

The idea behind this model is that the heterogeneity caused by the mass percentages corresponding to the different size fractions is not only a feature observed at a privileged scale, but also it occurs within a range of (smaller) scales in a similar manner. This is a simple hypothesis among all plausible ones in order to formulate a conjecture for the complex behaviour of the PSD at unobservable scales. Self-similarity is a natural assumption that squeezes the *a priori* poor information carried by the three single percentages of textural data to determine the entire PSD distribution.

The capability of the self-similar model to predict the PSD of a given real soil was discussed in detail in Taguas et al. (1999). In that work the descriptive power of the model was strongly evidenced from the analysis of a large class of soils, thus conferring validity on the self-similarity approach to be further exploited while keeping, at the same time, the need for additional information to a minimum. However, it is interesting to notice that, independently of the quality of the performance of the model for a given real soil PSD, it is useful to test how close (or how far) is the PSD from exact fractal self-similarity.

Fair direct computations of entropy dimension require textural data for a wide range of scales e.g., measurements of laser diffraction analysis, which are not usually available from standard data e.g., measurements of hydrometer and sieving analysis. In principle, this implies that entropy dimension, being theoretically well adapted to measure texture, would be useless for practical purposes.

However, the Martín and Taguas (1998) fractal model for PSD described in this section can be used to overcome the difficulty in computing entropy dimensions discussed in the preceding paragraph. Indeed, the assumption of the model plus theoretical results from fractal geometry shows that, for a self-similar PSD, the entropy dimension is given by the simple formula (Martín and Taguas 1998)

$$D = \frac{\sum_{i=1}^{3} p_i \log p_i}{\sum_{i=1}^{3} p_i \log r_i} \tag{8}$$

where the p_i is the fraction of each soil texture class (clay, silt, or sand) and $r_i = l_i/l$ in which l_i is the length of each soil texture class interval and l is the length of the total size interval i.e. $l = 2$. For example, in the USDA system $r_1 = 0.001$, $r_2 = 0.024$ and $r_3 = 0.975$. The formula above can be easily computed from conventional textural data, and it thus provides an efficient and straightforward way to evaluate the entropy dimension of the fractal model replicating PSD from limited textural data. The entropy dimension D takes values between 0 and 1: the higher the value of D, the more heterogeneous the PSD, and the richer the soil textural structure. Moreover, since D may take any value from 0 to 1, entropy dimension – supplying a continuum of textural classes – adds a further criterion of discrimination of soil textures in terms of heterogeneity, when compared with standard classifications. A thorough interpretation of the above formula and its theoretical properties in terms of texture analysis is given by Martín et al. (2001).

6.3 The Log-self-similar Model Based on Random Cascades

The self-similar mass distribution generated by the IFS and proposed by Martín and Taguas (1998) appeared to be a sensible approach to conceive self-similarity in soil PSDs. The IFS method is an iterative process by which relative mass proportions of size classes are spread over short subintervals that are smaller, linear copies of the three initial size intervals. It provides a useful method to test soil self-similarity behaviour and calculate self-similar PSDs from the limited information provided by standard textural data.

Below a model is described that produces a log-self-similar distribution via random cascades and replaces the deterministic nature of the self-similar distribution produced by IFS techniques. This model and its testing are given in Martín and García-Gutiérrez 2008 and García-Gutiérrez and Martín 2008 respectively.

Roughly speaking, a cascade is a process which fragments a given set (the size interval in our case) into increasingly smaller pieces according to a certain rule and simultaneously divides the measure of the set according to some (possibly random) mass fragmentation rule. The process defines a limit measure that is multifractal.

In precise terms, let I be the size interval, i.e., $I = [0, 2000]$, and $I_1 = [0, \alpha]$, $I_2 = [\alpha, \beta]$, $I_3 = [\beta, 2000]$ the subintervals of sizes corresponding to three soil texture classes. Also let $p_1 = \mu(I_1)$, $p_2 = \mu(I_2)$ and $p_3 = \mu(I_3)$ be the mass proportions for the intervals I_1, I_2 and I_3, respectively ($p_1 + p_2 + p_3 = 1$).

We shall consider the transformation given by $\Phi(x) = \log(1 + x)$ that transforms the texture interval $I = [0, 2000]$ into $I^* = \Phi(I) = [\Phi(0), \Phi(2000)] = [0, 7.601]$, where $0 = \log(1)$ and $7.601 = \log(2001)$. For $i = 1, 2, 3$, let $I_i^* = \Phi(I_i)$.

For $i = 1, 2, 3$, let φ_j be the linear transformation that transforms I^* en I_i^*, $I_i^* = \varphi_j(I^*)$. These three transformations are applied first to the interval I^* and then to any of the resulting intervals I_i^* ("subintervals") following the branching process *ad infinitum*.

One interval of the kth stage of the multiplicative cascade that results from the iterative application of a certain sequence of linear transformations is denoted by I_{kj}^*. The mass, which is supposed to be uniformly spread in a "son" (sub-subinterval) $I_{k+1,i}^* = \varphi_i(I_{kj}^*)$, is given by:

$$\mu(I_{k+1,i}^*) = \mu(I_{kj}^*)V \tag{9}$$

where V is a random variable that follows a normal distribution of mean p_i (see further details in the next section), and $\mu(I_{kj}^*)$ is the mass of the interval I_{kj}^*.

At the limit, the process defines a statistically self-similar mass distribution supported on I^* (see Falconer 1990, 1997).

Finally, let us define $I = \Phi^{-1}(I^*)$ and $\mu(J) = \mu(J^*)$, with $J \subset I$ and $J^* = \Phi(J)$, as a model for the PSD distribution.

Closer examination of this model led us to rethink self-similarity in soil PSD. Soil data are usually reported in terms of the contents in clay (soil particles smaller than 0.002 mm), silt (0.002 – 0.050 mm) and sand (0.050 – 2 mm). This assigns relatively similar importance to these textural separates. In contrast, the respective size intervals differ by several orders of magnitude (i.e. 0.002 mm, 0.048 mm and 1.95 mm, respectively). With the above-described self-similar modeling approaches, these proportions and size intervals would lead to vast amounts of soil mass accumulating in very small rescaled copies of the size interval (specifically, in the rescaled copies of the clay interval). This might be unrealistic in pedological terms. In fact, the analysis made by Martín and Montero (2001), via a multifractal spectrum computed from data obtained by laser diffraction analysis of textures, revealed an excellent scaling behaviour when a log-rescaled size interval was scaled down. Using the rescaled interval instead of the usual interval in scaling analyses is strongly supported by the nature of the data provided by texture analysis instruments (see also Martín et al. 2001, Martín and Montero 2001).

The previous facts suggest a different (self-similar) way of conceiving how the fine scale structure is echoed by global mass-size data based on elemental size fraction contents. The key idea is to view the PSD as the result of a fractal-like mass-spreading iterative process occurring in the log-rescaled size interval at any scale. The new regularity may be termed log-self-similarity. Just as normality is replaced with log-normality in order to explain fundamental aspects of natural order, log-self-similarity might help establish regularity laws for a wide range of scenarios where fractal features have been demonstrated.

7. Thermodynamic Approach to Model PSD

Theory of Information and thermodynamic formalism are well-known tools to study complex systems where one faces considerable lack of information. A new point of view presented in this section aims to explain PSD in granular media. The goal is giving an epistemological point of view by applying a primary principle, the Maximun Entropy Principle (Jaynes 1957), together with concepts and tools from Information Theory and Statistical Mechanics to a granular system of "many particles". A useful reference to see these kind of methods and the interconnection of disciplines is Beck and Schlögl (1993).

In the following, we present discrete and continuous approaches to model soil PSD. Any of them provide some support to respective PSD models previously discussed in section 6.

7.1 The Discrete Approach

Consider a granular system consisting of grains with similar shape (e.g., spheres, however there is no restriction) following an arbitrary grain-size distribution. Then, the total number of grains in the system is N distributed in k size classes i.e., $I_1, I_2,\ldots, I_k$. We first suppose that in each class $p_i = \frac{n_i}{N}$ in which n_i is the number of grains in the i-th class.

Again the information entropy to the probability distribution $p_1, p_2,\ldots, p_k$ is defined by

$$I = \sum_{i=1}^{k} p_i \log p_i \tag{10}$$

with the convention that $I = 0$ if $p_i = 0$ for some i.

The maximum entropy corresponds to the case $n_1 = n_2 = \ldots n_k$. If m_i is the "representative" mass of grains in the class I_i, and M is the total mass of the system, the application of the Maximun Entropy Principle (Jaynes 1957) leads to maximize the total entropy with the natural constrains $M = \Sigma n_i m_i$ and $N = \Sigma n_i$.

This method is analogous to the derivation of the canonical ensemble in statistical mechanics. Using the Lagrange multipliers technique by computing the extremes of the auxiliary function

$$\Phi = -\sum_{i=1}^{k} p_i \log p_i + \lambda \sum_{i=1}^{k} n_i m_i - M \tag{11}$$

one obtains that

$$-\log p_i - 1 - \lambda m_i = 0 \tag{12}$$

and then

$$m_i = -\frac{1+\log p_i}{\lambda} \tag{13}$$

Notice that the equation (13) may be interpreted as a qualitative result indicating that the representative mass m_i corresponding to an interval would correlate with $\log p_i = \frac{n_i}{N}$. Thus, $m_i \approx -\frac{\log n_i}{N}$, which means that the smaller the size of the grains in a certain class I_i the greater the number n_i of grains in that class. Although this is a quite plausible result in most natural granular media where it may be shown that they approximately obey such a rule, the numerical examination of such correlation may be somehow difficult, since determination of the number of particles n_i cannot be easily estimated in practice.

One may derive mass-based relations (instead of number) with extra assumptions. Geometric granular models, assuming that grain sizes follow a geometric sequence $d_k = cd_{k-1}$ (c is a constant) forming classes of size d_k, have been used in former models of granular systems (see e.g.,

Andreasen and Andersen 1930). Under such hypothesis if M_i is the total mass corresponding to the size fraction i, one obtains

$$M_i = n_i m_i \approx N e^{m_i} m_i = N e^{c m_{i\text{-}1}} c m_{i\text{-}1} = \alpha M_{i\text{-}1} \tag{14}$$

This means that fractions appear in quantities M_k that are a fixed proportion α of the quantity M_{k-1} corresponding to the previous finer fraction, when a constant increase in the log-scale of sizes is considered. This is in fact basically the hypothesis of the model postulated in Andreasen and Andersen (1930).

7.2 The Continuous Approach

Here we present an entropy based approach which in some way reinforces the PSD self-similar model proposed in Martín and Taguas (1998).

This approach is compatible with the grain size distribution of granular systems which could be the result of a long grinding process with random rules that do not differ, statistically speaking, through all grain sizes. Thus, our focus here does not include granular systems with anarchic order but those systems in which grain sizes are arranged in the same statistical manner for the smaller than for the greater sizes. We postulate that all grain sizes under a given upper limit may take place. No limitation on the grain shape is *a priori* established, although the statistical similarity between grains of different sizes is assumed.

Geometrical similarity refers to visual (geometric) perception and is related with the information received about grains (number/size or mass) in each fraction. In practical terms the statistical similarity property for a grain distribution would be expressed by the requirement that after sieving some amount of granular material with sieves of different sizes, the structure of the grain distribution of the material sieved is equivalent (in information terms) to the grain distribution of the initial one.

Let us suppose that the granular material is sieved, retaining grains of size greater than r (e.g., $r = ½$). If the fraction of material sieved is p_1, the new information received is

$$H_1 = -(p_1 \log p_1 + p_2 \log p_2) \tag{15}$$

being $p_2 = 1 - p_1$. In order to obtain the information content of the separated fractions new sievings are required.

So, in order to fix ideas let us suppose that the range of grain sizes available is divided into discrete intervals of length r^m, $m = 1, 2, 3, \ldots$, so that we have totally $N(m) = r^{-m}$ intervals. For the rest, we fix $r = 1/2$ and the interval of sizes the unit interval [0, 1].

We shall denote that $p_i = \dfrac{n_i}{N}$ the proportion of grains whose size (established in a given sense depending on the shape—diameter, side length, ...) is in a subinterval $I_i \subset [0, 1] = I$. Successive sieving of both parts (the fraction sieved and the fraction retained) provides new information. If the distribution structure is similar (recall we suppose the same statistical arrangement for the smaller than for greater sizes), the information

$$H_2 = -\sum p_{ij} \log p_{ij} \tag{16}$$

would duplicate (under any extra assumption the simplest hypothesis, *Ockham´s razor*). In the general case the amount of information that one would obtain after successive sieving increases at the same rate depending solely upon the scale.

The candidate distribution μ has to maximize the total entropy $H_k = -\Sigma p_{kj} \log p_{kj}$ at any level k with the constrains:

1. $H_k = -\Sigma p_{kj} \log p_{kj} = H \log r^k$
2. $\Sigma_j p_{kj} = 1$
3. $\int Cx^3 d\mu(x) = M$, M being the total mass (C constant determined by the grain shape).

It turns out that if $H_1 = -(p_1 \log p_1 + p_2 \log p_2)$, then at the level $k = 2$ one gets $p_{ij} = p_i p_j$ and in general at the level k the mass inside a dyadic interval $I_{j_1 j_2 \ldots j_k}$ is $p_{j_1 j_2 \ldots j_k} = p_{j_1} p_{j_2} \ldots p_{jk}$.

The distribution inside the intervals and thus the distribution μ is the limit of a multiplicative cascade (see Falconer 1990, 1997). Notably this measure obtained by the thermodynamic approach essentially agrees with that proposed in Martín and Taguas (1998) and tested in Taguas et al. (1999) as a model of mass size particle-size.

8. From the Self-similar PSD to Packing Parameters

The key factor that builds the bridge between the PSD and the pore space is the random packing arrangement of particles. The crucial influence of particle size distribution on the random packing and the corresponding pore structure increases the interest in relating both, either theoretically or by computational methods: it is an old dream among particulate scientists of different scientific fields. Even when the particles are modelled by hard spheres models of granular media, addressing this issue may be useful in predicting different properties of a wide number of natural and engineered systems such as soil, ceramics and porous materials. Packing of log-normally distributed spheres by means of computer simulations have been studied in Nolan and Kavanag (1993) and He et al. (1999) who used a Monte Carlo simulation for a random model of spherical particles of sizes obeying any given distribution.

The mass self-similar PSD described in previous section 6.2 provides an exact self-similar PSD governed by a single input parameter driving the heterogeneity of the PSD.

Suppose that the granular material is sieved, retaining grains of size greater than r (think $r = 1/2$). Assume further that the fraction of material sieved is $p_1 = p$ and the retained fraction $p_2 = 1 - p$. Regardless of how close the real PSD is to the ideal self-similar model, it is worth finding clear relationships between the parameter $p_1 = p$ and the resulting void fraction produced by the random packing of particles.

Figure 9 shows, as a mere illustrative example, a realization of the packing algorithm, used by Martin et al. (2015).

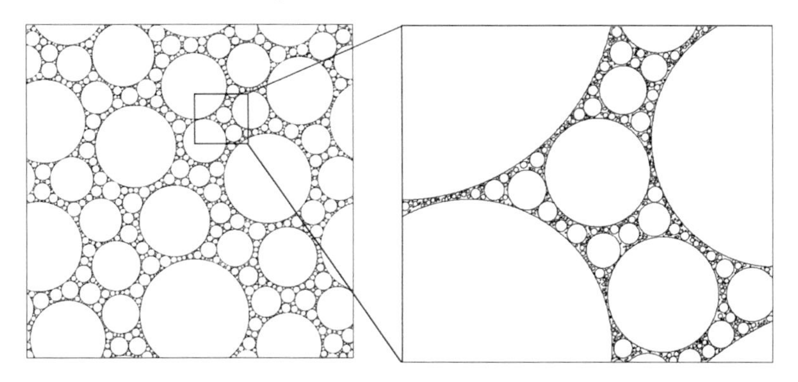

Figure 9: An example and its magnification of the result of the packing algorithm corresponding to a PSD obtained with $p = 0.6$.

As shown in Fig. 10, the porosity of the simulated porous medium depicted in Fig. 9 has a clear dependence on the value of p.

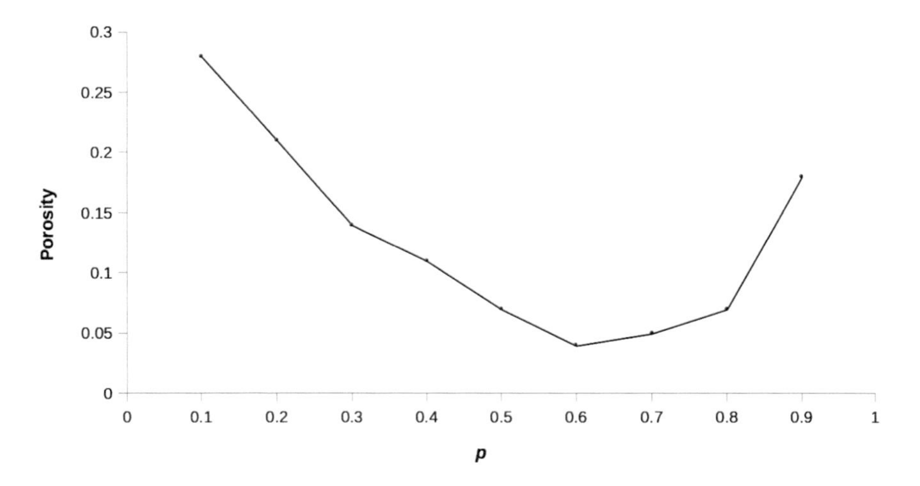

Figure 10: Mean value of porosity obtained for $p = 0.1$ to 0.9.

The value of p reflects the relative predominance of the lower grain sizes in relation to the greater ones (at every scale because of self-similarity). This might be because the intergranular space between grains of a certain size class is occupied by grains of a correlatively lower size class: the degree of this occupation would be affected by the mentioned predominance. It can be observed that the porosity does not present a monotonically increasing behaviour as a function of $p_1 = p$. This has a quite clear explanation: the occupation of the intergranular space should take an optimal degree for a certain value of the above mentioned predominance of the lower grain sizes with respect to the greater ones. Around this optimum point, which should be close to the Apolonian packing used as a model of dense granular media (see Anishchik and Medvedev 1995), random packings are expected to move away from that optimum scenario and to have a greater intergranular space.

9. Conclusion

Multi-particle granular systems share important features with complex systems. The Mathematics of Complexity (Theory of Information, Measure Theory and Fractal Geometry) provide concepts, theories and tools that may be useful to understand, to quantify, to model and to simulate those systems. Although those mathematics include somehow specialized elements not always easy to understand, a multidisciplinary approach may be useful for predictions in the field of soil and granular media.

In particular promising future investigations will be to obtain better results in the simulation of the structure of the pore space geometry from the single knowledge of the PSD. This would help to better understand the influence of PSD in processes that take place in porous media.

Acknowledgement

This work was funded by Spain's Plan Nacional de Investigación Científica, Desarrollo e Innovación Tecnológica (I+D+I), under ref. AGL2011-25175 and ref. AGL2015- 69697-P. It includes part of the research done by the Research Group of Fractals and Applications to Soil and Environmental Sciences (PEDOFRACT) of the Technical University of Madrid.

REFERENCES

Anderson, A.N., A.D. McBratney and J.W. Crawford. 1998. Applications of fractals to soil studies. *Advances in Agron.* 63: 1–76.

Andreasen, A. and J. Andersen. 1930. Ueber die Beziehung zwischen Kornabstufung und Zwischenraum in Produkten aus losen Ko¨rnern. *Kolloid-Zeitschrift* 50: 217–228.

Anishchik, S.V. and N.N. Medvedev. 1995. Three-Dimensional Apollonian Packing as a Model for Dense Granular Systems. *Phys. Rev. Lett.* 75: 4314–4317.

Bagnold, R.A. 1941. The Physics of Blown Sand and Desert Dunes. Chapman & Hall, London.

Bagnold, R.A. and O.E. Barndorff-Nielsen. 1980. The pattern of natural size distributions. *Sedimentology* 27(2): 199–207.

Barndorff-Nielsen, O.E. 1977. Exponentially decreasing distributions for the logarithm of particle size. *Proc. Roy. Soc. London*, A353: 401–419.

Barndorff-Nielsen, O.E. 1978. Hyperbolic distributions and distributions on hyperbolae. *Scandinavian Journal of Statistics* 5(3): 151–157.

Bertoin, J. and S. Martínez. 2005. Fragmentation Energy. *Adv. in Appl. Probab.* 37(2): 553–570.

Beck, C. and F. Schlögl. 1993. Thermodynamics of chaotic systems. Cambridge University Press, Cambridge.

Bittelli, M., G.S. Campbell and M. Flury. 1999. Characterization of particle-size distribution in soils with a fragmentation model. *Soil Sci. Soc. Am. J.* 63: 782–788.

Caniego, F.J., M.A. Martín and F. San José. 2001. Singularity features of pore-size soil distribution: Singularity strength analysis and entropy spectrum. *Fractals* 9(3): 305–316.

Caniego, F.J., M.A. Martín and F. San José. 2003. Renyi dimensions of soil pore size distributions. *Geoderma* 112: 205–216.

Chhabra, A. and R.V. Jensen. 1989. Direct determination of the $f(\alpha)$ singularity spectrum. *Phys. Rev. Lett.* 62: 1327–1330.

Englman, R., N. Rivier and Z. Jaeger. 1988. Size-distribution in sudden breakage by the use of entropy maximization. *J. Appl. Phys.* 63: 4766. http://dx.doi.org/10.1063/1.340114.

Evertsz, C.J.G. and B.B. Mandelbrot. 1992. Multifractal measures. In: Chaos and fractals. Peitgen, H.-O., H. Jürgens and D. Saupe (Eds). Springer-Verlag, N.Y. pp. 921–953.

Falconer, K. 1990. Fractal Geometry. John Wiley & Sons, New York.

Falconer, K. 1997. Techniques in Fractal Geometry. John Wiley & Sons, New York.

Fieller, N.R.J., D.D. Gilbertson and W. Olbricht. 1984. A new method for environmental analysis of particle size distribution data from shoreline sediments. *Nature* 311: 648–651.

Frish, U. and D. Sornette. 1997. Extreme deviations and applications. *J. Phys. I France* 7: 1155–1171.

García-Gutiérrez, C. and M.A. Martín. 2008. Testing log-selfsimilarity of soil particle size distribution simulation with minimum inputs. *Pure and Applied Geophysics* 165: 1117–1129.

He, D., N.N. Ekere and L. Cai. 1999. Computer simulation of random packing of unequal particles. *Phys. Rev.* E60: 7098–7104.

Hutchinson, J.E. 1981. Fractals and self similarity. *Indiana Univ. Math. J.* 30: 713–747.

Hwang, S.I. and S.I. Choi. 2006. Use of a lognormal distribution model for estimating soil water retention curves from particle-size distribution data. *Journal of Hydrology* 323(1): 325–334.

Ibáñez, J.J., S. de Alba, A. Lobo and V. Zucarello. 1998. Pedodiversity and global soil patterns at coarse scales (with Discussion). *Geoderma* 83: 71–192.

Jaynes, E.T. 1957. Information Theory and Statistical Mechanics. *Physical Review* 106(4): 620–630.

Khinchin, A.I. 1957. Mathematical Foundation of Information Theory. Dover Publications, New York.

Kolmogorov, A.N. 1941. On the logarithmic normal distribution of particle sizes under grinding. *Dokl. Akad. Nauk SSSR* 31: 99–101.

Kravchenko, A.N., C.W. Boast and D.G. Bullock. 1999. Multifractal analysis of soil spatial variability. *Agron. J.* 91: 1033–1041.

Lehmann, J., D. Solomon, J. Kinyangi, L. Dathe, S. Wirick and C. Jacobsen. 2008. Spatial complexity of soil organic matter forms at nanometre scales. *Nature Geoscience* 1: 238–242.

Mandelbrot, B.B. 1982. The fractal geometry of nature. W.H. Freeman, New York.

Margalef, R. 1958. Information theory in ecology. *Gen. Syst.* 3: 36–71.

Martín, M.A. and F.J. Taguas. 1998. Fractal modelling, characterization, and simulation of particle-size distribution in soil. *Proceedings of the Royal Society of London A. Mathematical, Physical and Engineering Sciences* 454: 1457–1468.

Martín, M.A. and J.M. Rey. 2000. On the role of Shannon entropy measuring heterogeneity. *Geoderma* 98: 1–3.

Martín, M.A. and E. Montero. 2001. Laser diffraction and multifractal analysis for the characterization of dry volume-size soil distributions. *Soil & Tillage Research* 64(1–2): 113–123.

Martín, M.A., J.M. Rey and F.J. Taguas. 2001. An entropy–based parametrization of soil texture via fractal modelling of particle–size distribution. *Proc. R. Soc. London Ser. A* 457: 937–948.

Martín, M.A., J.M. Rey and F.J. Taguas. 2005a. An entropy-based heterogeneity index for mass-size distributions in Earth science. *Ecological Modelling* 182: 221–228.

Martín, M.A., Y. Pachepsky, J.M. Rey, F.J. Taguas and W.J. Rawls. 2005b. Balanced entropy index to characterize soil texture for soil water retention estimation. *Soil Science* 170(10): 759–766.

Martín, M.A. and C. García-Gutiérrez. 2008. Log-selfsimilarity of continuous soil particle size distributions estimated using random selfsimilar cascades. *Clays & Clay Minerals* 56(3): 389–395.

Martín, M.A., C. García-Gutiérrez and M. Reyes. 2009. Modeling multifractal features of soil particle size distributions with Kolmogorov fragmentation algorithms. *Vadose Zone J.* 8: 202–208.

Martin, M.A., F.J. Ortega, M. Reyes and F.J. Taguas. 2015. Computer simulation of random packings for self-similar particle size distributions in soil and granular materials: Porosity and pore size distribution. *Fractals* 22(3).

Martín, M.A., F.J. Muñoz-Ortega, M. Reyes and F.J. Taguas. 2015a. Computer Simulation of Packing of Particles with Size Distributions Produced by Fragmentation Processes. *Pure and Applied Geophysics* 172(1): 141–148.

Martín, M.A., M. Reyes and F.J. Taguas. 2015b. Intermittent pluri-sink model and the emergence of complex heterogeneity patterns: A simple paradigm for explaining complexity in soil chemical distributions. *Journal of Chemistry* 2015. Article ID 138202, 5 pages.

Miller, B.A. and R.J. Schaetzl. 2012. Precision of soil particle size analysis using laser diffractometry. *Soil Science Society of America Journal* 76(5): 1719–1727.

Muller, J. and J.L. McCauley. 1992. Implication of fractal geometry for fluid flow properties of sedimentary rocks. *Transport in Porous Media* 8(2): 133–147.

Nolan, G.T. and P.E. Kavanag. 1993. Computer simulation of random packings of spheres with log-normal distributions. *Powder Technol.* 76: 309–316.

Peyton, R.L., C.J. Gantzer, S.H. Anderson, B.A. Haeffner and P. Pfeifer. 1994. Fractal dimension to describe soil macropore structure using X-ray computed tomography. *Water Resour. Res.*, 30(3): 691–700.

Posadas, A., D. Gimenez, R. Quiroz and R. Protz. 2003. Multifractal characterization of soil pore systems. *Soil Science Society of America Journal* 67: 1361–1369.

Prigogine, I. 1945. Modération et transformations irreversibles des systemes ouverts. Bulletin de la Classe des Sciences, Academie Royale de Belgique 31: 600–606.

Renyi, A. 1957. In Trans. 2nd Prague Conf. on Information Theory, Statistical Decision Functions and Random Processes, pp. 545–556.

San José Martínez, F., M.A. Martín, F.J. Caniego, M. Tuller, A. Guber, Y. Pachepsky and C. García-Gutiérrez. 2011. Multifractal analysis of discretized X-ray CT images for the characterization of soil macropore structures. *Geoderma* 156: 32–42.

Segal, E., P.J. Shouse, S.A. Bradford, T.H. Skaggs and D.L. Corwin. 2009. Measuring particle size distribution using laser diffraction: Implications for predicting soil hydraulic properties. *Soil Science* 174(12): 639–645.

Shannon, C.E. 1948. A mathematical theory of communication I. *Bell Syst. Tech. J.* 27: 379–423.

Sornette, D. 2000. Critical Phenomena in Natural Sciences. Chaos, Fractals, Self-organization and Disorder: Concepts and Tools. Springer Series in Synergetics, New York.

Taguas, F.J., M.A. Martín and E. Perfect. 1999. Simulation and testing of self-similar structures for soil particle-size distributions using iterated function systems. *Geoderma* 88: 191–203.

Turcotte, D.L. 1986. Fractals and fragmentation. *J. Geophys. Res.* 91: 1921–1926.

Tyler, S.W. and S.W. Wheatcraft. 1992. Fractal Scaling of Soil Particle-Size Distributions: Analysis and Limitations. *Soil Science Society American Journal* 56(2): 362–369.

Wu, Q., M. Borkovec and H. Sticher. 1993. On particle-size distributions in soils. *Soil Science Society of America Journal* 57(4): 883–890.

CHAPTER

5

The Fractals of Percolation Theory in the Geosciences

Allen G. Hunt[1]* **and Fang Yu**[2]

[1] Department of Physics, Wright State University, 3640 Colonel Glenn Hwy., Dayton OH 45435, United States

[2] Department of Earth & Environmental Sciences, Wright State University, 3640 Colonel Glenn Hwy., Dayton OH 45435, United States

1. Introduction

Wherever optimization of a flux in a disordered material or network is important, percolation theory has potential relevance. The optimization procedure can require construction of an optimal network in space, or it may simply reflect the selection of an optimal path through a disordered network. Both types of optimization are often relevant in the geosciences, in view of theoretical explanations of the formation of drainage networks in terms of "feasible optimality" (Rigon et al. 1998), or the continually refining optimizations within living organisms (Bejan 1997a, b). Thus, development of drainage networks in surface flow corresponds to an active process of connecting potential paths through a medium (Hunt 2016a), the second to the choice of the paths of tree roots in the subsurface (Hunt 2016b). For the former case, the exponents of percolation theory appear to describe the range of sinuosities in river networks (Gray 1961, Maritan et al. 1996), and for the latter, the tortuosity of flow, diffusion, or electrical conduction (Ghanbarian et al. 2013, Hunt and Ewing 2016) as well as (Hunt 2016b), the fractal dimensionality of roots (Levang-Brilz and Biondini 2002). In the latter case, percolation structures govern the mass flux over the network (e.g., that determine net primary productivity), or the flow of nutrients or solutes (limiting vegetation growth and soil formation) through the medium. The reasons for this relate most simply to the physical foundation of percolation theory.

*Corresponding author: allen.hunt@wright.edu

Percolation theory has many additional applications in the geosciences beyond those relating to fluxes described here, such as in the viscosity (Campbell and Forgacs 1990, Vigneresse et al. 1996) and yield strength (Hoover et al. 2001) of magmatic suspensions, distribution of forest fire sizes (Mackay and Jan 1984, Cox and Durrett 1988, Hunt 2008), explosive eruption fragments (Gaonac'h et al. 2003), crystallization in melts (Avramov et al. 2000), rock fracture properties (Guéguen et al. 1997), connectivity of landscapes and corridors relative to speciation (Keitt et al. 1997, Wiens et al. 1997), pre-seismic electromagnetic phenomena (Hunt 2005, Hunt et al. 2007) and so forth (Sahimi 1994), but the potential in these possibilities is too vast and incompletely explored to describe here. In fact, the focus of this chapter will be on applying percolation theory to soil formation, vegetation growth, natural and intensively managed, and net primary productivity. These applications, though comparatively new, are also highly topical, with relevance to climate change and the global carbon cycle.

Percolation is a theory of connectivity (Stauffer and Aharony 1994). Connectivity is relevant to a number of processes that involve flow or transport of material from one place to another, or to whether a rigid portion of a medium connects across a system. The fundamental conclusions of percolation theory are strongly dependent on the dimensionality of the system. For example, both rigidity and fluidity can percolate simultaneously in three dimensions, but not in two. For scientists constructing models of porous media, no flow is possible in a two-dimensional grain-supported medium, while simultaneous flow through both wetting and non-wetting phases is possible in three dimensions, even if the medium is grain-supported (i.e., the medium maintains structural rigidity through the grain contacts).

In order to fully understand the many applications of percolation theory in the geosciences, it is necessary first to be aware of two distinct applications in the case of flow. For example, in a poorly connected medium, the flow paths may be highly tortuous, fractally branching, and rare. The description of these paths may be dominated by characteristics of the medium, but the paths tend to conform to the fractal geometry of percolation. In a well-connected medium, however, if the medium is also sufficiently heterogeneous, the dominant flow paths generally have the characteristics of percolation. This result is a property of what is called critical path analysis, which identifies the dominant flow paths by equating a fractional volume associated with the most permeable portions of the medium with the critical threshold for percolation. As stated by Muhammad Sahimi (Sahimi 1994), when flow is largely confined to the critical network, the properties of solute transport are described by the scaling exponents in percolation theory.

Although such generality might not be anticipated in solute transport, the relevance of percolation extends further than mere transport properties. In fact, the scaling properties of solute transport then govern such diverse properties as chemical weathering, soil formation, or the deposition of calcium carbonate layers in arid-land soils, whenever such phenomena are solute transport-limited, which appears to be nearly always the case, at least at longer time scales. This particular subject is now employed as an introduction to soils and vegetation.

2. Relevant Percolation Results

A few quantities of interest as well as some terminology are now reviewed.

In the following, the abbreviation 2D (3D) for two-dimensional (three-dimensional) will be used. A system is at the percolation threshold when an interconnected cluster of infinite size comes into existence. For bond percolation, for example, an infinite cluster of bonds appears when the probability of connecting a bond between two nearest neighbor sites exceeds the percolation threshold, p_c, a value which depends on both the dimensionality of the system and on the particularities of the connections of the bonds between the sites (most strongly on the number of possible connections, or coordination number, Z, at an arbitrary site or bond). The linear dimension, R_s, of the largest cluster of interconnected sites diverges according to a power of $p - p_c$, with $R_s \approx (p - p_c)^{-\nu}$, with $\nu = 0.88$ in 3D, and $\nu = 4/3$ in 2D. The volume concentration of clusters with s connected bonds (or sites, for site percolation), n_s, is proportional to $s^{-\tau}$ at the percolation threshold, with $\tau = 2.18$ in 3D and 2.05 in 2D. Away from the percolation threshold, $n_s \approx s^{-\tau} \exp\{-[s^{\sigma}(p - p_c)]^2\}$, with $\sigma\nu = 1/d_f$, with d_f the mass fractal dimensionality of large clusters near the percolation threshold. The actual value of σ will not be relevant in the following, but the fact that $(\tau - 1)/\sigma\nu = d$, with d the dimensionality of the network (2D on a surface, 3D in the bulk) is important, as this relationship allows the transformation of the cluster statistics to a form that gives the number of clusters of a given maximum resistance value within a region of volume L^3 (or area L^2).

Large clusters near the percolation threshold are characterized by large holes, loops and many dead ends, which can be composed of interconnections of bonds (or sites), but which connect to the remainder of the cluster only at one point. The backbone, which determines the scaling of solute transport times, is formed by pruning all the dead ends from the cluster. The backbone mass fractal dimensionality, D_b, can take on a much wider range of values than the universal values of ν and τ above, but in the absence of certain long-range correlations (Sahimi and Mukhopadhyay 1996), D_b values are restricted to four cases, which can be summarized as 2D random or invasion (1.64 and 1.217), or 3D random or invasion

(1.87 and 1.46). Invasion percolation values are obtained for either wetting or drying conditions in 2D, but only during drying conditions in 3D (Sheppard et al. 1999). Here, the typical time required for solute to traverse a cluster of Euclidean length x scales as x^{Db}. Other important exponent values describe a tortuosity that diverges at the percolation threshold when the system size tends to infinity, and include those for optimal paths, (D_{opt} = 1.21 in 2D, 1.43 in 3D) and the minimum separation, or chemical path length (D_{min} = 1.13 or 1.21 in 2D, 1.37 or 1.43 in 3D). The chemical path length is the shortest distance along an interconnected path across a cluster. The variability in D_{min} values is associated with the form of percolation theory, invasion or random, and bond or site. As pointed out e.g., in Sheppard et al. (1999), the wetting process is invasion site percolation (with trapping) and the drying process is invasion bond percolation (with trapping).

Optimal paths are defined for networks with a very wide range of local bond strengths, or resistances, as the paths with the minimal total resistance. Finally, finite-size scaling (Fisher 1971) relates quantities that diverge (or vanish) at the percolation threshold with exponent μ (i.e., as $(p - p_c)^{\mu}$) to their dependence as a function of system size, x, at the threshold as $x^{-\mu/\nu}$. While this result will not be used directly here, it is of sufficient generality and importance to require mentioning.

3. Solute Transport

The following discussion is based on theory for solute transport presented by Lee et al. (1999) and on exponent values published by Sheppard et al. (1999). The theoretical discussion was developed for steady-state, incompressible flow and advective solute transport (Lee et al. 1999). Kirchoff's laws, equivalent to Laplace's equation, were solved numerically for disordered networks, and the times of solute transport through links were calculated as inversely proportional to the link velocity, with the probability of exiting the link proportional to the fluid velocity. Particle tracking was then employed as a means to generate solute arrival time distributions. The theoretical foundations, though not the methods, were shared by analytical treatments (Hunt and Skinner 2008, Hunt et al. 2011, Ghanbarian-Alavijeh et al. 2012) based on the results of the authors given. While such a theoretical description is thus well-grounded for solute transport, here the same results will also be employed to generate scaling relationships for non-steady-state water transport, and literature comparison with problems in infiltration and evapotranspiration. In those cases, the theoretical justification is less certain.

Lee et al. (1999) showed that the distance and time of solute transport through a porous medium near but above the percolation threshold are distinct, non-trivial, properties of a system size. These authors studied the ability of fluid flow to advect particles from one side of a system to the other. In particular, for solute sources on the left side of a system of Euclidean length x, the typical path length for particles exiting on the right hand side was proportional to $x^{D_{min}}$, whereas the most likely time for particles to exit on right hand side was x^{D_b}. The first exponent D_{min} is the chemical paths exponent, while the second D_b is the fractal dimensionality of the percolation backbone. Thus, the minimum path length, L, across such a system scales as (exponent values from Sheppard et al. 1999):

$$L = x_0\left(\frac{x}{x_0}\right)^{1.46} \text{ (3D drainage) } L = x_0\left(\frac{x}{x_0}\right)^{1.37} \text{ (3D saturated or wetting) (1a)}$$

$$L = x_0\left(\frac{x}{x_0}\right)^{1.13} \text{ (2D saturated) } L = x_0\left(\frac{x}{x_0}\right)^{1.21} \text{ (2D wetting or drainage) (1b)}$$

Here, x_0 is a bond length. In the case of porous media, such a bond length should correspond to the distance between pore bodies, or a typical grain size. The most likely time required for the particles to traverse the system scales as (exponent values from Sheppard et al. 1999)

$$t = t_0\left(\frac{x}{x_0}\right)^{1.87} \text{ (3D saturated or wetting) } t = t_0\left(\frac{x}{x_0}\right)^{1.46} \text{ (3D drainage) (2a)}$$

$$t = t_0\left(\frac{x}{x_0}\right)^{1.64} \text{ (2D saturated) } t = t_0\left(\frac{x}{x_0}\right)^{1.22} \text{ (2D unsaturated) (2b)}$$

Here t_0 is the time for fluid to traverse the distance x_0, equal to a pore separation; in other words, x_0/t_0 is the fluid flow rate at the pore scale. These results help to explain why paths that do not appear to be extraordinarily tortuous (20 m in a 1 m column) can still retard the arrival of solutes by an enormous factor (over 1000) (Hunt et al. 2015). Under common relevant conditions of 3D wetting or full saturation, the transport distance x is proportional to the transport time, t, to the 0.53 power, close enough to 0.5, characteristic of diffusion, to allow considerable confusion in the literature.

In media that are strongly disordered, it is possible to define connected paths through the system that produce the minimum possible total resistance. These paths are called optimal paths. Their tortuosity and backbone dimensions are essentially equal (1.43 and 1.42 in 3D, and both 1.21 in 2D) (Sheppard et al. 1999). This means that the time for solutes to traverse these paths scales the same way with the Euclidean length across the system as does the length of the paths.

The following addresses only the time of solute transport, as that dependence generates the typical solute velocity (flux) as a function either of time or solute transport distance. For $t = t_0(x/x_0)^{Db}$, one finds, apart from numerical constants of order unity,

$$\frac{dx}{dt} = \left(\frac{x_0}{t_0}\right)\left(\frac{t}{t_0}\right)^{\frac{1}{D_b}-1} = \left(\frac{x_0}{t_0}\right)\left(\frac{x}{x_0}\right)^{1-D_b} \tag{3}$$

In most cases, the value of D_b extracted from experiments has been 1.87, as expected for saturated conditions and fully 3D flow (Hunt et al. 2015). In one case, the value was 1.21, associated with 2D unsaturated conditions. An important distinction between the two forms of the right-hand-side of Eq. (3) is that the second is directly proportional to the flow rate, x_0/t_0, while the first has a more complicated dependence, and is very nearly proportional to the square root of the flow rate.

In Eq. (3) the ratio x_0/t_0 is a fluid flow rate, which in experiments simulating vertical flow (White and Brantley 2003, Salehikhoo et al. 2013) is typically proportional to the hydraulic conductivity, but which in situ is the net infiltration rate (Hunt and Ghanbarian 2016). Infiltration rates are typically smaller, since they are limited by precipitation, and thus more likely to be consistent with transport-limitations on weathering. The solute transport velocity is therefore a negative power of time as well as of distance. The implication of such a result is that at long enough times or large enough length scales, solute transport rates must decline sufficiently to limit the rate of chemical reactions, including chemical weathering, in porous media. As a consequence, the actual value of the solute velocity, which is related to the fluid velocity, takes on a rather large importance; at the time at which an associated solute advection time exceeds a reaction time for well-mixed solutions, the reaction becomes transport-limited, and the further time dependence is that of the solute velocity. Such a cross-over to transport-limited conditions has been argued by Maher (2010) to occur at a time scale of roughly a day, or a length scale of about 0.5 mm (ca. 1.8 m/yr flow rates). However, it can occur at much smaller length scales, depending on the fluid flow rate. In arid regions, a typical length scale for transport-limitations to become dominant appears to be a single pore (30 μm, see below), though because of the correspondingly slower flow (infiltration rates of 100 mm/yr or less), the time scale is still roughly a day.

The full theoretical treatment of solute transport is somewhat more complicated, and does not lead to a precise power law solute velocity. But it is only required for quantitative agreement with data in the case that more than five orders of magnitude of elapsed time (or about three of space) are relevant. Since it is frequently the case that a smaller range of data is considered, the simple scaling laws often suffice.

4. Applying Scaling

4.1 Soil Development

For soil development, it was proposed (Hunt 2016b, Hunt and Ghanbarian 2016) that the total solute transport distance at time t should be equal to the soil depth, x, at the same time. Since this solute transport defines weathering depths at the bottom of the B horizon, the exponent chosen was 1.87, considered with 3D flow under wetting or saturated conditions. Thus,

$$x = x_0 \left(\frac{t}{t_0} \right)^{\frac{1}{1.87}} = x_0 \left(\frac{t}{t_0} \right)^{0.53} \quad (4)$$

In order to make the scaling laws predictive, it is necessary to choose values for x_0, the characteristic grain size, and $v_0 = x_0/t_0$. In the case of soil formation, v_0 is proportional to the infiltration rate. The infiltration rate is, by water conservation equal to the precipitation less the evapotranspiration, and can vary locally in accord with surface water routing (run-on less run-off). These values are not available generally, though particle sizes are sometimes reported. Precipitation is reported more commonly, but evapotranspiration is less well-known, and local values of run-on and run-off are almost never available. This means that some alternate strategies for testing the hypothesis must be developed. Sources of data are preferred with as much detail as possible; thus $P - ET$ values are better than merely P. When x_0 values are not available, we can either use a universal value, which may hopefully be typical, or we can use x_0 as a fit parameter and check against qualitative information, such as soil texture classes (e.g., sandy loam), or bedrock substrate characteristics (silty sandstone, for example). Nevertheless, there exists evidence to suggest a choice of x_0 of roughly 30 μm may be appropriate across a fairly wide range of conditions. This evidence consists of data for particle sizes in arid regions, where soil calcic horizon development can be compared with theory (below), some evidence from soil production (Tennessee Valley, CA, site), typical soil hydraulic conductivity values, as well as the geometric mean value of the silt particle range in soils.

It can be shown (Hunt and Manzoni 2016) that a uniform medium with pore diameters about 10 μm produces a saturated hydraulic conductivity of 1 μm/s. By definition, soil particle sizes range from clay through silt to sand. The arithmetic mean silt particle diameter (3.9 μm $< d_s <$ 62.5 μm) is about 32 μm. Given the typical assumption that pore diameters are about 30% as large as particle diameters (Gvitzmann and Roberts 1991), the hydraulic conductivity information would suggest that a typical particle size, or pore separation, should be about 33 μm, or very close to the mean diameter of the middle particle class (clay, silt, sand). Precipitation values can range over nearly four orders of magnitude, from barely 0.002 m/yr in the Atacama Desert to about 10 m/yr in a number of places, including the New Zealand Alps. The formation of soil, however, is a process that scarcely applies in the driest Atacama sites, and the vast majority of the data for soil depths as a function of time comes from sites which exhibit a range of net infiltration values that extends over only about three orders of magnitude. Thus, $x_0 = 30$ μm is a good general value to use, but t_0 values corresponding to the time for water to cross a pore can vary over about three orders of magnitude, from 30 s to 30,000 s.

4.2 Vegetation Growth Scaling

It will be seen that vertical solute transport times for 100 m tend to be about 100 Myr, a vast time scale compared with vegetation growth over similar length scales, including root elongation in the subsurface. In particular, Sequoia species and Eucalyptus species can reach heights of nearly 100 m in 100 years. In the case that nutrient transport is considered key to limiting vegetation growth rates, relating passive solute transport times to vegetation growth will thus be impossible. Nevertheless, it has been suggested that "suboptimal" nutrient availability is nearly universal (Lynch 1995). This limitation is then proposed to lead to particular strategies of root growth, with fractal dimensionality dependent on species type and needs (Lynch 1995, Fitter and Stickland 1992). Hunt (2016b) offered that it supports a more general hierarchical strategy of root growth, allowing optimal rates of finding nutrients distributed heterogeneously. In the context of this hierarchy, roots would then be expected to grow preferentially towards nutrient sources, signals of which will be transmitted along the dominant (optimal) advective flow paths, along which minimal total resistance to root extension would be encountered. Thus it was hypothesized that root growth should follow the optimal paths from percolation. Since (Lynch 1996) such nutrients are primarily confined to the top meter or less of the soil, the appropriate exponent to choose is the 2D optimal paths exponent, equal to 1.21, and

$$L = x_0 \left(\frac{t}{t_0} \right)^{1/D_{opt}} = 0.00003\,\mathrm{m} \left(\frac{t}{30} \right)^{0.82} \tag{5}$$

The same values of x_0 and t_0 are applied in Eq. (5) as in soil depth scaling (Eq. 4). In Eq. (5), however, x_0 may be considered to be either (1) a geometric mean xylem diameter (or, equivalently, a pore size diameter that controls flow rates), in which case the appropriate value would be 10 μm (Hunt and Manzoni 2016), or (2) a typical particle diameter (a distance between neighboring pores), which yields closer to 30 μm. x_0/t_0 is then a pore-scale flow rate, as in Eq. (4); however, at larger scales, at least, it should relate to an evapotranspiration rate.

For 30 μm pore sizes, a pore size flow rate of 32 m/yr (0.000001 m/s) yields a time scale of 30 s. A 20 m/yr flow rate yields a time scale of about 45 s. Consider what results from a change in perspective associated with an increase in spatial scale to ecosystem size. If one expresses L as proportional to a depth of water drawn from the ground over a growing season (AET(t_{grow})), the equation is transformed to

$$L = \text{AET}\,(t_{grow})\left(\frac{t}{t_{grow}}\right)^{\frac{1}{D_{opt}}} = 1.62\,\text{m}\left(\frac{t}{6\,\text{mos.}}\right)^{0.82} \tag{6}$$

The value 1.62 m derives from requiring the equivalence of Eq. (6) and Eq. (5). But 1.62 m in a growing season is nearly the maximum reported yearly biome evapotranspiration (ca. 1650 mm) (Box et al. 1989), the bulk of which is assumed to take place during the growing season. Thus, a change of scale and perspective leads to the ability to represent the same equation either in terms of a pore-scale flow rate, or a seasonal evapotranspiration value.

When nutrients do not limit growth, as can be the case either when they are uniformly distributed, or distributed in sufficient concentrations, L has been suggested to be linear in time,

$$L = x_0 \frac{t}{t_0} \tag{7}$$

A second way to arrive at Eq. (7), if the nutrient distribution is spatially uniform, is to assume the relevance of the advection-dispersion equation ADE at all length scales. In Eq. (7), x_0/t_0 produces the same velocity at any time scale; thus it is a flow rate at the pore scale, and at the scale of a growing season, it is the AET. If x_0/t_0 = AET, and if Eq. (7) is consistent with the same flow rate as Eqs (4) and (5), then $L/t = x_0/t_0 = 32$ m/yr. Thus, the maximum value of L after a growing season, about 16m or so, is within about 35% of the tallest corn plant ever recorded (almost 12 m), while the largest tomato plant reached a length of about 20m in a year (Hunt 2016b). This explains the necessity to increase water and nutrients simultaneously, if the objective is to achieve maximum plant growth, as well as the rather extreme values of water volume required for such growth. It is possible that the ratio of 1.6 m/16 m (ratio of growing season AET to pore scale flow distance) = 1/10, may provide as well as a realistic estimate of the ratio of the unsaturated value of the hydraulic conductivity at typical growing season soil conditions to its value under saturated conditions.

4.3 Productivity

The net primary productivity is equal to the mass added to a plant, or community of plants, over a year, though, in fact this mass increase occurs primarily over the growing season. Then, the root mass of an adequately fertilized plant and watered plant, whose root radial extent reaches L after time t, will be bounded by

$$M \sim L^{D_m} = K^{D_m} t^{D_m} \tag{8}$$

In this equation, D_m is the mass fractal dimensionality of the root system. Since plant roots in each species tend to seek out the nutrients near the surface of the soil, it is proposed that the appropriate choice for D_m is that of the 2D mass fractal dimensionality of percolation, 1.896. A collection of data for root mass as a function of root radial extent, in other words values for D_m, in 55 grass and forb species (Fig. 1a) is in general accord with $D_m = 1.9$. In particular, if one excludes the four examples with $D_m > 5$, and the single example with $D_m < 0.75$ (different from the predicted value by more than a factor 2.5), the remaining values average to 1.90 in both cases, forbs and grasses. If, instead, one compiles the exponents for the power-law that relates root biomass with root length, one should expect a result $D_m/D_{opt} = 1.9/1.21 = 1.57$, a value generally consistent with the data (Fig. 1b).

If t corresponds to a growing season, then M should be proportional to the net transpiration during the growing season to the power D_m, and in this case M is approximately the net primary productivity, NPP, since above- and below-ground productivity are roughly equal over a wide range of plant species.

When translated to an entire ecosystem and scaled up in time to a growing season, the AET value should be limited to the water mass that is regionally withdrawn from the soil, rather than by a given plant. If deep water sources can be neglected, this value is likely to be considerably smaller than the maximum that can be accessed by an advantageously located large individual.

In the case that growth is retarded by nutrient limitations, proceeding in the same fashion leads to the relationship,

$$M \sim L^{D_m} = \left[\mathrm{AET}(t_{grow}) \right]^{D_m} \tag{9}$$

where AET (t_{grow}) is the depth of evapotranspiration over a growing season, not an instantaneous rate of evapotranspiration. Note the high degree of universality of Eq. (9); however, the AET is strongly climate-limited. Equation (9) is consistent with a total plant height, or root radial extent, at the end of a growing season equal to the actual evapotranspiration of the growing season. In most natural ecosystems the growing season AET is constrained to be less than about 1650 mm, which provides a pretty reasonable upper limit on natural tree growth in the first half a year, and is much smaller than the maximum corn stalk height above. If Eq. (6) for L is accurate, however, one also generates a close relationship between a germination time, t_{germ} and the growing season length, t_{grow},

$$t_{grow} = \left| \frac{\mathrm{AET}(t_{grow})}{L_{germ}} \right|^{D_{opt}} t_{germ} \tag{10}$$

In this equation, L_{germ} represents seed length at germination. Equation (10) is a new, untested, prediction in this chapter.

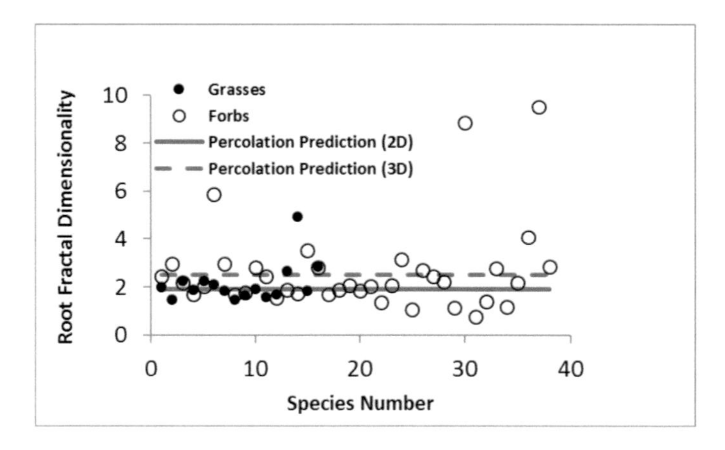

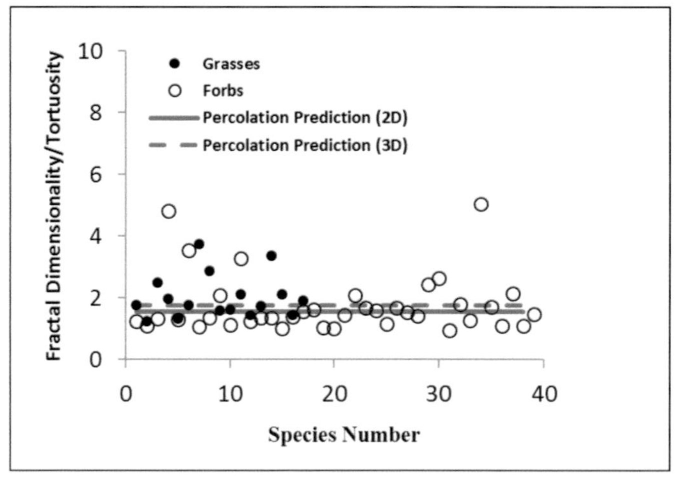

Figure 1: (a) Root mass fractal dimensionalities (in terms of root radial extent) for 55 forb and grass species (data from Levang-Brilz and Biondini 2002), (b) Same except that the dimensionality is calculated with respect to the actual root length, rather than the root radial extent.

5. Comparison with Scaling Data

5.1 Calcic and Gypsic Horizons

For first comparisons with data, consider the hypothesis that calcic and gypsic horizon depths in the subsurface, like soil depths, are equal to total solute transport distances.

Consider first data from Jenny (1941) regarding the depths of calcic horizons in the arid Mojave Desert and semi-arid Great Plains of the western USA as a function of precipitation. According to the author, the ages of the soils were constrained to be no more than 15,000 years. In order to make a quantitative prediction of the calcic horizon depth using $x = x_0(t/t_0)^{1/Db}$ and t = 15,000 years, a typical particle size diameter, which gives x_0, but P – AET, which gives x_0/t_0, are both required. Although local evapotranspiration values for Jenny's sites cannot be found now, we accessed regional evapotranspiration rates to use in the calculation of P – AET. These are approximately 100 mm/yr throughout the Mojave Desert and 350 mm/yr along the Great Plains (Sanford and Selnick 2013). It is equally impossible to find local values for the particle size distribution. However, it is possible to estimate typical particle-sizes on the Great Plains by considering data from dust storms (Hagen and Woodruff 1973). Those authors considered data from 39 stations from southern Texas to Montana over a period of 10 years (approximately 4700 dusty days) and found a median particle size of 50 μm. Since the winds on these days were predominantly southerly, sources of the dust particles were also predominantly from the same provenance. Pewe et al. (1981) investigated desert dust in Arizona, and found that the particle sizes ranged from 5 μm to 50 μm, with arithmetic mean 27.5 μm, but geometric mean value 15.8 μm. Wells et al. (1985) investigated soil mantles in the Cima basalt field (eastern Mojave) and found a very similar particle size distribution to that of Pewe et al. (1981), but with diameters twice as large, i.e., from 10 μm to 100 μm (the authors mention a maximum size of 125 μm). A range of 10 μm to 100 μm gives a geometric mean diameter of 31 μm, but the estimate of 125 μm as the upper bound yields 39 μm. The similarity of the Av horizon across the Mojave Desert indicates a consistent source of particles in eolian dust deposition (Reheis and Kihl 1995), meaning that the particle diameters from Wells et al. (1985) are reasonably characteristic of the region, though they would appear to be a little large for the Sonoran desert in Arizona.

Using values of 39 μm and 50 μm for x_0 for the Mojave Desert and the Great Plains, respectively, leads to the predictions for calcic horizon depths as a function P of – AET (Fig. 2). In this comparison we emphasize that zero adjustable parameters were used; every parameter in the equation is independently predicted from surveys. Residual scatter is presumably due largely to the absence of limits on the minimum age of the deposits, which are constrained only to be no more than 15,000 years old. However, local variability in AET and in particle size distributions will also contribute.

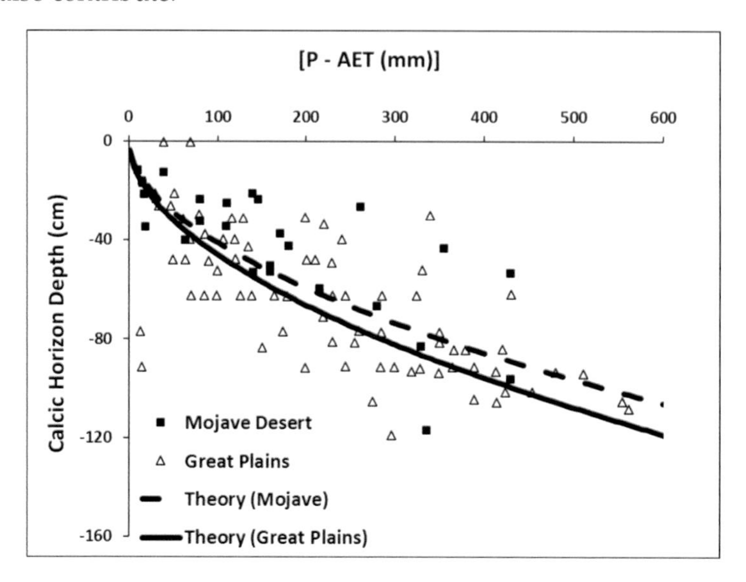

Figure 2: Calcic horizon depths in post-glacial soils as a function of P - AET in the Mojave Desert and on the Great Plains. Theory is the solute transport distance from Eq. (4) using values for x_0 taken from references listed in the text, the regional AET values, which define the ratio of x_0/t_0 from Sanford and Selnick (2013).

Retallack and Huang (2010) give depths to gypsic horizons around the world with the same age constraints as Jenny (1941), namely Holocene, though these are stated to be 13,000 years maximum in age. Two difficulties present themselves, however. The sources are more widely spread, and documentation for ET is more difficult to find, whereas knowledge of particle sizes is impossible. Without specific knowledge of the particle sizes, a reasonable guess for x_0 would be the arithmetic mean value of the particle sizes for the Great Plains and Mojave Desert, covering two distinct ecotones and substrates that are nevertheless susceptible to calcic horizon development. A conservative estimate for ET would be the regional ET of the Mojave Desert, 100 mm/yr. However, even this small value for ET leads to more negative than positive values of P – ET at zero horizon depth. Constraining the average P – ET value to be zero when gypsic horizon depths are zero, leads to a choice of ET = 60 mm/yr, as shown in Fig. 2, while leaving an uncertainty (standard deviation) of about 35 mm/yr, meaning that at the lowest precipitation values, our results are consistent with a range of ET values between about 25 mm/yr and 95 mm/yr.

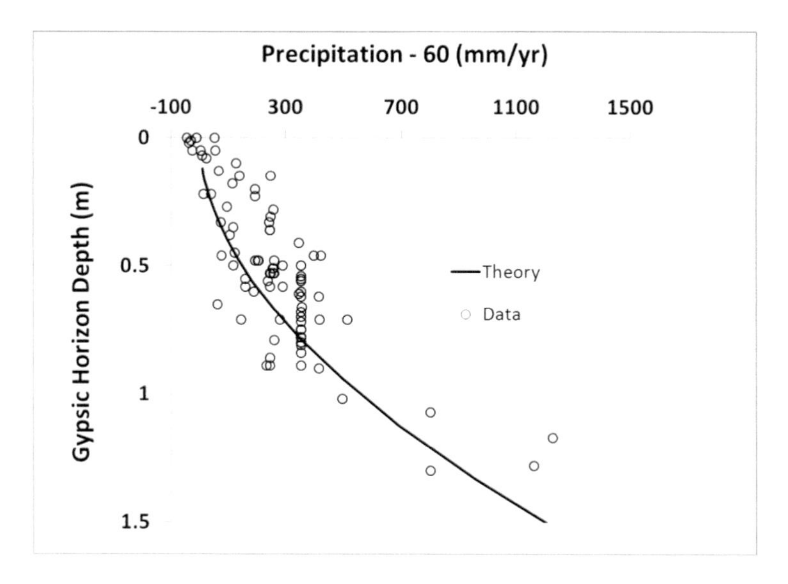

Figure 3: Gypsic horizon depths for post-glacial soils around the world as a function of precipitation less a minimum AET value of 60 mm/yr. The theory is taken again from the solute transport distance given in Eq. (4). The particle size, x_0, was chosen to be the mean values of the Great Plains and Mojave Desert soils above.

5.2 Soil Depth

Soils are complex biological entities, which have often been likened to organisms (Hillel 2005). The existence, or non-existence, of an organism in a colony of living entities is properly a topic in complexity theory. Nevertheless, the classification of soils has traditionally been heavily dominated by physical characteristics, such as color. In particular, the topsoil, or A horizon, is often identified as the brown layer, and the subsoil, or B horizon, in terms of its redness. The brown color at the top is a sign of carbon content, while the red is typically an indication of oxidation of, e.g., iron. The depth to the bottom of the subsoil can thus reasonably be equated to a chemical weathering depth. Chemical weathering rates are known to decline by up to five or six orders of magnitude over time scales from weeks to 6 Ma (White and Brantley 2003). Similarly, soil formation rates decline by the same amount over the same time frame (Egli et al. 2014). Thus, the general success in defining calcic and gypsic horizon depths in terms of solute transport distances invites an interpretation of soil depths within the same framework. How well does this work? From climatic variability the range of yearly precipitation rates ranges from 0.002 m to 10 m. But a maximum flow rate is about three times larger. Accounting for run-on and run-off might lead to a somewhat larger maximum infiltration rate than 10 m/yr. If the range of pore-scale flow rates in porous media from around 0.02 m/yr to approximately 20 m/yr is applied to the scaling relationship with 30 μm soil particle size (a rather typical silt particle diameter), one generates Fig. 4.

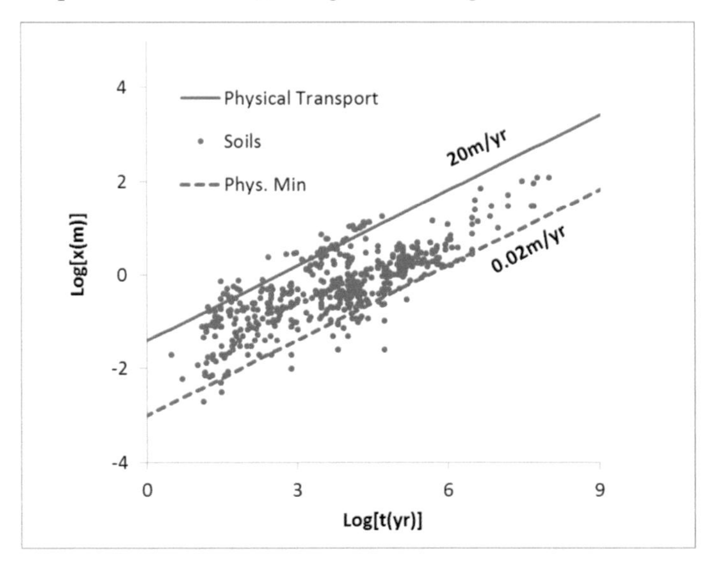

Figure 4: Depths and ages of approximately 50 of the world's soils (over 600 data points). References given mainly in Hunt (2016b). Comparison is with Eq. (4) using 30 μm for x_0, and 20 m/yr as the maximum pore-scale flow rate and 0.02 m/yr as a minimum. In the Atacama Desert, flow rates and order of magnitude smaller are noted (Owen et al. 2013), but soils in the traditional sense do not really exist. Such a large value of t_0 (a factor 10 larger), would generate one additional half order of magnitude variation; nevertheless, the soils accessed are all from wetter regions, and the additional unexplained variability is more likely due to occasional relevance of smaller particle sizes.

Although the scaling relationship for typical x_0 and known range of t_0 values generates the envelope of soil depths as a function of time over an extraordinarily wide range of time scales, not all soils follow the prediction of Eq. (4) individually. Several reasons for such discrepancies can be found. The site-to-site variability in x_0 is not accounted for, for example. At longer time intervals, particularly, soil erosion can remove soil as fast as it is produced, in which case a steady-state soil depth is achieved, with no further increase. When erosion can be neglected, i.e., when soil development does not approach steady-state, Eq. (4) should be applicable. Figure 4 shows a collection of soils with time dependent depths that follow the scaling relationship of Eq. (4).

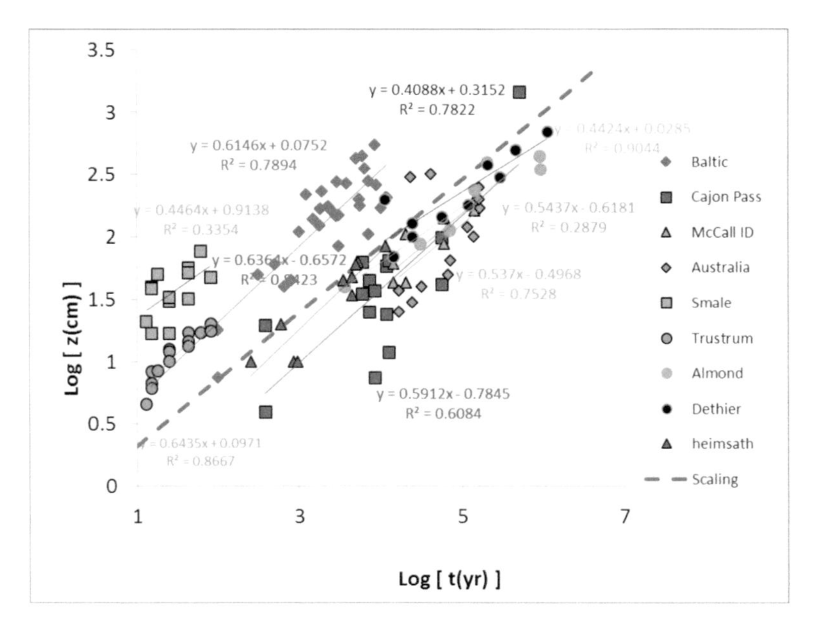

Figure 5: Soil depths as function of ages for nine soils (Heimsath et al. 2006, 2010, Trustum and de Rose 1988, Alexandrovskiy 2007, Smale et al. 1997, Colman and Pierce 1986, McFadden et al. 1987, Dethier et al. 1988, Almond et al. 2007) showing substantial individual agreement with Eq. (4). While generally compatible with the interpretation in terms of the role of the water flux q, because the New Zealand soils (Smale et al., 1997) and Trustum and de Rose (1988), and the Baltic and Russian soils (Alexandrovskiy 2007), all in humid regions, form more rapidly, while the Australian (Heimsath et al. 2001a; 2006; 2009; 2010) and Cajon pass (McFadden and Weldon 1987) soils in arid regions form more slowly, the predicted proportionality to $q^{1-1/Db} \approx q^{1/2}$ needs to be tested directly. In particular, for precise agreement with the predictions of Eq. (4), the Baltic soils would require the use of x_0 of approximately 150 μm. Since a set of these soils is referred to as "sandy" (Alexandrovskiy 2007), this is possible, though not confirmed.

5.3 Soil Production and Relationship with Chemical Weathering

Equation (3), using $D_b = 1.87$, gives the soil production rate as the temporal derivative of the soil depth. When represented as a function of soil depth, the soil production rate is independent of the age t of the soil, but is proportional to the –0.87 power of the depth, and directly proportional to the infiltration rate. Both of these predictions are validated separately (Figs 6 and 7). Although the scatter in the data as a function of depth is quite large, making it somewhat difficult to distinguish between our power-law prediction and the Heimsath group's exponential phenomenology, when other data are analyzed as a function of time, the scatter will be less.

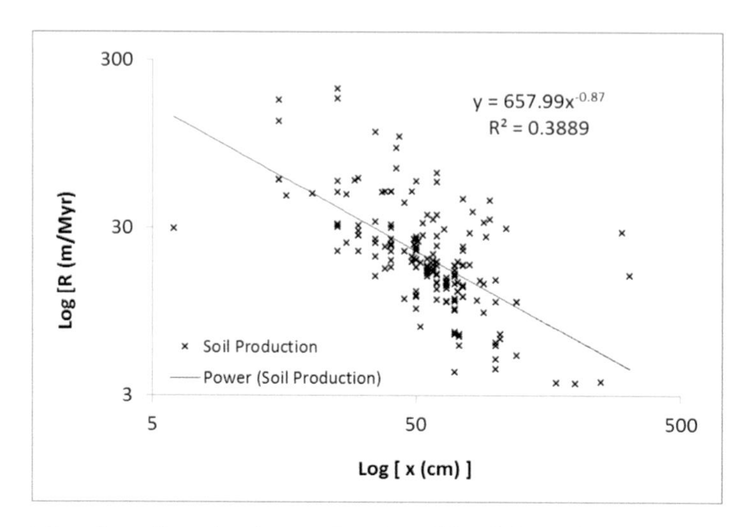

Figure 6: Data for soil production as a function of depth from all 10 published studies of the Heimsath group (Burke et al. 2006; 2009, Dixon et al. 2009a, b, Heimsath et al. 1997, Heimsath et al. 1999, Heimsath et al. 2001a, b, Heimsath et al. 2002, Heimsath et al. 2006, Heimsath et al. 2009). The extracted power agrees with prediction, but scatter is large. Intra-site scatter is presumably due to various heterogeneities; inter-site scatter is due mostly to infiltration rate differences (see Fig. 7).

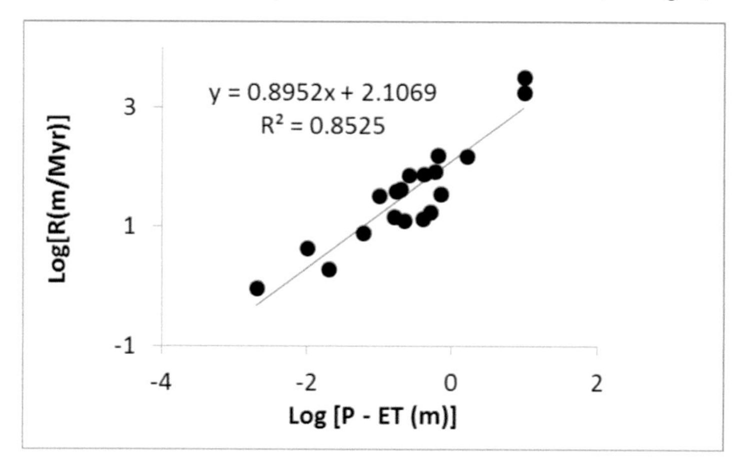

Figure 7: Data from Amundson et al. (2015) for the soil production rate as a function of precipitation. The reported values for soil production are parameters extracted from exponential fits to data such as shown in Fig. 6, thus eliminating the intra-site variability in that figure. In those sites where it was possible to find the AET value (from the Australian Bureau of Meteorology, or from Sanford and Selnick 2013), these values were used for the ordinate. Note, however, that in the two extreme cases of the Atacama Desert (P = 0.002 m/yr) and the New Zealand Alps (P = 10 m/yr) AET data were not available, and P was used as a substitute.

Represented as a function of soil depth, it is possible to modify Eq. (3) for soil production to incorporate a constant soil erosion rate, A, and then to find a general solution for the soil depth as a function of both erosion and soil production. A steady-state solution exists in which the soil production and erosion rates are identical. More generally, the soil depth increases initially in accord with Eq. (4), but reaches an asymptote in the large time limit.

Some of the data for soil depth as a function of time suggest such an asymptotic approach to a steady-state situation. The question is whether published (or reasonable) values of A_0 explain the discrepancies between observed and predicted soil depths at larger times for e.g., the Norwegian, Canadian, Franz Josef, Peruvian, Merced River alluvial terraces, and Himalayan soils (Schülli-Maurer et al. 2007, VandenBygaart and Protz 1995, Stevens 1968, Goodman et al. 2001, White et al. 1996, Jacobson et al. 2002). In these comparisons landscape denudation rates will be applied as the constant A_0.

For the particular case of the Merced terraces, ages ranging from 200 years to 3×10^6 years, mean particle sizes are (White et al. 1996) 170 μm, 320 μm, 650 μm, and 720 μm for the 10 kA Modesto, 40 kA Modesto, 330 kA Riverbank, and 600 kA Turlock Lake examples, respectively, with overall mean = 465 μm. The precipitation is 300 mm (White et al. 1996), and on the western slopes of the Sierra a typical AET fraction is 50% (Sanford and

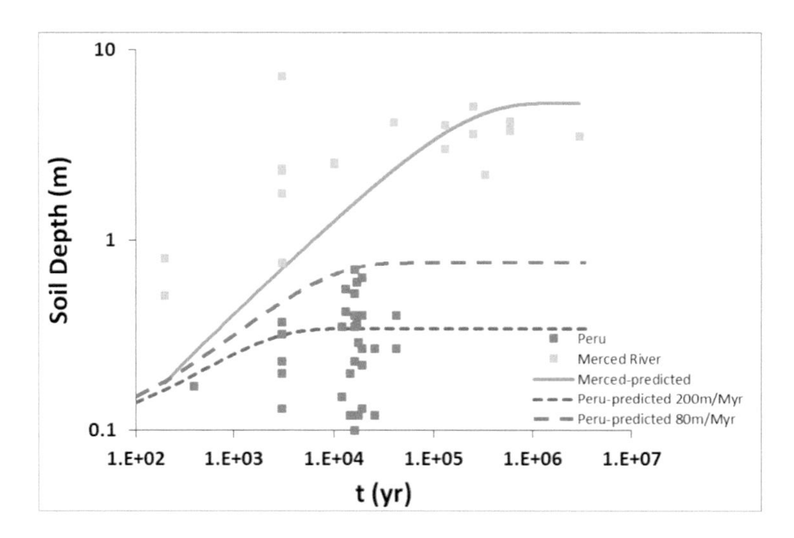

Figure 8: Estimation of erosion effects on soil depth using, for Merced terraces, particle sizes 465 μm and infiltration rate 0.15 m/yr (White et al. 1996), denudation rate 24 m/Myr (Riebe et al. 2001) and for Peru, particle size 20 μm flow rate 1.8 m/yr Peru (Goodman et al. 2001) (known precipitation is 1.5 m), denudation rates 80 m/Ma and 200 m/Ma (Abbühl et al. 2010).

Selnick, 2013), leaving ca. 150 mm as infiltration. White et al. (1996) give an erosion rate of only 2 m/Myr, but this value is closer to erosion rates in the Namibian desert (Bierman and Nichols, 2004). Riebe et al. (2001) give denudation rates across the northern Sierra Nevada that range from 24 m/Myr to 61 m/Myr, with little dependence on longitude or elevation. Stock et al. (2005) report a denudation rate of 20 m/Myr in the last 3.5×10^6 years for the Sierra Nevada province. We choose the minimum value from Riebe et al. (2001), since it is roughly in accord with the value given by Stock et al. (2005). Interestingly, Bullen et al. (1997) report for the same region that chemical weathering rates decline by a factor of 30 over the times 10 kyr to 30 Myr, in accord with the $t^{-0.62}$ dependence of chemical weathering rates reported by White and Brantley (2003), which yields a factor of 34. The topic of chemical weathering is considered in detail later.

5.4 Steady-state Conditions and Landscape Equilibrium

In principle steady-state conditions, with equal soil erosion and production rates, can exist. However, this is a rarity. In order to address this issue, it was necessary to expand the tabulated data sets from published sources using surface age dating methods. Soil ages, t, are determined using (Ivy-Ochs and Kober 2008),

$$C_{(t)} = \frac{P_{(0)}}{\lambda}(1 - e^{-\lambda t}) \tag{11}$$

In Eq. (11), $C_{(t)}$(atoms/g) can be the concentrations of Be^{10} or Al^{26}, $P_{(0)}$ the nuclide production rate at the sampling site: $P_{(0)} = 0$ for Be^{10} and $P_{(0)} = 36.8$ for Al^{26} at sea level and high latitude (>60°) according to Heimsath et al. (1999, 2001a, b). $\lambda = \ln(2)/t_{1/2}$ is the radioactive decay constant and $t_{1/2} = 1500000$ yr for Be^{10} and $t_{1/2} = 701000$ yr for Al^{26}.

For an exponential soil production function favored by the Heimsath group, soil production tends much more rapidly to zero at greater depths, allowing an easier approach to equilibrium. But with a power-law decay in soil production, equilibrium is not so easily attained. In order to address the question as to whether steady-state conditions arise, the following differential equation must be solved (which we have done numerically for $A(t)$ = constant A_0),

$$\frac{dx}{dt} = \frac{1}{1.87}\left(\frac{x_0}{t_0}\right)\left(\frac{x}{x_0}\right)^{1-D_b} - A(t) \tag{12}$$

where $A(t)$ is a time-dependent soil removal rate. In the simplest case, $A = A_0$, but if changes in climate are to be considered, the more general possibility must be addressed. Setting $dx/dt = 0$ generates an equilibrium depth, which can be calculated in terms of the erosion rate; in fact, the equilibrium depth is very simply represented as the product of the particle size with the ratio of the infiltration rate to the erosion rate raised to the power 1/0.87 = 1.14. Such a formula for equilibrium soil depths (and infiltration rates of 1 m/yr, particle size of 30 μm) yields about a 5 cm soil depth for 1000 m/Myr erosion rate, increasing to about 10 m for 10 m/Myr erosion rates.

Using x_0 = 30 μm and t_0 consistent with infiltration rates bounded by 0.31 m/yr (high = Snug) and 0.11 m/yr (low = Nunatock River) generates Fig. 9a. While, at larger erosion rates the prediction is accurate, there is a failure to predict soil depths at small erosion rates. The implication is that at such small erosion rates, the time required to reach steady-state depth is very long. In particular, if the soil ages derived from Eq. (11) are applied,

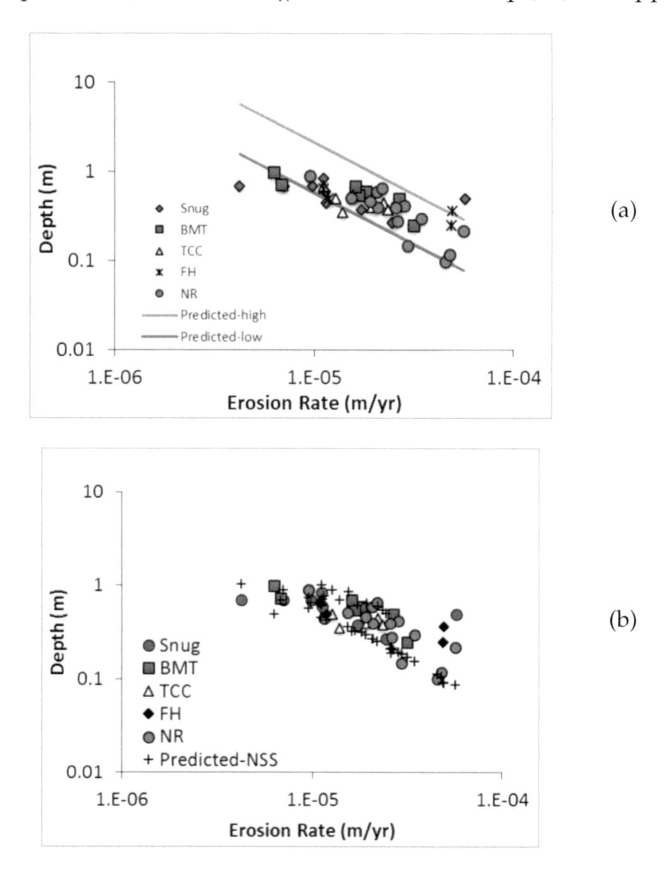

Figure 9: (a) Predicted dependence of soil depth on erosion rate assuming equality of soil erosion and production rates. SGM or San Gabriel Mountains and TV, or Tennessee Valley are located in California, NR Nunnock River, FH Frog's Hollow, BM Brown Mountain, and Snug are all located in southeast Australia. **(b)** Calculated values of soil depths from Eq. (5) using surface ages determined by cosmogenic dating (Eq. (11)).

the predictions match the median observations for all erosion rates in Fig. 9b. But, now a large scatter in depths for high erosion rates reflects a more discontinuous erosion process due, e.g., to landsliding (San Gabriel Mountains, So. Cal., USA).

5.6 Chemical Weathering and More Advanced Solute Transport Treatment

The process of chemical weathering has often been suggested to be limited by advective solute transport rates (Maher 2010, Salehikhoo et al. 2012, Hunt et al. 2015, to name a few). More specifically, Hunt et al. (2015) proposed that chemical weathering rates could, under a wide range of conditions, be scaled using solute transport velocities. In other words, the decline in chemical weathering rates of time follows the same functional dependence as that of the solute transport velocity. Such identification is precisely compatible with equating the soil depth to the solute transport distance, as long as the soil depth is considered equal to the chemical weathering depth, i.e., the depth to the bottom of the B horizon.

A potentially important distinction between the temporal evolution of soil depths with the discussion of gypsic and calcic horizon depths is that the latter may be taken to be forming in a pre-existing medium. Soils may form in a variety of ways, both in pre-existing porous media, such as alluvial deposits, or directly from weathering of bedrock, though sometimes highly fractured, such as saprolite. In the case of soil formation, since the bedrock and the soil are in series for vertical flow, the bedrock flow properties and particle size distributions may be the appropriate choices for the scaling function parameters. In particular, very low bedrock hydraulic conductivity values may suppress the relevant flow rate significantly compared with P – ET leading to significant run-off. Run-off may contribute more to soil erosion than to soil formation. The grain sizes of the bedrock may also be more suitable in the parameter x_0, since weathering may be focused along the boundary of the bedrock and the soil. These questions are, as yet, not answered.

While gypsic and calcic horizon depths above were studied as functions of precipitation at (nominally) a single time scale, the following will address chemical weathering rates and soil formation rates over a wide range of time scales, from days to 100 million years, but neglecting precipitation or particle size dependence. Over such a wide range of time scales, the predictions of solute transport derived from predictions of the full solute arrival time distribution, and those from the simple scaling function approach do differ, making it necessary to include a discussion of the full solute transport theoretical approach. Notably, the distinction in the predictions of the solute velocity is also associated with complex behavior in the dispersivity, which itself depends on the solute velocity.

The theoretical description centers around two problems: (1) finding the distribution of such system-crossing paths with given flow-rate limitations, and (2) finding the influence of the topology of these flow paths on the arrival time distribution. The first portion combines techniques of critical path analysis and the cluster statistics of percolation, and is also relevant to the determination of an upscaled hydraulic conductivity, K (Hunt 1998). It is the second portion, which addresses the fractal topology of the flow paths, where the scaling arguments of percolation theory enter. Note that the first part of this framework was already put in place in Hunt (1998). There it was also pointed out that, in heterogeneous media, the upscaling must be based on the resistance value distribution, rather than resistivity. Given that the predicted hydraulic conductivity of the medium is thus based on the critical conductance value from percolation theory (Ambegaokar et al. 1971, Pollak 1972, Friedman and Seaton 1998, Bernabe and Bruderer 1998, Hunt 2001), this rate-controlling, or bottleneck, conductance value defines a solution of Kirchoff's laws on a network representing the porous medium. This makes the upscaling technique a solution of Laplace's equation over the medium.

The solute arrival time distributions calculated have approximate power-law (heavy) tails, similar to those modeled in Continuous Time Random Walk (CTRW), or the Fractional Advection-Dispersion Equation (FADE). The chief difference is that the power relates directly to a known quantity from percolation theory, namely the fractal dimensionality of the backbone, D_b. The fact that, under a wide range of conditions, D_b takes on only four distinct values (Sheppard et al. 1999) associated with differing conditions of saturation and the dimensionality of the connections between the flow paths, makes the theoretical approach predictive, as the results of the calculation do not depend strongly on medium characteristics. Even more restrictively, the results described here are *all* generated by solute transport under conditions of three dimensional connectivity and saturated flow conditions for which a single value, $D_b = 1.87$, is relevant.

The likelihood that a path with a given limiting resistance (or, equivalently, conductance) value can be found, that physically spans a system of a given length, depends ultimately on what local resistance distribution is chosen. In all our comparisons with experiment, a fractal distribution of pore sizes has been used (Rieu and Sposito 1991); note, however, that the power of the power-law solute arrival time distributions does not relate to the pore size distribution, instead it relates to D_b. Indeed the pore-size distribution has an influence on the arrival time distribution only in a relatively narrow range of times near the distribution peak (Hunt et al. 2011), at least in sufficiently large systems (hundreds of pore separations or larger).

For clarity, the continuum analogue to the discrete random fractal model of Rieu and Sposito (1991) (RS) assumes a pdf, $f(r) = r^{2-D}$ (3-D)/

r_m^{3-D} for pore radii, r, within the range $r_0 \leq r \leq r_m$ for pore-space fractal dimensionality, D. This particular result has been shown to generate the known RS porosity and water retention results (Hunt 2001).

Hunt (1998) and Hunt and Skinner (2008) used the Gaussian form of the cluster statistics (Stauffer 1979) given in terms of the bond fraction, p, its critical value, p_c, and the volume of the cluster, s, to derive the relative probability, $W_p(g \mid x)$, that a system of length x could be spanned by an interconnected cluster with smallest conductance value, g. The requirement, to generate an expression in terms of the useful parameters, g and x, necessitated the transformation of variables from p to g, based on the cumulative pore size distribution from the RS model, and a transformation from s to x, based on the mass fractal dimensionality of the percolation cluster. The latter derivation required application of the dimensionally-dependent scaling law for percolation exponents (Hunt 1998). The result for $W_p(g \mid x)$, applying the RS pore-size function, was,

$$W_p(g|x) \propto Ei\left[\left|1-\left(\frac{g}{g_c}\right)^{\frac{3-D}{3}}\right|^2 \left(\frac{x}{L}\right)^{\frac{2}{v}}\right], \qquad g_{\min} < g < g_{\max} \tag{13}$$

Here, g_c is the critical value of g for percolation (the maximum possible value of the bounding g, for which a spatial representation of all the local conductance values, $g > g_c$ forms an interconnected, system-spanning cluster), Ei is the exponential integral, L is the heterogeneity length, i.e., pore scale, and $\nu = 0.88$ (in three dimensions) is the critical exponent for the correlation length. The choices $\nu = 0.88$ and $D_b = 1.87$ are thus both required by saturation conditions and three dimensional flow connectivity. The values $g_{\min}$ and $g_{\max}$ are generated from r_0 and r_m, (Ghanbarian-Alavijeh et al. 2012):

$$g_{\min} = g_c\left[\frac{1-\varphi}{1-\theta_1}\right]^{3/(3-D)} \tag{14}$$

$$g_{\max} = g_c\left[\frac{1}{1-\theta_1}\right]^{3/(3-D)} \tag{15}$$

with θ_t the percolation threshold, expressed as a critical moisture content, and φ the porosity.

Next, the most likely arrival time, t_g, for solute transported across a cluster with limiting conductance value, g, is required. In this calculation, the contributions to t_g from the resistance distribution on the cluster and the topological slowing from the path tortuosity and connectivity are multiplied together. This strategy is, in principle, a generalization of the treatment of resistance R in fundamental physics texts as a product of resistivity, ρ, and geometrical factors, $R = \rho l/A$, with l the system length and A the cross-sectional area. Summing the times along a path of resistances connected in series is consistent with the formulation for the equivalent resistance of such a path in terms of its cumulative resistance. The result was (Hunt and Skinner 2008):

$$t_g = t_0\left(\frac{x}{L}\right)^{D_b} h(g) \tag{16}$$

where L is again the heterogeneity length scale, which for microscopic applications, such as mineral weathering and soil production, can be considered a pore scale, t_0 is an advective pore crossing time, and

$$h(g) = \frac{D}{3-D}\frac{1}{(1-\theta_t)^{v(D_b-1)}}\left[\left(1+\frac{\theta_1}{1-\theta_1}\right)\left(\frac{g_c}{g}\right)^{1-D/3}-1\right]\left[\frac{1}{\left(\frac{g}{g_c}\right)^{1-D/3}-1}\right]^{v(D_b-1)} \tag{17}$$

The arrival time distribution is then found from a probabilistic identity with the constraint that the solute flux must always be proportional to the water flux. Since using the rate-limiting conductance, g, to define the cluster flow, is consistent with using critical path analysis (Ambegaokar et al. 1971, Pollak 1972, Friedman and Seaton 1998, Bernabe and Bruderer 1998, Hunt, 2001) to solve Laplace's equation, the same conceptual link between flow and transport is established in the percolation theoretical formulation as is noted in stochastic treatments (Dagan 1987, 1991, Dagan and Neuman 1997, Freeze 1975, Gelhar 1986, Gelhar and Axness 1983, Chao et al. 2000, Neuman and di Federico 2003, Winter et al. 2003). Practically speaking, this result underlies the explicit decision (Hunt 2016) to make the solute transport rate proportional to the flow rate. The proportionality of solute transport to flow requires a proportionality of the solute flux to the limiting resistance value, g, leading to multiplication by this factor in the probabilistic identity in Eq. (18) below (Hunt and Skinner 2008, 2010):

$$gW_p(g\,|\,x) = W_p(t\,|\,x)dt/dg \tag{18}$$

and making the factor t_0 in Eq. (16) inversely proportional to the flow rate (Ghanbarian-Alavijeh et al. 2012). Numerical solution of Eq. (18) was complicated by the necessity to invert the equation for $t(g)$ to solve for $g(t)$, which has multiple solutions (between 1 and 3 for monomodal distributions, but up to 7 for bimodal distributions), and secondarily on account of the need to approach the singularity at $g = g_c$ closely in order to find the functional form of the heavy-tail of the distribution.

Figure 10 demonstrates the relationship between the predicted values of solute velocity as obtained from the simple scaling relationship (Eq. 3), and from Eq. (18) above. Note that, particularly in view of experimental scatter, it would be difficult to distinguish the relevance of the predictions of Eq. (3) and Eq. (8) without having data that cover five or more orders of magnitude of time scales.

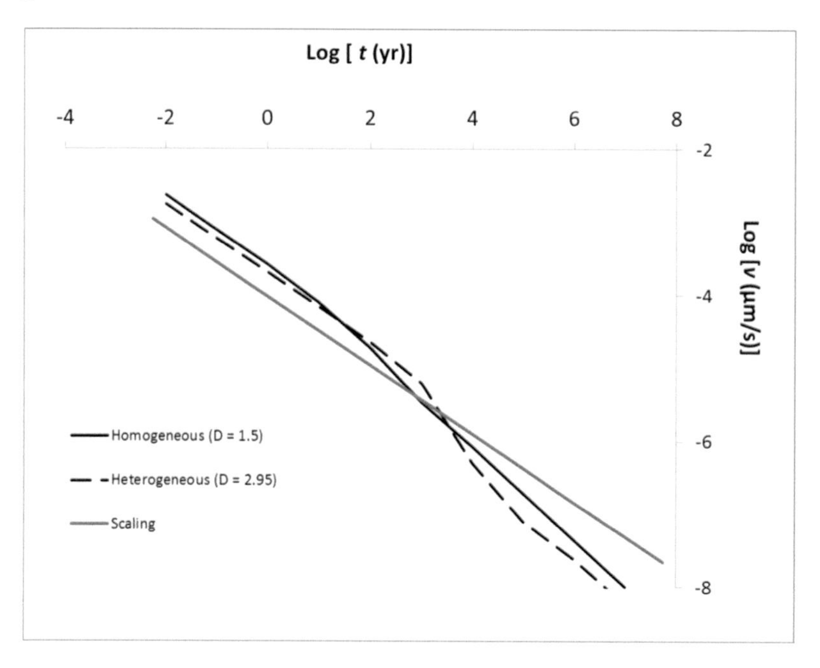

Figure 10: Comparison of the predictions of the velocity, v, of the solute centroid from simple scaling theory and from the full theoretical machinery for calculating the spatial solute distribution as a function of time. Interpretation is simplified with the scaling relationship, but it does sacrifice some accuracy at the later time scales if substantially more than five decades of time are investigated.

In the next figure is shown how successful the prediction of chemical weathering rates, soil production rates, and C and N sequestration rates as a function of time is using a typical infiltration rate and approximately one order of magnitude variation in either direction (see Fig. 11).

5.7 Vegetation Growth

Comparison with several dozen individual tree studies confirmed the validity of the scaling relationship (Eq. (5)) for vegetation growth rates. One (modified) bilogarithmic plot (Fig. 12) and one linear plot (Fig. 13) are shown here. Figure 13 employs data from logged trees, for which it is possible by coring to construct the history of the tree height as a function

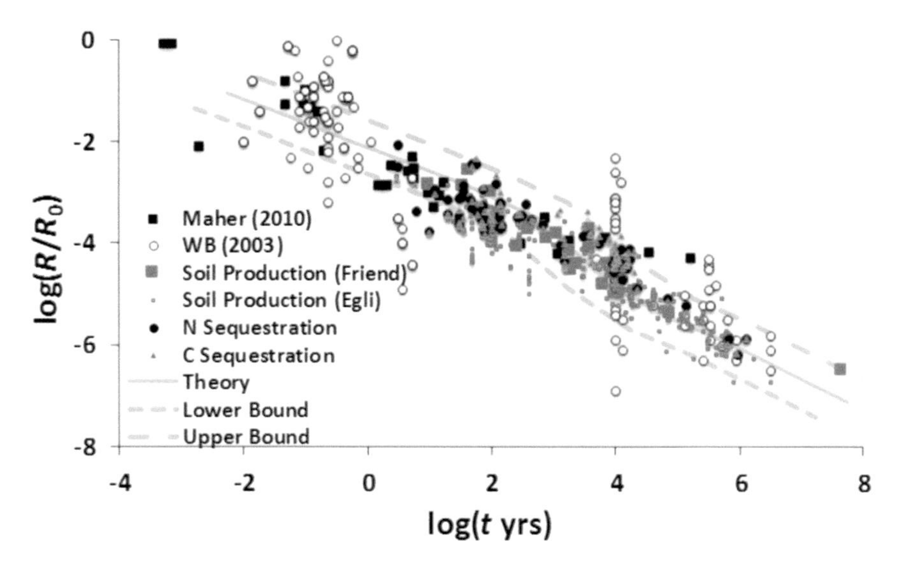

Figure 11: Comparison of the predicted chemical weathering, soil production, and soil N and C sequestration rates with theory. The original figure, from Hunt et al. (2015), compared the theoretical solute transport velocity with chemical weathering rates. The same functional form is now shown to describe the remaining properties as well. WB (2003) refers to White and Brantley (2003), Egli refers to Egli et al. (2014), while the N and C sequestration data are from Egli et al. (2012).

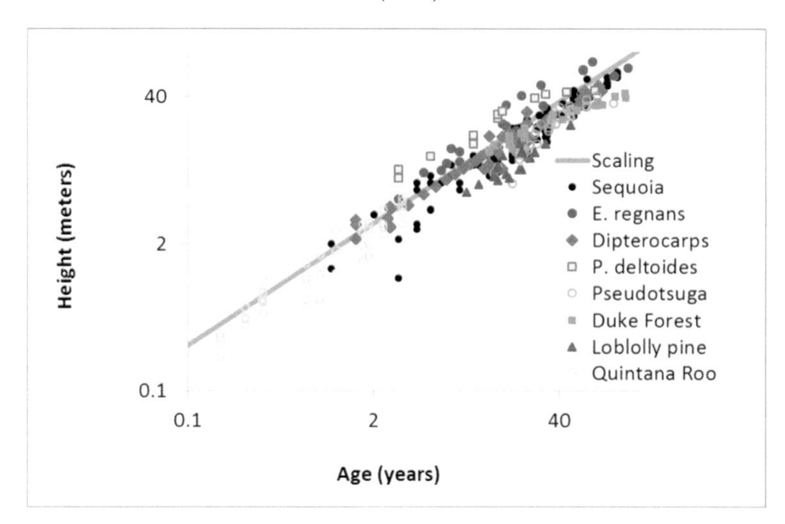

Figure 12: Data for the growth curves of a number of fast-growing tree species, from the tropics to the temperate zones. The data from Quintana Roo (Allen et al. 2003) are for trees three years of age or less. Remaining data sources are given in Hunt and Manzoni (2016).

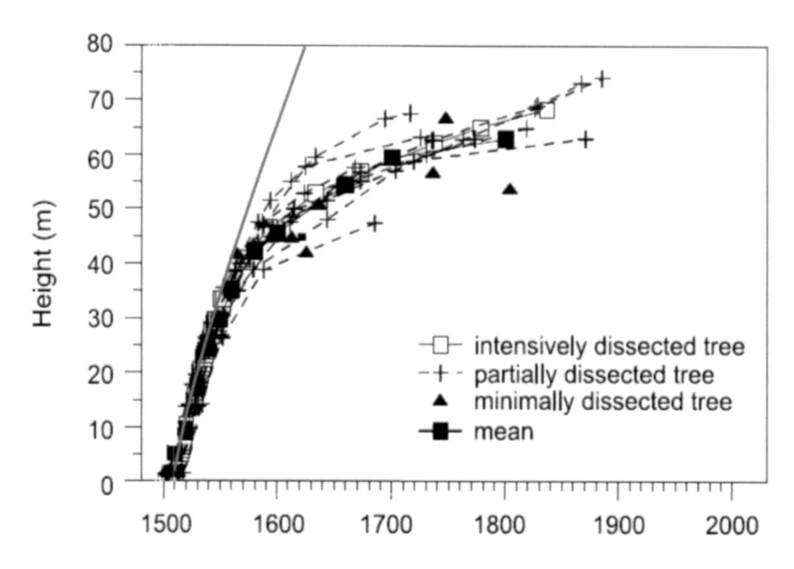

Figure 13: Data for logged *Pseudotsuga menziesii* from the Pacific Northwest (Winter et al. 2002), showing the life growth curves of a large number of individual trees. Our predicted scaling function describes their growth curves almost exactly in the first approximately 40 years without use of adjustable parameters.

of age (Winter et al. 2002). In Hunt and Manzoni (2016), the comparison (Fig. 12) used the parameters of 1 μm and 32 m/yr as length scale and maximum flow rate. Hunt and Manzoni (2016), however, argued that 10 μm is a better length scale for pore or xylem diameter (compatible with 30 μm for particle sizes above), and the above comparison for soil development used 20 m/yr as a maximum flow rate. Thus, in order to be completely compatible with Fig. 4 for soil development, these parameters are used here as well. Except for the trees of Quintana Roo (Yucatan) (Allen et al. 2003), the remaining data sources were used in Hunt and Manzoni (2016). Since a dozen references were required for the *Sequoiadendron giganteum* data alone, the interested reader is referred to Hunt and Manzoni (2016) for the data sources. It is important that the species and individuals shown are noted for their rapid growth rates and large xylem diameters. The adaptive advantage of their rapid growth rates fits their ecological niche, namely in their early response to fire (Sequoia, Douglas fir), flood (*Populus deltoides*), or canopy gap formation (Dipterocarps) (Hunt and Manzoni, 2016, and references therein). Further, *deltoides* is noted to prefer sandier overbank deposits with high water contents throughout the growing season.

Although data for fast-growing woody plants were collected from a large number of sources, one large database containing 6650 plant height–plant age data pairs (Falster et al. 2015) was accessed as well. The latter data set did not select for large size, but for reliability. Equation (5) generates a prediction that is approximately a maximum growth rate, but it is important to generate as well a minimum growth rate in order to constrain the data. Later, it is shown that ecosystem AET measurements range from 20 mm/yr in the Namibian desert (Seeley 1978) to 1650 mm/yr in tropical evergreen rainforests (Box et al. 1989). Such a ratio of maximum to minimum flow rate (a factor 82) constrains an analogous range of pore scale flow rates, from 20 m/yr to 0.24 m/yr, as shown in Fig. 14. Note that scaled up to one half year, the pore-scale flow rates of 20 m/yr to 0.24 m/yr yields the range 1650 mm to 20 mm (in six months), thus reproducing exactly the range of ecosystem AET values shown in a later figure.

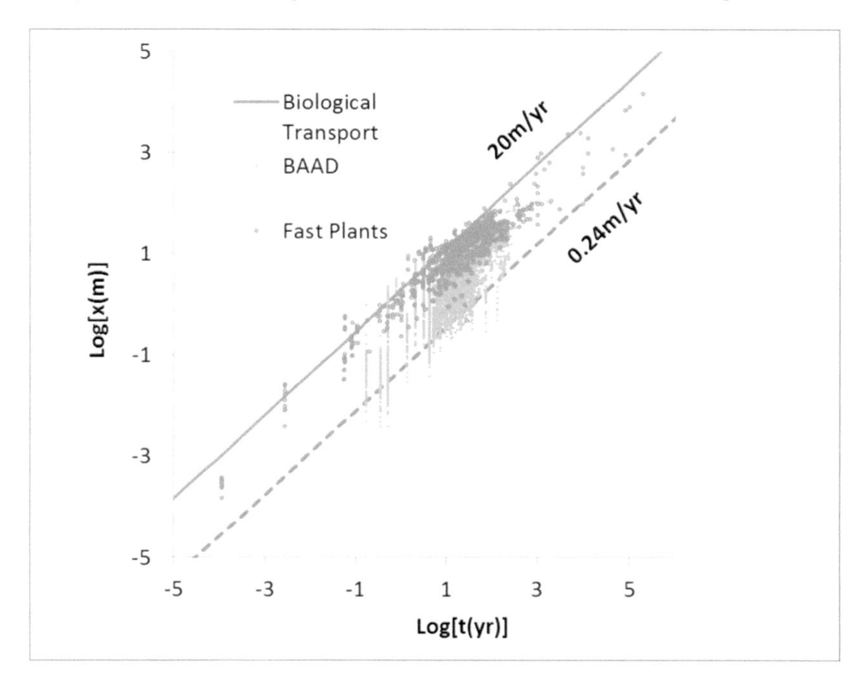

Figure 14: Comparison of the scaling relationship for vegetation growth using known limiting values of evapotranspiration with "fast-growing plants," gleaned from dozens of studies (Hunt and Manzoni 2016) and the single BAAD database (Falster et al. 2015) with 6650 data pairs for woody plant height and age.

5.8 Summary of Spatio-temporal Scaling

When the predictions of soil development, vegetation root radial extent (or height), natural and intensively managed, are put together, Fig. 15 results.

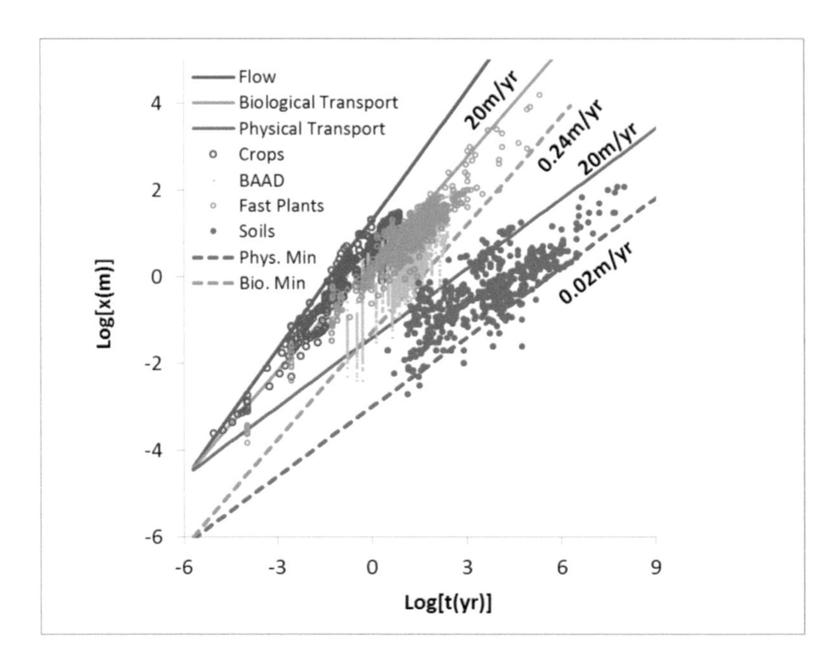

Figure 15: Simultaneous representation of the predictions for a maximum flow distance, a maximum and minimum plant size, and a maximum and minimum soil depth, all as functions of time, over almost fifteen orders of magnitude of time scales. Since the predictions of the bounding growth functions are associated with specific values of transpiration (e.g., 0.02 m to 1.65 m), and that of the soil are associated with the same pore scale flow rates as generate the range of transpiration values, and all three upper bounds correspond to a blend of the typical soil flow rate under saturated conditions (ca. 30 m/yr) and a maximum infiltration rate (10 m/yr), as well as typical fundamental length scales of 10 μm for pores, 30 µm for particles, this representation corresponds to a zero parameter prediction of all three scaling relationships simultaneously. Although the minimum flow rate chosen for soil production is an order of magnitude less than the value chosen for vegetation growth, soil production has been defined even for slower flow rates in the Atacama Desert. But soil depths for such slow flow rates are not available.

5.9 Productivity

Now consider the relationship between net primary production, NPP and AET. Although a scaling prediction is available, it is not yet possible to generate an exact numerical coefficient (prefactor) in the prediction. Yet, it may be possible to account for the magnitude of the *variability* in the coefficient. In particular, one can address the ratio of length scales at six months provided by the distinct predictions for crop heights and natural vegetation heights. At the largest flow rates (20 m/yr, or AET 1650 mm/yr), this ratio is approximately ten; raised to the 1.9 power, this yields 79. At the smallest flow rates (0.24 m/yr), the ratio is about 4.4; raised to the

1.9 power this yields 17. Using these particular values for the predicted range of coefficients, and a one adjustable parameter prediction for the lowest value of the coefficient generates in Fig. 16 an envelope of predicted values of NPP as a function of AET. Note that an alternate means to generate a variability in predictions can utilize a known correlation of N and C values in the soil (Cleveland and Liptzin 2007, Stevenson 1994, Tian et al. 2010a) assumed linear dependences of productivity on nutrients, in particular N (Bouman et al. 1996, van Ittersum et al. 2003, van Dam et al. 1997), and a known variation of organic C contents among the world's biomes (Cao and Woodward, 1998) or soil types (Batjes, 1996) of about 40, to generate a maximum NPP that is 40 times the minimum value for all AET values. The ratio 40 is bounded by the geometric mean (37) and the arithmetic mean (48) of the theoretical ratios (17 and 79) at low and high AET, respectively. The following 17 data sources are acknowledged: Al-Jamal et al. (2000), Bai et al. (2008), Barrett and Skogerboe (1980), Box et al. (1989), Chen et al. (2013), Hillel and Guron (1973), Jensen and Sletten (1965), McNulty et al. (1996), Mogensen et al. (1985), Muldvin et al. (2008), Payero et al. (2008), Rosenzweig (1968), Seeley (1978), Sun et al. (2004), Tian et al. (2010b), Webb et al. (1978) and Zhang and Oweis (1999).

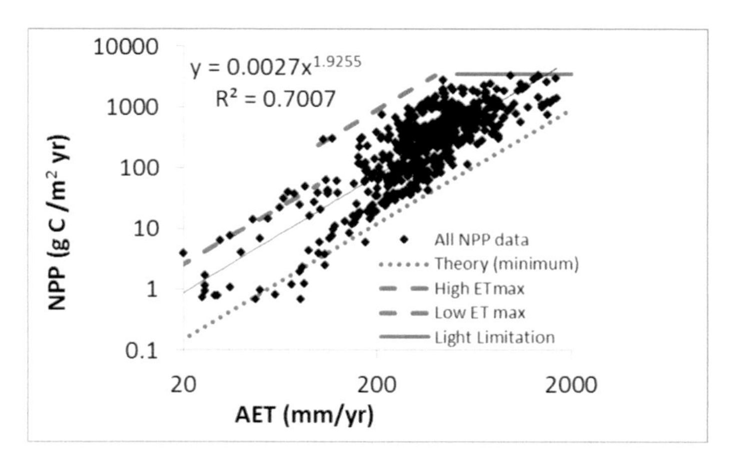

Figure 16: Net primary productivity as a function of actual evapotranspiration for the 17 studies listed in the text. Light limitation is from de Wit (1965). Because the theoretical description is a power-law in form, a single adjustable parameter, which is a numerical coefficient, can generate a theoretical minimum value. The maximum functions at high and low ET (High ET max, Low ET max) are generated from the ratios (79 and 17) of fertilized to unfertilized predictions for maximum root radial extent, as discussed in the text. A ratio of maximum to minimum productivity equal to a factor 40, as deduced from a linear dependence on N content, would lead to a single maximum function nearly halfway between High ET max and Low ET max, in accord with the geometric mean of these ratios.

Note that the slope extracted from the data differs from the predicted value by only 1.5%. The light limitation at about 3500 g C/m^2yr is given in (de Wit 1965), and is very close to the maximum observed NPP in a range of biomes, including tropical savannah and tropical forests. Breaking down the data into natural and agricultural ecosystems does not lead to different mean values of the exponent, but it does add uncertainty to the extracted values.

6. Conclusions

Fractal exponents of percolation theory, which describe such quantities as the backbone of the percolation cluster, tortuosity of optimal flow paths, and the mass of randomly connected objects, are relevant in the biogeosciences. These dimensionalities define rates of soil formation and chemical weathering, plant growth, and give the seasonal net primary productivity, respectively. The results for soil formation and plant growth are conceptually linked through the relevance of the net infiltration rate and the evapotranspiration, respectively. The specific predictions for the scaling of soil depth and vegetation growth use known values of typical pore and particle sizes, and limits on flow rates derived from the literature. The range of transpiration rates available, from tropical evergreen forests to the Namibian and Chihuahuan deserts, describes not only the range of growth rates for natural plants, but also the range of productivities of ecosystems and biomes. The seasonal transpiration rates are compatible with the pore-scale flow rates that are used to predict the evolution of the soil depth. The proposed functional forms of the scaling relationships are verified, as well as the numerical constants present and the relationship with flow and transport conditions (2D, 3D, wetting or saturated), making the entire suite of predictions self-consistent and accurate.

REFERENCES

Abbühl, L.M., K.P. Norton, F. Schlunegger, O. Kracht, A. Aldahan and G. Possnert. 2010. El Niño forcing on ^{10}Be-based surface denudation rates in the northwestern Peruvian Andes? *Geomorphology* 123: 257–268.

Al-Jamal, M.S., T.W. Sammis, S. Ball and D. Smeal. 2000. Computing the crop water production function for onion. *Agricultural Water Management* 46: 29–41.

Alexandrovskiy, A.L. 2007. Rates of soil-forming processes in three main models of pedogenesis. *Revista Mexicana de Ciencias Geológicas* 24(2): 283–292.

Allen, E.G., H.A. Violi, M.F. Allen and A. Gomez-Pampa. 2003. Restoration of tropical seasonal forest in Quintana Roo. In: A. Gomez-Pompa, M.F. Allen, S.L. Fedick, and J. Jimenez-Osornio (eds). The lowland Maya area: Three millenia at the human–wildland interface. Haworth Press, Binghamton, New York, USA. Pp. 587–598.

Almond, P., J. Roering and T.C. Hales. 2007. Using soil residence time to delineate spatial and temporal patterns of transient landscape response. *Journal of Geophysical Research* 112: F03S17, doi:10.

Ambegaokar, V.N., B.I. Halperin and J.S. Langer. 1971. Hopping conductivity in disordered systems. *Phys Rev B*, 4: 2612.

Amundson, R., A. Heimsath, J. Owen, K. Yo and W.E. Dietrich. 2015. Hillslope soils and vegetation. *Geomorphology* 234: 122–132.

Avramov, I., R. Keding and C. Rüssel. 2000. Crystallization kinetics and rigidity percolation in glass-forming melts. *J. Non-Cryst. Solids* 272: 147–153.

Australian Bureau of Meteorology. 2015. http://www.bom.gov.au/jsp/ncc/climate_averages/evapotranspiration/index.jsp. Commonwealth of Australia, Bureau of Meteorology (accessed most recently, Feb. 19, 2016)

Bai, Y., J. Wu, Q. Xing, Q. Pan, J. Huang, D. Yang and X. Han. 2008. Primary production and rain use efficiency across a precipitation gradient on the Mongolia Plateau. *Ecology* 89: 2140–2153.

Barrett, H. and G. Skogerboe. 1980. Crop production functions and the allocation and use of irrigation water. *Agricultural Water Management* 3: 53–64.

Batjes, N.H. 1996. Total carbon and nitrogen in the soils of the world. *European Journal of Soil Science* 47: 151–163.

Bejan, A. 1997a. Constructal-theory network of conducting paths for cooling a heat generating volume. *Int. J. Heat Mass Transfer* 40: 799–816.

Bejan, A. 1997b. Advanced Engineering Thermodynamics, 2nd ed. Wiley, New York.

Bernabé, Y. and C. Bruderer. 1998. Effect of the variance of pore size distribution on the transport properties of heterogeneous networks. *J. Geophys. Res.* 103: 513–525.

Bierman, P.R. and K.K. Nichols. 2004. *Annu. Rev. Earth Planet. Sci.* 32: 215–255.

Bouman, B.A.M., H. van Keuelen, H.H. van Laar and R. Rabbinge. 1996. The school of de Wit, crop growth simulation models: A pedigree and historical overview. *Agricultural Systems*, 52: 171–198.

Box, E.O., B.N. Holben and V. Kalb. 1989. Accuracy of the AVHRR vegetation index as a predictor of biomass, primary productivity, and net CO_2 flux. *Vegetatio* 80: 71–89.

Bullen, T., A. White, A. Blum, J. Harden and M. Schulz. 1997. Chemical weathering of a soil chronosequence on granitoid alluvium: II. Mineralogic and isotopic constraints on the behavior of strontium. *Geochimica et Cosmochimica Acta* 61: 291–306.

Burke, B.C., A.M. Heimsath and A.F. White. 2006. Coupling chemical weathering with soil production across soil-mantled landscapes. *Earth Surf. Proc. and Landforms* DOI: 10.1002/esp.1443.

Burke, B., A.M. Heimsath, J. Chappell and K. Yoo. 2009. Weathering the escarpment: Chemical and physical rates and processes, southeastern Australia. *Earth Surface Processes and Landforms* DOI: 10.1002/esp.1764.

Campbell, G.A. and G. Forgacs. 1990. Viscosity of concentrated suspensions: An approach based on percolation theory. *Physics Review* 41A: 4570–4573.

Cao, M. and F.I. Woodward. 1998. Net primary and ecosystem production and carbon stocks of terrestrial ecosystems and their responses to climate change. *Global Change Biology* 4: 185–198.

Chao, H., H. Rajaram and T. Illangasekare. 2000. Intermediate-scale experiments and numerical simulations of transport under radial flow in a two-dimensional heterogeneous porous medium. *Water Resour. Res.* 36: 2869.

Chen, X., J. Bai, X. Li, G. Luo, J. Li and B. Larry Li. 2013. Changes in land use/land cover and ecosystem services in Central Asia during 1990-2009. *Current Opinions in Environmental Sustainability* 5: 116–127.

Clapp and G. Hornberger. 1978. Empirical equations for some hydraulic properties. *Water Resour. Res.* 14: 601–604.

Cleveland, C.C. and D. Liptzin. 2007. C:N:P stoichiometry in soil: Is there a ''Redfield ratio'' for the microbial biomass? *Biogeochemistry* 85: 235–252.

Colman, S.M. and K.L. Pierce. 1986. Glacial sequence near McCall, Idaho: Weathering rinds, soil development, morphology, and other relative-age criteria. *Quaternary Research* 25: 25–42.

Cox, J.T. and R. Durrett. 1988. Limit theorems for the spread of epidemics and forest fires. *Stochastic Processes and their applications*, 30: 171–191.

Dagan, G. 1987. Theory of solute transport by groundwater. *Ann. Rev. Fluid Mech.* 19: 183–215.

Dagan, G. 1991. Dispersion of a passive solute in nonergodic transport by steady velocity-fields in heterogeneous formations. *J. Fluid Mech.* 233: 197–210.

Dagan, G. and S.P. Neuman (Eds). 1997. Subsurface Flow and Transport: A Stochastic Approach. Cambridge University Press, Cambridge, U.K.

Dethier, D.P. 1988. The soil chronosequence along the Cowlitz River. Washington, US Geological Survey Report 1590F.

de Wit. 1965. Photosynthesis of leaf canopies. Agricultural Research Report no. 663 PUDOC Wageningen.

Egli, M., F. Favilli, R. Krebs, B. Pichler and D. Dahms. 2012. Soil organic carbon and nitrogen accumulation rates in cold and alpine environments over 1 Ma. *Geoderma* 183–184, 109–123.

Egli, M., D. Dahms and K. Norton. 2014. Soil formation rates on silicate parent material in alpine environments: Different approaches–different results? *Geoderma* 213: 320–333.

Falster, D.S., R.A. Duursma, M.I. Ishihara, D.R. Barneche, R.G. Fitzjohn, A. Varhammar, M. Aiba, M. Ando, N. Anten, M.J. Aspinwall, J.L. Baltzer, C. Baraloto, M. Battaglia, J.J. Battles, B. Bond-Lamberty, M. Van Breugel, J. Camac, Y. Claveau, L. Coll, M. Dannoura, S. Delagrange, J.-C. Domec, F. Fatemi, W. Feng, V. Gargaglione, Y. Goto, A. Hagihara, J.S. Hall, S. Hamilton, D. Harja, T. Hiura, R. Holdaway, L.S. Hutley, T. Ichie, E.J. Jokela, A. Kantola, J.W.G. Kelly, T. Kenzo, D. King, B.D. Kloeppel, T. Kohyama, A. Komiyama, J.-P. Laclau, C.H. Lusk, D.A. Maguire, G. Lemaire, A. Mäkela, L. Markesteijn, J. Marshall, K. Mcculloh, I. Miyata, K. Mokany, S. Mori, R.L.W. Myster, M. Nagano, S.L. Naidu, Y. Nouvellon, A.P. O'grady, K.L. O'hara, T. Ohtsuka, N. Osada, O.O. Osunkoya, P.L. Peri, A.M. Petritan, L. Poorter, A. Portsmuth, C. Potvin, J. Ransijn, D. Reid, S.C. Ribeiro, S.D. Roberts,

R. Rodriguez, A. Saldana-Acosta, I. Santa-Regina, K. Sasa, N.G. Selaya, S.C. Sillett, F. Sterck, K. Takagi, T. Tange, H. Tanouchi, D. Tissue, T. Umehara, H. Utsugi, M.A. Vadeboncoeur, F. Valladares, P. Vanninen, J.R. Wang, E. Wenk, R. Williams, F. De Aquino Ximenes, A. Yamaba, T. Yamada, T. Yamakura, R. Yanai and R.A. York. 2015. BAAD, a biomass and allometry database for woody plants. *Ecological Archives* HO96-128. http://esapubs.org/archive.

Fisher, M.E. 1971. The theory of critical point singularities. In: Critical Phenomena (M.S Green, ed.), Proc. 1970 Enrico Fermi Int'l. Sch. Phys., Course No. 51. Varenna, Italy. Academic Press, New York, pp. 1–99.

Fitter, A.H. and T.R. Stickland. 1992. Fractal characterization of root system architecture. *Functional Ecology* 6(6): 632–635.

Freeze, R.A. 1975. A stochastic-conceptual analysis of one-dimensional groundwater flow in nonuniform homogeneous media. *Water Resour. Res.* 11: 725–741.

Freeze, R.A. and J.A. Cherry. 1979. Groundwater. Prentice-Hall, Englewood Cliffs, N.J.

Friedman, S.P. and N.A. Seaton. 1998. Critical path analysis of the relationship between permeability and electrical conductivity of three-dimensional pore networks. *Water Resour. Res.* 34: 1703–1710.

Friend, J.A. 1992. Achieving soil sustainability. *Journal of Soil and Water Conservation* 47: 156–157.

Gaonac'h, H., S. Lovejoy and D. Schertzer. 2003. Percolating magmas and explosive volcanism. *Geophysical Research Letters*, DOI: 10.1029/2002GL016022.

Gelhar, L.W. 1986. Stochastic subsurface hydrology from theory to applications. *Water Resour. Res.* 22: 1358–1458.

Gelhar, L.W. and C.L. Axness. 1983. Three-dimensional stochastic analysis of macrodispersion in aquifers. *Water Resour. Res.* 19: 161–180.

Ghanbarian-Alavijeh, B., T.E. Skinner and A.G. Hunt. 2012. Saturation dependence of dispersion in porous media. *Phys. Rev.* E86: 066316.

Ghanbarian, B., A.G. Hunt, M. Sahimi, R.P. Ewing and T.E. Skinner. 2013. Percolation theory generates a physically based description of tortuosity in saturated and unsaturated porous media. *Soil Science Society of America Journal* 77(6): 1920–1929.

Goodman, A.Y. and D.T. Rodbell, G.O. Seltzer and B.G. Mark. 2001. Subdivision of Glacial Deposits in Southeastern Peru Based on Pedogenic Development and Radiometric Ages. *Quaternary Research* 56: 31–50. doi:10.1006/qres.2001.2221, available online at http://www.idealibrary.com

Gray, D.M. 1961. Interrelationships of watershed characteristics. *J. Geophys. Res.* 66: 1215–1223.

Guéguen, Y., T. Chelidze and M. Le Ravalec. 1997. Microstructures, percolation thresholds, and rock physical properties. *Tectonophysics* 279: 23–35.

Gvirtzman, H. and Roberts, P.V. 1991. Pore scale spatial analysis of two immiscible fluids in porous media. *Water Resour Res* 27: 1165–1176.

Hagen, L.J. and N.P. Woodruff. 1973. Air pollution from duststorms in the Great Plains. *Atmospheric Environment* 7: 323–332.

Heimsath, A.M., W.E. Dietrich, K. Nishiizumi and R.C. Finkel. 1997. The soil production function and landscape equilibrium. *Nature* 388: 358–361.

Heimsath, A.M., W.E. Dietrich, K. Nishiizumi and R.C. Finkel. 1999. Cosmogenic nuclides, topography, and the spatial variation of soil depth. *Geomorphology* 27: 151–172.

Heimsath, A.M., J. Chappell, W.E. Dietrich, K. Nishiizumi and R.C. Finkel. 2001a. Late Quaternary erosion in southeastern Australia: A field example using cosmogenic nuclides. *Quaternary International* 83–85: 169–185.

Heimsath, A.M., W.E. Dietrich, K. Nishiizumi and R.C. Finkel. 2001b. Stochastic processes of soil production and transport: Erosion rates, topographic variation and cosmogenic nuclides in the Oregon coast range. *Earth Surf. Process. Landforms*, 26: 531–552.

Heimsath, A.M., J.C. Spooner and D.G. Questiaux. 2002. Creeping soil. *Geology* 30(2): 111–114.

Heimsath, A.M., J. Chappell, R.C. Finkel, K. Fifield and A. Alimano. 2006. Escarpment erosion and landscape evolution in southeastern Australia. Geological Society of America Special Paper 398. Boulder, CO.

Heimsath, A.M., D. Fink and G.R. Hancock. 2009. The 'humped' soil production function: Eroding Arnhem Land, Australia. *Earth Surf. Process. Landforms* 34: 1674–1684, doi: 10.1002/esp.1859.

Heimsath, A.M., J. Chappell and K. Fifield. 2010. Eroding Australia: Rates and processes from Bega Valley to Arnhem Land. In: P. Bishop and B. Pillans (eds), Australian Landscapes. Geological Society, London. Special Publications, 346: 225–241. DOI: 10.1144/SP346.12.

Hillel, D. and Y. Guron. 1973. Relation between evapotranspiration and rate and maize yield. *Water Resources Research* 9: 743–748.

Hillel, D. 2005. *J. Nat. Resour. Life Sci. Educ.* 34: 60–61.

Hoover, S.R., K.V. Cashman and M. Manga. 2001. The yield strength of subliquidus basalts. *J. Volcanology and Geothermal Research* 107: 1–18.

Hunt, A.G. 1998. Upscaling in subsurface transport using cluster statistics of percolation. *Transport in Porous Media* 30(2): 177–198.

Hunt, A.G. 2001. Applications of percolation theory to porous media with distributed local conductances. *Advances in Water Resources* 24(3, 4): 279–307.

Hunt, A.G. 2005. Comment on "Modeling low-frequency magnetic-field precursors to the Loma Prieta Earthquake with a precursory increase in fault-zone conductivity," by M. Merzer and S.L. Klemperer. *Pure and Applied Geophysics*, DOI: 10.1007/s00024-005-2776-6.

Hunt, A.G. 2008. A new conceptual model for forest fires based on percolation theory. *Complexity* 13(3): 12–17.

Hunt, A.G. 2016a. Explanation of the values of Hack's drainage basin, river length scaling exponent. *Non-linear Processes in Geophysics* 23: 91–93.

Hunt, A.G. 2016b. Spatio-temporal scaling of vegetation growth and soil formation from percolation theory. *Vadose Zone J.* 15(2), doi:10.2136/ vzj2015.01.0013.

Hunt, A., N. Gershenzon and G. Bambakidis. 2007. Pre-seismic electromagnetic phenomena in the framework of percolation and fractal theories. *Tectonophysics* 431(1–4): 23–32.

Hunt, A.G. and T.E. Skinner. 2008. Longitudinal dispersion of solutes in porous media solely by advection. *Philosophical Magazine* 88(22): 2921–2944.

Hunt, A.G. and T.E. Skinner. 2010. Incorporation of effects of diffusion into advection-mediated dispersion in porous media. *J. Stat. Phys.*, 140: 544–564.

Hunt, A.G., T.E. Skinner, R.P. Ewing and B. Ghanbarian-Alavijeh. 2011. Dispersion of solutes in porous media. *European Physical Journal* B 80(4): 411–432.

Hunt, A.G., B. Ghanbarian-Alavijeh, T.E. Skinner and R.P. Ewing. 2015. Scaling of Geochemical Reaction Rates via Advective Solute Transport. *Chaos* 25(7): 075403. DOI: 10.163/1.4913257.

Hunt, A.G. and R.P. Ewing. 2016. Scaling. In: Handbook of Groundwater Engineering, 3rd edition. J.H. Cushman and D. Tartakovsky (eds). CRC Press, Boca Raton, FL.

Hunt, A.G. and B. Ghanbarian. 2016. Percolation Theory for Solute Transport in Porous Media: Geochemistry, Geomorphology, and Carbon Cycling. *Water Resour. Res.* DOI: 10.1002/2016WR019289.

Hunt, A.G. and S. Manzoni. 2016. Networks on Networks. Morgan and Claypool Publication. Institute of Physics. CA, USA.

Ivy-Ochs, S. and F. Kober. 2008. Surface exposure dating with cosmogenic nuclides. *Eiszeitalter und Gegenwart Quaternary Science Journal* 57/1–2: 179–209.

Jacobson, A.D., J.D. Blum, C.P. Chamberlain, M.A. Poage and V.F. Sloan. 2002. Ca/Sr and Sr isotope systematics of a Himalayan glacial chronosequence: Carbonate versus silicate weathering rates as a function of landscape surface age. *Geochimica et Cosmochimica Acta* 66(1): 13–27.

Jenny, H. 1941. Factors of Soil Formation: A System of Quantitative Pedology. Dover, N.Y.

Jensen, M.E. and W.H. Sletten. 1965. Evapotranspiration and soil moisture-fertilizer interrelations with irrigated grain sorghum in the southern Great Plains. Agricultural Research Service. US Department of Agriculture.

Keitt, T.H., D.L. Urban and B.T. Milne. 1997. Detecting critical scales in fragmented landscapes. *Conservation Ecology* 1: 4.

Lee, Y., J.S. Andrade, S.V. Buldyrevm, N.V. Dokholoya, S. Havlin, P.R. King, G. Paul and H.E. Stanley. 1999. Traveling time and traveling length in critical percolation clusters. *Phys. Rev. E* 60(3): 3425–3428.

Levang-Brilz, N. and M.E. Biondini. 2002. Growth rate, root development and nutrient uptake of 55 plant species from the Great Plains Grasslands, USA. *Plant Ecology* 165: 117–144.

Lynch, J. 1995. Root architecture and plant productivity. *Plant Physiology* 109: 7–13.

Maher, K. 2010. The dependence of chemical weathering rates on fluid residence time. *Earth Plan. Sci. Lett.* 294: 101–110.

MacKay, G. and N. Jan. 1984. Forest fires as critical phenomena. *J. Phys.* A 17: L757–L760.

Maritan, A., A. Rinaldo, R. Rigon, A. Giacometti and I. Rodriguez-Iturbe. 1996. Scaling laws for river networks. *Phys. Rev.* E 53(2): 1510–1515.

McFadden, L.D. and R.J. Weldon. 1987. Rates and processes of soil development on Quaternary terraces in Cajon Pass, California. *Geological Society of America Bulletin* 98: 280–293.

McNulty, S.G., J.M. Vose and W.T. Swank. 1996. Loblolly pine hydrology and productivity across the southern United States. *Forest Ecology and Management* 86: 241–251.

Mogensen, V.O., H.E. Jensen and Md. Abdur Rab. 1985. Grain yield, yield components, drought sensitivity and water use efficiency of spring wheat subjected to water stress at various growth stages. *Irrigation Science* 6: 131–140.

Muldavin, E.H., D.I. Moore, S.L. Collins, K.R. Wetherill and D.C. Lightfoot. 2008. Aboveground net primary production dynamics in a northern Chihuahuan desert ecosystem. *Oecologia* 155: 123–132.

Neuman, S.P. and V. Di Federico. 2003. Multifaceted nature of hydrogeologic scaling and its interpretation. *Rev. Geophys*. 41(3): 1014.

Owen, J.J., R. Amundson, W.E. Dietrich, K. Nishiizumi, B. Sutter and G. Chong. 2010. The sensitivity of hillslope bedrock erosion to precipitation. *Earth Surface Processes and Landforms* 36: 117–135.

Owen, J.J., W.E. Dietrich, K. Nishiizumi, G. Chong and R. Amundson. 2013. Zebra stripes in the Atacama desert. *Geomorphology* 182: 157–172.

Payero, J.O., D.D. Tarkalson, S. Irmak, D. Davison and J.L. Petersen. 2008. Effect of irrigation amounts applied with subsurface drip irrigation on corn evapotranspiration, yield, water use efficiency, and dry matter production in a semiarid climate. *Agricultural Water Management* doi:10.1016/j. agwat.2008.02.015.

Pewe, T.L., E.A. Pewe and R.H. Pewe, A. Journaux and R.M. Slatt. 1981. Desert dust: Characteristics and rates of deposition in central Arizona. *Geological Society of America*, Special Paper 186: 169–190.

Pollak, M. 1972. A percolation treatment of dc hopping conduction. *J. Non Cryst. Solids* 11: 1–24, doi:10.1016/0022-3093(72)90304-3.

Reheis, M.C. and R. Kihl. 1995. Dust deposition in southern Nevada and California, 1984–1989: Relations to climate, source area and source lithology. *J. Geophys. Res.* 100, 8893–8918.

Retallack, G.J. and C. Huang. 2010. Depth to gypsic horizon as a proxy for paleoprecipitation in paleosols of sedimentary environments. *Geology* 38: 403–406.

Riebe, C.S., J.W. Kirchner, D.E. Granger and R.C. Finkel. 2001. Minimal climatic control on erosion rates in the Sierra Nevada, California. *Geology* 29: 447–450.

Rieu, M. and G. Sposito. 1991. Fractal fragmentation, soil porosity, and soil water properties. I. Theory. *Soil Sci. Soc. Am. J.* 55: 1231–1238.

Rigon, R., I. Rodriguez-Iturbe and A. Rinaldo. 1998. Feasible optimality implies Hack's law. *Water Resour. Res.* 34(11): 3181–3189.

Rosenzweig, M.L. 1968. Net primary productivity of terrestrial communities: Prediction from climatological data. *The American Naturalist*, 102: 67–74.

Sahimi, M. 1994. Applications of Percolation Theory. Taylor & Francis, London.

Sahimi, M. and S. Mukhopadhyay. 1996. Scaling properties of a percolation model with long-range correlations. *Phys. Rev.* E 54: 3870, doi: 10.1103/ PhysRevE.54.3870.

Salehikhoo, F., Li, L. and Brantley, S. 2013. Magnesite dissolution rates at different spatial scales: The role of mineral spatial distribution and flow velocity. *Geochimica et Cosmochimica Acta* 108: 91–106.

Sanford, W.E. and D.L. Selnick. 2013. Estimation of evapotranspiration across the conterminous United States using a regression with climate and land-cover data. *J. Am. Water Res. Assoc.* 49: 217–230.

Schülli-Maurer, I., D. Sauer, K. Stahr, R. Sperstaad and R. Sorensen. 2007. Soil formation in marine sediments and beach deposits of southern Norway: Investigations of soil chronosequences in the Oslofjord region. *Revista Mexicana de Ciencias Geológicas* 24(2): 237–246.

Seeley, M.K. 1978. Grassland productivity: The desert end of the curve. *South African Journal of Science* 74: 295–297.

Sheppard, A.P., M.A. Knackstedt, W.V. Pinczewski and M. Sahimi. 1999. Invasion percolation: New algorithms and universality classes. *J. Phys. A: Math. Gen.* 32: L521–L529.

Smale, M.C., M. McLeod and P.N. Smale. 1997. Vegetation and soil recovery on shallow landslide scars in tertiary hill country, East Cape region, New Zealand. *New Zealand Journal of Ecology* 21(1): 31–41.

Stauffer, D. 1979. Scaling theory of percolation clusters. *Physics Reports* 54: 1–74.

Stauffer, D. and A. Aharony. 1994. Introduction to Percolation Theory. Taylor and Francis, London.

Stevens, P.R. 1968. A chronosequence of soils near the Franz Josef Glacier. Dissertation. Lincoln College Canterbury NZ.

Stevenson, F. 1994. Humus Chemistry – Genesis, Composition Reactions. Wiley, New York.

Stock, G.M., R.S. Anderson and R.C. Finkel. 2005. Rates of erosion and topographic evolution of the Sierra Nevada, California, inferred from cosmogenic ^{26}Al and ^{10}Be concentrations. *Earth Surface Processes and Landforms* 30: 985–1006.

Sun, R., J.M. Chen, Q. Zhu, J. Liu, J. Li, S. Liu, G. Yan and S. Tang. 2004. Spatial distribution of net primary productivity and evapotranspiration in Changbaishan Natural Reserve, China, using Landsat ETM+ data. *Can. J. Remote Sens.* 30: 731–742.

Tian, H., G. Chen, C. Zhang, J.M. Melillo and C.A.S. Hall. 2010a. Pattern and variation of C:N:P ratios in China's soils: A synthesis of observational data. *Biogeochemistry* 98: 139–151.

Tian, H., G. Chen, M. Liu, C. Zhang, G. Sun, C. Lu, X. Xu, W. Ren, S. Pan and A. Chappelka. 2010b. Model estimates of net primary productivity, evapotranspiration, and water use efficiency in the terrestrial ecosystems of the southern United States during 1895–2007. *Forest Ecology and Management* 259: 1311–1327.

Trustum, N.A. and R.C. de Rose. 1988. Soil depth-age relationships of landslides on deforested hillsides, Taranaki, New Zealand. *Geomorphology* 1: 143–160.

Van Dam, J.C., J. Huygen, J.G. Wesseling, R.A. Feddes, P. Kabat, P.E.V. van Walsum, P. Groenendijk and C.A. van Diepen. 1997. Theory of SWAP version 2.0 Simulation of water flow, solute transport and plant growth in the Soil-Water-Atmosphere-Plant environment, Report 71. Department of Water Resources, Wageningen Agricultural University. Technical Document 45, DLO Winand Staring Centre, Wageningen.

van Ittersum, M., P. Leffelaar, H. van Keulen, M. Kropff, L. Bastiaans and J. Goudriaan. 2003. On approaches and applications of the Wageningen crop models. *Eur. J. Agron.* 18: 201–234.

VandenBygaart, A.J. and R. Protz. 1995. Soil genesis on a chronosequence, Pinery Provincial Park, Ontario. *Canadian Journal of Soil Science* 75: 63–72.

Vigneresse, J.L., P. Barbey and M. Cuney. 1996. Rheological Transitions During Partial Melting and Crystallization with Application to Felsic Magma Segregation and Transfer. *J. Petrol.* 37: 1579–1600.

Webb, W., S. Szarek, W. Lauenroth, R. Kinerson and M. Smith. 1978. Primary productivity and water use in native forest, grassland, and desert ecosystems. *Ecology* 59: 1239–1247.

Wells, S.G., J.C. Dohrenwend, L.D. McFadden, B.D. Turrin and K.D. Mahrer. 1985. Late Cenezoic landscape evolution on lava flow surfaces of the Cima volcanic field, Mojave Desert, California. *GSA Bulletin* 96: 1518–1529.

White, A.G., A.E. Blum, M.S. Schulz, T.D. Bullen, J.W. Harden and M.L. Peterson. 1996. Chemical weathering rates of a soil chronosequence on granitic alluvium: I. Quantification of mineralogical and surface area changes and calculation of primary silicate reaction rates. *Geochimica et Cosmochimica Acta* 60(14): 2533–2550, PII S0016-7037(01)00755-4.

White A.F. and S.L. Brantley. 2003. The effect of time on the weathering rates of silicate minerals. Why do weathering rates differ in the lab and in the field? *Chem. Geol.* 202: 479–506.

Wiens, J.A., R.L. Schooley and R.D. Weeks Jr. 1997. Patchy landscapes and animal movements: Do beetles percolate? *Oikos*, 78: 257–264.

Winter, L.E., L.B. Brubaker, J.F. Franklin, E.A. Miller and D.Q. DeWitt. 2002. Initiation of an old-growth Douglas fir stand in the Pacific Northwest: A reconstruction from tree-ring records. *Can. J. Forest Res.* 32: 1039–1056.

Winter, C.L., D.M. Tartakovsky and A. Guadagnini. 2003. Moment differential equations for flow in highly heterogeneous porous media. *Surv. Geophys.* 24(1): 81–106.

Zhang, H. and T. Oweis. 1999. Water-yield relations and optimal irrigation scheduling of wheat in the Mediterranean region. *Agricultural Water Management* 38: 195–211.

Zhou, G., Y. Wag, Y. Jiang and Z. Yang. 2002. Estimating biomass and net primary production from forest inventory data: A case study of China's Larix forests. *Forest Ecology and Management* 169: 149–157.

CHAPTER
6

Nonlocal Models for Transport in Fractal Media

Daniel O'Malley[1] **and John H. Cushman**[2*]
[1]Computational Earth Science, Los Alamos National Laboratory
MS T003, Los Alamos, NM 87545, USA
[2] Department of Earth, Atmospheric and Planetary Sciences,
Department of Mathematics, Purdue University, 550 Stadium Mall
Dr West Lafayette, IN 47907, USA

1. Introduction

Two approaches to the subject of transport in fractal porous media have been derived using minimal assumptions. One is derived from a thermodynamics and statistical mechanics (Cushman 1991, Cushman and Ginn 1993, Cushman et al. 1994), and the other utilizes a continuum mechanics approach where the divergence of the velocity is a random function (Neuman 1993). Despite their disparate origins the two approaches result in transport equations that are remarkably similar, but not identical. We begin our discussion by providing an overview of these two approaches. Subsequently, we explore three additional models of transport in fractal media. There are many such models and these three are by no means meant to be exhaustive. However, we hope that they provide some indication of the flavors of models that are used in this domain.

2. A Continuum Approach

Neuman (1993) derived a continuum approach that utilizes an observational volume, ω, in the sense of Cushman (1984) and assumes that a dispersion-free version of Fick's law applies on this scale. That is,

$$\mathbf{q}(\mathbf{x}, t) = \mathbf{v}(\mathbf{x}, t)\, c(\mathbf{x}, t) \tag{1}$$

*Corresponding author: jcushman@purdue.edu

$$\frac{\partial c(\mathbf{x},t)}{\partial t} = -\nabla \cdot \mathbf{q}(\mathbf{x},t) + g(\mathbf{x},t) \tag{2}$$

holds where $\mathbf{q}(\mathbf{x}, t)$ is the mass flux, $\mathbf{v}(\mathbf{x}, t)$ is the velocity, $c(\mathbf{x}, t)$ is the mass concentration, and $g(\mathbf{x}, t)$ is a mass source or sink. In this context, the quantities in Eq. (1) are considered observable on a support volume, ω, centered at x at time t. While Fick's laws are assumed to hold on this local scale, they are deemed unsuitable for predicting concentrations due to uncertainty in the velocity field. An alternative to Eq. (1) and Eq. (2) for the expected value of $c(\mathbf{x}, t)$ conditioned on available data is derived where the right hand side of Eq. (1) is nonlocal in space and time (i.e., it contains integral terms).

In addition to assuming that Fick's law holds on the local scale given by ω, it is assumed that the divergence of the velocity is given by a random function. That is,

$$\nabla \cdot \mathbf{v}(\mathbf{x},t) = f(\mathbf{x},t) \tag{3}$$

In the context of hydrologic applications, $f(\mathbf{x}, t)$ represents fluid sources. It is usually impractical to obtain $\mathbf{v}(\mathbf{x}, t)$ at all points in space and time through any means, but it is plausible to obtain a smooth, unbiased estimate, $\langle \mathbf{v}(\mathbf{x},t) \rangle_d$, of $\mathbf{v}(\mathbf{x}, t)$. Here $\langle \cdot \rangle_d$ denotes the mean of a quantity that has been conditioned on available data. This data can come in many forms. For example, the data may consist of velocity measurements at a number of points in space and time, and/or other measurements that provide information about the velocity such as hydraulic conductivity or pressure measurements.

The main quantity of interest here is $\langle c(\mathbf{x},t) \rangle_d$, which is the mean concentration conditioned on the data used to obtain $\langle \mathbf{v}(\mathbf{x},t) \rangle_d$. Averaging over Eq. (2) results in

$$\frac{\partial \langle c(\mathbf{x},t) \rangle_d}{\partial t} = -\nabla \cdot \langle \mathbf{q}(\mathbf{x},t) \rangle_d + \langle g(\mathbf{x},t) \rangle_d \tag{4}$$

where

$$\langle \mathbf{q}(\mathbf{x},t) \rangle_d = \langle \mathbf{v}(\mathbf{x},t) \rangle_d \langle c(\mathbf{x},t) \rangle_d + \langle \mathbf{v}'(\mathbf{x},t) c'(\mathbf{x},t) \rangle_d \tag{5}$$

$$\mathbf{v}'(\mathbf{x},t) = \mathbf{v}(\mathbf{x},t) - \langle \mathbf{v}(\mathbf{x},t) \rangle_d \tag{6}$$

$$c'(\mathbf{x},t) = c(\mathbf{x},t) - \langle c(\mathbf{x},t) \rangle_d \tag{7}$$

The main result is the integral equation for $\langle \mathbf{v}'(\mathbf{x},t)c'(\mathbf{x},t) \rangle_d$ given by

$$\langle \mathbf{v}'(\mathbf{x},t)c'(\mathbf{x},t) \rangle_d = \int_{R^d} \int_0^t \alpha(\mathbf{x},t,\mathbf{y},s) \nabla_y \cdot \langle \mathbf{v}'(\mathbf{y},s)c'(\mathbf{y},s) \rangle_d \, ds \, d\mathbf{y} \tag{8}$$

$$-\int_{R^d} \int_0^t \beta(\mathbf{x},t,\mathbf{y},s) \nabla_y \langle c(\mathbf{y},s) \rangle_d \, ds \, d\mathbf{y}$$

$$-\int_{R^d}\int_0^t \gamma(\mathbf{x},t,\mathbf{y},s)\langle c(\mathbf{y},s)\rangle_d \, ds \, d\mathbf{y}$$

where

$$\alpha(\mathbf{x},t,\mathbf{y},s) = \langle c(\mathbf{x},t \mid \mathbf{y},s)\mathbf{v}'(\mathbf{x},t)\rangle_d \tag{9}$$

$$\beta(\mathbf{x},t,\mathbf{y},s) = \left\langle c(\mathbf{x},t \mid \mathbf{y},s)\mathbf{v}'(\mathbf{x},t)\mathbf{v}'^T(\mathbf{y},s)\right\rangle_d \tag{10}$$

$$\gamma(\mathbf{x},t,\mathbf{y},s) = \langle c(\mathbf{x},t \mid \mathbf{y},s)\mathbf{v}'(\mathbf{x},t)f(\mathbf{y},s)\rangle_d \tag{11}$$

with $c(\mathbf{x}, t \mid \mathbf{y}, s)$ the concentration of a delta function source originating at location $\mathbf{y}$ at time t. Some care must be used in interpreting Eqs. (9-11) because the $\mathbf{v}'(\mathbf{x}, t)$ terms are to be treated effectively as constants that are not subject to differentiation and integration with respect to $\mathbf{x}$ or t. Equations (9-11) can be estimated via closure approximations or sampling methods. However, use of a sampling method would largely defeat the purpose of deriving Eq. (4) and Eq. (8) since the sampling would make it trivial to evaluate $\langle c(\mathbf{x},t)\rangle_d$ and other quantities of interest.

Combining Eq. (4), Eq. (5), and Eq. (8) results in a space/time nonlocal equation for the conditioned mean concentration, $\mathbf{v}'(\mathbf{x}, t)$. In this setting, the nonlocality is essentially due to uncertainty in the velocity induced by Eq. (3) and data that is insufficient to constrain the velocity field. The dependence of Eq. (4) on the available data is an important point, and stands against the idea that a dispersion coefficient or tensor is something that can be measured in an absolute sense. Dispersion parameters often do more to quantify our ignorance, than they do to quantify fundamental physical properties of the system. When the velocity field is represented with greater accuracy, the apparent dispersion is smaller. On the other hand, when the velocity field is represented with less accuracy (e.g., with a constant velocity), the apparent dispersion is greater.

3. A Statistical-mechanical Approach

This approach, which is derived in Cushman et al. (1994), utilizes a statistical-mechanical framework. Suppose that a fluid and a conservative tracer are embedded in the pore space, V_p, of some solid volume, V_s (i.e., V_p is the complement of V_s). Let M be the number of particles comprising fluid and N be the number of particles comprising the conservative tracer, resulting in a total of $J = N + M$ particles. The phase space, containing the positions and momenta of these particles, is given by

$$\Omega = (V_p)^J \times R^{3J} \tag{12}$$

with the $(V_p)^J$ part containing the particle positions and the R^{3J} part containing the momenta. The solid phase is not represented with particles, but its impact is represented through a force field acting on the particles (which can be random). Let $f(\mathbf{x}, \mathbf{p}; t)$ be the probability density of the system being at the point $(\mathbf{x}, \mathbf{p})$ in phase space at time t.

The basic physical assumption underlying this approach is that the system is Hamiltonian and hence (McQuarrie 1976) the probability density function, $f(\mathbf{x}, \mathbf{p}; t)$, satisfies a Liouville equation

$$\frac{\partial f}{\partial t} = -\sum_{j=1}^{J} \left[\frac{\mathbf{p}_j}{m_j} . \nabla_{x_j} + \mathbf{F}_j \cdot \nabla_{p_j} \right] f \tag{13}$$

where $\mathbf{F}_j$ is the force acting on the jth particle. This equation states that the time rate of change of the probability density function is equal to the convective derivative in phase space. For notational convenience, two shorthand versions of the operator on the right hand side of Eq. (13) are used. These are

$$L = -i \sum_{j=1}^{J} \left[\frac{\mathbf{p}_j}{m_j} \cdot \nabla_{x_j} + \mathbf{F}_j \cdot \nabla_{p_j} \right] \tag{14}$$

and

$$\mathbf{V} \cdot \nabla_{\omega} \equiv \sum_{j=1}^{J} \left[\frac{\mathbf{p}_j}{m_j} \cdot \nabla_{\mathbf{x}_j} + \mathbf{F}_j \cdot \nabla_{\mathbf{p}_j} \right] \tag{15}$$

where

$$\mathbf{V} = \left(\frac{\mathbf{p}_1}{m_1}, \ldots, \frac{\mathbf{p}_J}{m_j}, \mathbf{F}_1, \ldots, \mathbf{F}_J \right) \tag{16}$$

and

$$\nabla_{\omega} = (\nabla_{x_1}, \ldots, \nabla_{x_J}, \nabla_{p_1}, \ldots, \nabla_{p_J}) \tag{17}$$

Using these equations combined with a projection operator formalism, two main results can be obtained for the distribution of a tagged particle (Cushman et al. 1994), one for the equilibrium case where $\frac{\partial f}{\partial t} = 0$ and one for the general nonequilibrium case.

Let $G(\mathbf{x}, t)$ denote the probability density of finding the tagged particle at position $\mathbf{x}$ at time t. In the equilibrium case, $G(\mathbf{x}, t)$ satisfies

$$\frac{\partial G}{\partial t} = \nabla_x \cdot \int_0^t \int_{R^3} \mathbf{D}(y, \tau) \cdot \nabla_{x-y} G(\mathbf{x} - y, t - \tau) d\mathbf{y}\, d\tau \tag{18}$$

where

$$\mathbf{D}(\mathbf{y}, \tau) = \Im^{-1} \left\{ \left(\exp[iQ_0 L\tau] \mathbf{v}_0 e^{i\mathbf{k}\cdot\mathbf{x}_0}, e^{i\mathbf{k}\cdot\mathbf{x}_0} \right) \right\} \tag{19}$$

$\Im^{-1}$ denotes the inverse Fourier transform, $\mathbf{k}$ is the Fourier variable, $(a, b) \equiv \int_{\Omega} ab^* f(\mathbf{x}, \mathbf{p}; t)\, d\mathbf{x}\, d\mathbf{p}$, b^* denotes the complex conjugate of b, $\mathbf{x}_0$ is the initial position of the tagged particle, $\mathbf{v}_0$ is the initial velocity of the tagged particle, and Q_0 is a projection operator that acts on dynamic variables (i.e., functions of the form $\alpha(t) = \alpha(\mathbf{x}(t), \mathbf{p}(t))$). The action of Q_0 on $\alpha(t)$ is

$$Q_0\alpha(t) = \alpha(t) - e^{i\mathbf{k}\cdot\mathbf{x}_0}\left(\alpha(t), e^{i\mathbf{k}\cdot\mathbf{x}_0}\right) \tag{20}$$

Q_0 is a projection operator that projects onto a subspace that is ortogonal to $e^{i\mathbf{k}\cdot\mathbf{x}_0}$. Note that $G(\mathbf{x}, t)$ is a dynamic variable,

$$G(\mathbf{x}, t) = \left\langle \delta\left[\mathbf{x} - \left(\mathbf{x}_j(t) - \mathbf{x}_j(0)\right)\right]\right\rangle \tag{21}$$

where j is the index of the tagged particle and

$$\langle \alpha(t)\rangle \equiv \int_\Omega \alpha(\mathbf{x}(t), \mathbf{p}(t)) f(\mathbf{x}, \mathbf{p}; t) d\mathbf{x}\, d\mathbf{p} \tag{22}$$

The equation for $G(\mathbf{x}, t)$ in the nonequilibrium case is more complex, involving two integral terms (which are not convolutions) as well as a term for the mean velocity,

$$\begin{aligned}\frac{\partial G}{\partial t} &= -\nabla_x \cdot \left[\langle \mathbf{v}(t)\rangle G(\mathbf{x}, t)\right] \\ &+ \nabla_x \cdot \int_0^t \int_{R^3} \mathbf{D}_1(\mathbf{y}, t, \tau)\, G(\mathbf{x} - \mathbf{y}, t - \tau)\, d\mathbf{y}\, d\tau \\ &+ \nabla_x \cdot \int_0^t \int_{R^3} \mathbf{D}_2(\mathbf{y}, t, \tau) \cdot \nabla_{\mathbf{x}-\mathbf{y}}\, G(\mathbf{x} - \mathbf{y}, t - \tau)\, d\mathbf{y}\, d\tau\end{aligned} \tag{23}$$

where

$$\mathbf{D}_1(\mathbf{y}, t, \tau) = \aleph^{-1}\Im^{-1}\left\{\mathbf{Y}_1(k, s)\left[1 - \frac{i\mathbf{k}\cdot\mathbf{Y}_1(\mathbf{k}, s) + i\mathbf{k}\cdot\mathbf{Y}_2(\mathbf{k}, s)\cdot i\mathbf{k}}{s}\right]^{-1}\right\} \tag{24}$$

$$\mathbf{D}_2(\mathbf{y}, t, \tau) = \aleph^{-1}\Im^{-1}\left\{\mathbf{Y}_2(k, s)\left[1 - \frac{i\mathbf{k}\cdot\mathbf{Y}_1(\mathbf{k}, s) + i\mathbf{k}\cdot\mathbf{Y}_2(\mathbf{k}, s)\cdot i\mathbf{k}}{s}\right]^{-1}\right\} \tag{25}$$

$$\mathbf{Y}_1(\mathbf{k}, s) = -\aleph\left\langle \mathbf{a}'(t)\exp\left[i\mathbf{k}\cdot\mathbf{x}'(t)\right]\exp\left[-i\mathbf{k}\cdot\mathbf{x}_0\right]\right\rangle \tag{26}$$

$$\mathbf{Y}_2(\mathbf{k}, s) = -\aleph\left\langle \mathbf{v}'(t)\exp\left[i\mathbf{k}\cdot\mathbf{x}'(t)\right]\exp\left[-i\mathbf{k}\cdot\mathbf{x}_0\right]\mathbf{v}'(t)\right\rangle \tag{27}$$

$\aleph$ and $\aleph^{-1}$ denote the Laplace and inverse Laplace transforms, respectively; and $\mathbf{x}'(t)$, $\mathbf{v}'(t)$, and $\mathbf{a}'(t)$ are the fluctuating part of the tagged particles position, velocity, and acceleration, respectively. Formulating Eq. (23) in terms of a flux, as was done for the continuum model, we obtain

$$\frac{\partial G}{\partial t} = -\nabla \cdot q \tag{28}$$

$$q = \left[\langle \mathbf{v}(t)\rangle G(\mathbf{x},t)\right] - \int_0^t \int_{R^3} \mathbf{D}_1(\mathbf{y},t,\tau)\, G(\mathbf{x}-\mathbf{y},t-\tau)\, d\mathbf{y}\, d\tau - \int_0^t \int_{R^3} \mathbf{D}_2(\mathbf{y},t,\tau)\cdot \nabla_{\mathbf{x}-\mathbf{y}} G(\mathbf{x}-\mathbf{y},t-\tau)\, d\mathbf{y}\, d\tau \tag{29}$$

It has been shown (Cushman et al. 2011) that a number of commonly used dispersive processes including Brownian motion/Fickian dispersion, Levy motion/fractional dispersion (Benson et al. 2000a, b), continuous-time random walk (Metzler and Klafter 2000), processes with Levy velocities (Cushman et al. 2005), and Brownian motion run with a nonlinear clock (Cushman et al. 2009a, b) are special cases of Eqs. (28) and (29).

If a further assumption is added that the velocity is small in comparison to the scale of the heterogeneity of the medium (called a local equilibrium assumption), Eq. (29) simplifies into a form where the second term goes away. In a medium that exhibits fractal characteristics on a large scale, this would occur below a length scale where the medium no longer exhibits a fractal character, and is approximately uniform. If the medium exhibits fractal behavior on sufficiently small scales (relative to the velocity and the relaxation time), this approximation would not be appropriate and the full form of Eq. (29) would be necessary.

An experimental study of this nonlocal dispersion was carried out in (Moroni and Cushman, 2001) using a simplified version of the theory (Cushman and Moroni, 2001). The experiments performed 3D particle tracking velocimetry in a "matched index" porous medium. The phrase "matched index" refers to the fact that the fluid and solid in the media have the same refractive index making the solids appear transparent with no refraction. The solids in this case consisted of a packing of Pyrex beads with the fluid being glycerol. This makes it easy to track particles (air bubbles, in the case of this experiment) flowing through the medium. They estimated the generalized dispersion coefficient, $\mathbf{D}_2(\mathbf{y}, t, \tau)$, in Fourier-Laplace space where it was assumed that $\mathbf{D}_2(\mathbf{y}, t, \tau)$ does not depend on t, making the temporal integral in Eq. (29) a convolution. The focus was on the transverse dispersion, so $\langle \mathbf{v}(t)\rangle$ and $\mathbf{D}_1(\mathbf{y}, t, \tau)$ were both zero in these directions.

Obtaining the generalized dispersion coefficient from the particle trajectories presents numerical difficulties, but several methods were used and tests were performed to check the consistency and validity of the results. Quantitative results can be found in Moroni and Cushman (2001), but some interesting qualitative results were observed that we will recall here.

First, the Fourier-Laplace transform of the generalized dispersion coefficient tended to zero as the Fourier variable became large in magnitude. This indicates that the generalized dispersion coefficient has little power in high frequencies (small spatial scales), which makes sense given the uniform size of the beads. Another noteworthy observation is that the generalized transverse dispersion tensor appeared to be converging toward the transverse velocity covariance, as would be expected for Fickian dispersion. However, over the course of the experiments, it had not yet fully converged. This indicates that the dispersion remained non-Fickian throughout the course of the experiment.

4. Comparison between the Continuum and Statistical Mechanical Approaches

The two fluxes given by Eqs. (5-8) and (29) have remarkable similarities. They both contain a mean advective velocity term as well as integral terms which imply that the advection and dispersion are nonlocal. The second term in Eq. (8) is similar in form to the third term in Eq. (29) – both involving a spatiotemporal integral of a kernel matrix times the gradient of the mean concentration (Eq. (8)) or probability density (Eq. (29)). Similarly, the third term in Eq. (8) is similar to the second term in Eq. (29) – both involving a spatiotemporal of a vector kernel times the mean concentration (Eq. (8)) or probability density (Eq. (29)).

It is problematic to make a complete comparison between the two approaches because they are derived from different perspectives, under different assumptions, and are going after (slightly) different results. We enumerate some of the differences here.

1. The continuum mechanical approach produces an equation for the mean concentration. In contrast, the statistical mechanical approach produces an equation for the probability density of the displacement of a tagged particle.
2. Randomness is introduced into the continuum approach via a random fluid source/sink function, but this is antithetical to the statistical mechanical approach where the number of fluid particles is fixed at M. Randomness in the statistical mechanical approach is introduced via the initial positions and momenta of the particles.
3. The kernels of the integrals in Eq. (8) involves the location, $\mathbf{x}$, whereas the kernels in Eq. (29) depend on $\mathbf{x}_0$, the initial location of the tagged particle.
4. The first term in Eq. (8) does not have a clear analogue in Eq. (29).
5. The kernels in Eq. (8) are defined in terms of continuum mechanical concepts, whereas the kernels in Eq. (29) are defined in terms of statistical mechanical concepts.

The first two differences listed indicate that we should not expect the two equations to be the same, so we will not attempt to reconcile the other differences.

Despite the differences, the similarities highlight an important point: given the vast uncertainties that are present when modeling transport in highly-heterogeneous, fractal, porous media, it is challenging to produce an accurate model that utilizes a local version of the mass flux. They also indicate that solving transport problems in fractal porous media from first principles is not generally practical. For example, determining the kernels in Eqs. (24) and (25) would be a significant challenge in a real-world problem. If these kernels were available, solving Eq. (23) would present another significant challenge, but it can be performed in some cases (Hassan et al. 1997). Because of these challenges, alternative approaches to modeling transport in fractal porous media have been developed. These approaches are less rigorous, but simplify the problem and have appealing properties that ideally capture the essence of the transport problem at hand. While there are many, we explore three of them in the subsequent sections.

5. The Fractional Advection-dispersion Equation

The primary purpose of the dispersive component in a transport model is to represent the unresolved portion of the velocity. That is, when modeling transport in a highly-heterogeneous, fractal velocity field, some approximation of the velocity field is available to the model and the discrepancy between the approximation and the true velocity field is modeled via dispersion. The classical dispersive model treats the dispersion as if it is a (potentially very fast) form of molecular diffusion. From an Eulerian perspective, this is represented via the partial differential equation

$$\frac{\partial c}{\partial t} = -\nabla \cdot (\mathbf{v}c) + D\nabla^2 c \tag{30}$$

where c represents the concentration and $D\nabla^2 c$ is the dispersive component of the equation. From a Lagrangian perspective, this means that particles move by following the velocity field plus a displacement given by a white noise. This is because Eq. (30) is the Fokker-Planck equation for a stochastic differential equation driven by Brownian motion subject to drift. The dispersive portion here comes from the integral of the white noise – a Brownian motion.

The fractional advection-dispersion (Benson et al. 2000a, b, 2001) equation can be understood in much the same way as the classical advection-dispersion equation. However, it treats the dispersion differently. In some cases (Boggs et al. 1992, Adams and Gelhar 1992), the classical dispersive model fails to accurately represent the potential for the unresolved portion of the velocity field to contain transport paths that are very fast or slow. The fractional advection-dispersion equation provides a means of resolving this issue. The word "fractional" in "fractional advection-dispersion equation" refers to the fact that the Laplace operator in Eq. (30) is replaced with a fractional derivative,

$$\frac{\partial c}{\partial t} = -\nabla \cdot (\mathbf{v}c) + D\nabla_M^\alpha c \tag{31}$$

where $0 < \alpha \leq 2$ and ∇_M^α is a fractional "derivative" (really an integral operator) that is most readily expressed in Fourier space (Meerschaert et al. 1999),

$$\Im\left(\nabla_M^\alpha c\right) = \left[\int_{\|h\|=1} (i\mathbf{k} \cdot \mathbf{h})^\alpha M(d\mathbf{h})\right] \Im(c) \tag{32}$$

with **k** being the Fourier variable, and M is a mixing measure on the unit sphere. The inclusion of the measure M allows the dispersion to be biased in different directions, whereas the classical dispersion model is equally likely to move mass forward and backward. When $\alpha = 2$, Eq. (31) reduces to Eq. (30), but when $\alpha < 2$, ∇_M^α is a nonlocal operator. This nonlocality is what allows the dispersion model to move mass over long distances in short times.

Just as the classical dispersion model has a rich Lagrangian interpretation via Brownian motion, the fractional dispersion model has a rich Lagrangian interpretation via Levy motion. Brownian motion can be seen as the integral of a white noise where the noise has finite variance. Levy motion drops the assumption of finite variance and uses a noise that follows a stable distribution (Samoradnitsky and Taqqu 1994). The displacements of a Levy motion have heavy tails. That is, the probability of a Levy motion making a large displacement in a fixed time is high compared to the probability of a Brownian motion making a large displacement. As α decreases from 2, the probability of making very large jumps increases. Quantitatively, the probability of making a large jump whose magnitude is greater than Δx is proportional to $(\Delta x)^{-\alpha}$ when $\alpha < 2$. The Lagrangian perspective elucidates the ability for this dispersion model to capture the effect of unresolved transport paths that are very fast or slow. Fast unresolved paths can be represented, to some extent, via large jumps in the direction of the advection. On the other hand, slow unresolved paths can be represented, to some extent, via large jumps in the opposite direction of the advection (thereby nullifying some or all of the advective transport).

Equation (31) cannot generally be solved analytically in a closed form. However, semi analytical solutions can be written in terms of the probability density and cumulative density functions of the stable distributions (Benson et al. 2000a, O'Malley and Vesselinov 2014). Fig. 1 shows several semi-analytical, one-dimensional solutions of Eq. (31) with a symmetric M, $\mathbf{v} = 0$ when $Dt = 1$ (i.e., when the product of D from Eq.

(31) with the time is one). The initial condition is a delta function at the origin. The solution was obtained using numerical methods (Nolan 1997) to evaluate the probability density function for the stable distribution.

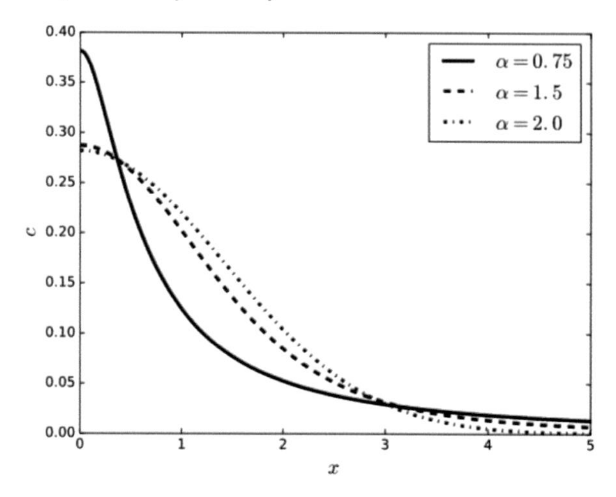

Figure 1: Symmetric, one-dimensional solutions of the fractional advection-dispersion equation with zero velocity as a function of the spatial coordinate are shown. Note that as α decreases from 2, the tails become heavier and the centroid becomes more peaked.

6. Microbial Dynamics in Fractal Porous Media

Here we describe a three-scale model (Park et al. 2005a, b, Park and Cushman 2006) of microbial transport where the mesoscale exhibits a fractal character (see Fig. 2). The microscale represents the motion of the microbe as the solution of a stochastic differential equation consisting of a drift process plus a Levy motion,

$$\mathbf{x}^0(t) = \mathbf{x}^0(0) + \int_0^t \mathbf{v}^0(\mathbf{x}^0(s))ds + \rho^0 L^0(t) \tag{33}$$

where the 0 in the superscript denotes the microscale (the superscripts 1 and 2 will be used to denote the mesoscale and macroscale respectively), ρ^0 is a constant, $L^0(t)$ is a Levy motion with stability index $\frac{1}{2} < \alpha^0 \leq 2$ and drift μ^0 (the other parameters of the Levy motion are not relevant at the macroscale), and $\mathbf{v}^0$ is a velocity that is assumed to be stationary, ergodic and Markovian. The drift μ^0 is separated from the velocity $\mathbf{v}^0$, because the velocity $\mathbf{v}^0$ is the velocity experienced by a passive tracer while the drift μ^0 is induced by the motile microbe (perhaps preferentially moving towards a nutrient). The Levy portion of the stochastic differential equation accounts for the motile behavior of the microbes (Berg 2000).

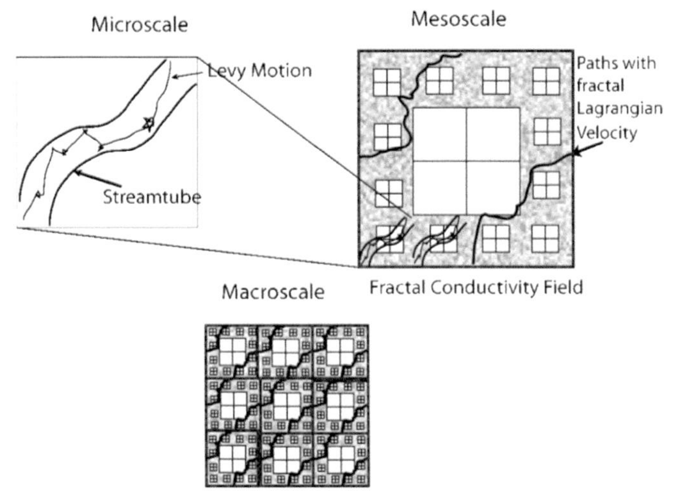

Figure 2: The three scales in the model of microbial transport in fractal porous media are shown. The microscale represents the motion of the microbe following a streamline, but pertubed by a Levy motion. The mesoscale represents the motion through a fractal porous media. The macroscale represents a periodic upscaling from the mesoscale.

An integrated Levy velocity is added on the mesoscale to represent advection through the fractal medium, resulting in an equation similar to Eq. (33), but with the velocity process being Levy. That is, we have

$$\mathbf{x}^1(t) = \mathbf{x}^1(0) + \int_0^t \mathbf{v}^1(\mathbf{x}^1(s))ds + \rho^1 L^1(t) \tag{34}$$

where $\mathbf{v}^1$ is a Levy process with stability index $1 < \alpha^1 \leq 2$. Upscaling from the mesoscale to the macroscale requires a generalized central limit theorem (Park and Cushman 2006), and results in the Fokker-Planck equation

$$\frac{\partial c}{\partial t} = -\mathbf{v}^2 \cdot \nabla c + D^2(t)\nabla_M^{\alpha^1} c \tag{35}$$

where

$$D^2(t) = \frac{-t^{\alpha^1}}{\cos\left(\dfrac{\sigma^1 \pi}{2}\right)} \tag{36}$$

and σ is the scale parameter of the Levy motion of the velocity process. Equation (35) holds provided that $1 < \alpha^1 \leq \dfrac{\alpha^0}{1-\alpha^0}$ when $\dfrac{1}{2} < \alpha^0 \leq \dfrac{2}{3}$. Note that the fractional derivative in Eq. (35) is associated with the stability index of the velocity process at the mesoscale rather than the stability index of the Levy motion representing the microbial motility at the microscale. This indicates that the dispersion associated with transport through the fractal flow field at the mesoscale dominates the dispersion associated with the microbial motility. The velocity involves two terms – one which averages over the unit cell on the mesoscale and the other is μ^0, the drift from the microbial dynamics.

7. Fractional Brownian Motion with a Nonlinear Clock

Fractional Brownian motion (Mandelbrot and Van Ness, 1968) is a Gaussian stochastic process that generalizes Brownian motion in a different way than Levy motions do. Levy motions generalize Brownian motion by replacing the finite variance displacements of a Brownian motion with displacements that have infinite variance. In contrast, fractional Brownian motion does away with Brownian motion's assumption that displacements over disjoint time intervals are independent, and replaces it an an assumption that the displacements are correlated over all time scales. Mandelbrot's early interest in the process seems to derive from hydrologic observations of the Nile (Mandelbrot and Wallis 1968). Subsequently, fractional Brownian motion has been used to study a variety of transport phenomena (e.g., Magdziarz et al. 2009, Panja 2011, Regner et al. 2013, O'Malley et al. 2014) including transport in fractal media (e.g., Park 2013). From a Lagrangian perspective, fractional Brownian motion is described via stochastic integral-differential equation,

$$B_H(t) = B_H(t) + \frac{1}{\Gamma\left(H+\frac{1}{2}\right)}\left[\int_{-\infty}^{0}(t-s)^{H-\frac{1}{2}} - (-s)^{H-\frac{1}{2}} dB(s) + \int_0^t (t-s)^{H-\frac{1}{2}} dB(s)\right] \tag{37}$$

where H, called the Hurst exponent, controls both the scaling behavior (O'Malley and Cushman 2012a, b, O'Malley et al. 2014) of the process and the fractal dimension of the trajectory (Falconer 2004). One of the key properties of fractional Brownian motions is that they can exhibit either persistence ($H > 1/2$) or anti-persistence ($H < 1/2$). Persistence is the property that the dispersion tends to continue in the same direction. Anti-persistence is the property that the dispersion tends to turn back on itself, so that displacements over one time interval tend to be cancelled out by displacements over another time interval.

Here, we focus on a generalization of fractional Brownian motion called fractional Brownian motion with a nonlinear clock (O'Malley and Cushman 2010, O'Malley et al. 2011). Fractional Brownian motion with a nonlinear clock can be described in a Lagrangian framework as

$$X(t) = B_H(F(t)) \tag{38}$$

where $X(t)$ is the fractional Brownian motion with a nonlinear clock, $B_H(t)$ is a Brownian motion and $F(t)$ is a deterministic, nondecreasing function called the clock. Fractional Brownian motion with a nonlinear clock inherits the persistence/anti-persistence from the fractional Brownian motion that it is based on. The clock transforms the time coordinate and makes it possible to decouple the scaling behavior of $X(t)$ from its fractal dimension. Under modest assumptions on the clock, the fractal dimension of the trajectory is determined by H,

$$\dim[graph(X)] = 2 - H \tag{39}$$

where dim denotes the Hausdorff fractal dimension,

$$graph(X) = [(t, x) : X(t) = x, t \in [a, b]] \tag{40}$$

and $[a, b]$ denotes an arbitrary time interval. Similarly, in a multidimensional context, the path of $X(t)$ has the same fractal dimension as the path of $B_H(t)$ – indeed the same paths. However, the scaling behavior is determined by both H and the clock. For example, if $F(t) = t^p$, $X(t)$ is pH-self-similar and hence pH-diffusive on both the short and long time scales (O'Malley and Cushman 2012b). This implies that when modeling transport in a medium with fractal transport paths, fractional Brownian motion with a nonlinear clock can be used to match the fractal dimension of the transport path and the scaling behavior of the dispersive component of the transport.

The Lagrangian perspective on fractional Brownian motion with a nonlinear clock can be clearly understood from the perspective of fractional Brownian motion combined with Eq. (38). However, like fractional Brownian motion, the Eulerian perspective on fractional Brownian motion with a nonlinear clock is murkier. To our knowledge, there is no known analogue of Eqs. (4), (18), (23), (31) or (35) for fractional Brownian motion or fractional Brownian motion with a nonlinear clock. Even though there is no Eulerian equation, an Eulerian "solution" in an infinite domain can be obtained from the Lagrangian perspective with the initial condition being a point source,

$$G(x, t) = \frac{1}{\sqrt{2\pi F(t)^{2H}}} \exp\left[-\frac{(x - x_0 - vt)^2}{2F(t)^{2H}} \right] \tag{39}$$

where x_0 is the location of the point source and v is the velocity. This essentially follows from the fact that fractional Brownian motion with a nonlinear clock is a Gaussian process. The solution is determined by the mean ($x_0 - vt$ in this case) and variance ($F(t)^{2H}$ in this case).

8. Discussion

We have presented a number of approaches for modeling transport in highly-heterogeneous, fractal velocity fields. The first two approaches we mentioned proceeded from basic physical principles with few assumptions. One utilized a continuum approach while the other utilized a statistical physics approach. The results obtained from these two approaches had similar qualities. They showed that the Eulerian equations governing transport were nonlocal in both space and time. The main drawback of these two approaches is that the transport equations are complex and involve integral kernels that cannot be readily estimated in many applications except through curve fitting.

We then mentioned three simpler models which lacked the full rigor of the first two approaches but are more readily applicable. The first of these was the fractional-advection dispersion equation, which makes it possible to model transport phenomena where there are very fast or very slow paths. The second was a three-scale model of microbial transport with a fractal medium at the mesoscale. The third was fractional Brownian motion with a nonlinear clock which makes it possible to model the fractal dimension of the paths and the scaling behavior of the transport phenomena. Like fractional Brownian motion, it maintains the ability to model dispersive transport processes where the dispersion is either persistent or anti-persistent.

These latter three models are meant to give a sample of the many diverse models of transport in fractal velocity fields. While the first two models we explored largely settled the question of transport in highly-heterogeneous, fractal velocity fields from a theoretical perspective, there is still considerable work to be done from a practical perspective in modeling these transport processes. For a given transport problem, there is often considerable disagreement among experts about which of the many models should be applied. This is not surprising because the simpler, more practical models typically have shortcomings when applied to a complex transport problem. For these reasons, transport modeling in fractal media remains a significant challenge.

Acknowledgements

DO wishes to acknowledge the support of a Los Alamos National Laboratory Director's Postdoctoral Fellowship. JHC wishes to acknowledge the support of NSF Grant No. EAR1314828.

REFERENCES

Adams, E.E. and L.W. Gelhar. 1992. Field study of dispersion in a heterogeneous aquifer: 2. Spatial moments analysis. *Water Resources Research* 28(12): 3293–3307.

Benson, D.A., S.W. Wheatcraft and M.M. Meerschaert. 2000a. The fractional-order governing equation of Lévy motion. *Water Resources Research* 36(6): 1413–1423.

Benson, D.A., S.W. Wheatcraft and M.M. Meerschaert. 2000b. Application of a fractional advection-dispersion equation. *Water Resources Research* 36(6): 1403–1412.

Benson, D.A., R. Schumer, M.M. Meerschaert and S.W. Wheatcraft. 2001. Fractional dispersion, Lévy motion, and the MADE tracer tests. In: Dispersion in Heterogeneous Geological Formations (pp. 211–240). Springer Netherlands.

Berg, H.C. 2000. Motile behavior of bacteria. *Physics Today* 53(1): 24–30.

Boggs, J.M., S.C. Young, L.M. Beard, L.W. Gelhar, K.R. Rehfeldt and E.E. Adams. 1992. Field study of dispersion in a heterogeneous aquifer: 1. Overview and site description. *Water Resources Research* 28(12): 3281–3291.

Cushman, J.H. 1984. On unifying the concepts of scale, instrumentation and stochastics in the development of multiphase transport theory. *Water Resources Research* 20(11): 1668–1676.

Cushman J.H. 1991. On diffusion in fractal porous media. *Water Resources Research* 27(4): 643–644.

Cushman J.H. and T. Ginn. 1993. Nonlocal dispersion in media with continuously evolving scales of heterogeneity. *Transport in Porous Media* 13(1): 123–138.

Cushman, J.H., X. Hu and T.R. Ginn. 1994. Nonequilibrium statistical mechanics of preasymptotic dispersion. *Journal of Statistical Physics* 75(5-6): 859–878.

Cushman, J.H. and M. Moroni. 2001. Statistical mechanics with three-dimensional particle tracking velocimetry experiments in the study of anomalous dispersion. I. Theory. *Physics of Fluids* 13(1): 75–80.

Cushman, J.H., M. Park, N. Kleinfelter and M. Moroni. 2005. Super-diffusion via Lévy lagrangian velocity processes. *Geophysical Research Letters* 32(19).

Cushman, J.H., D. O'Malley and M. Park. 2009a. Anomalous diffusion as modeled by a nonstationary extension of Brownian motion. *Physical Review* E79(3): 032101.

Cushman, J.H., M. Park and D. O'Malley. 2009b. Chaotic dynamics of super-diffusion revisited. *Geophysical Research Letters* 36(8).

Cushman, J.H., M. Park, M. Moroni, N. Kleinfelter-Domelle and D. O'Malley. 2011. A universal field equation for dispersive processes in heterogeneous media. *Stochastic Environmental Research and Risk Assessment* 25(1): 1–10.

Falconer, K. 2004. Fractal geometry: mathematical foundations and applications. John Wiley and Sons.

Hassan, A.E., J.H. Cushman and J.W. Delleur. 1997. Monte Carlo studies of flow and transport in fractal conductivity fields: Comparison with stochastic perturbation theory. *Water Resources Research* 33(11): 2519–2534.

Magdziarz, M., A. Weron, K. Burnecki and J. Klafter. 2009. Fractional Brownian motion versus the continuous-time random walk: A simple test for subdiffusive dynamics. *Physical Review Letters* 103(18): 180602.

Mandelbrot, B.B. and J.W. Van Ness. 1968. Fractional Brownian motions, fractional noises and applications. *SIAM Review* 10(4): 422–437.

Mandelbrot, B.B. and J.R. Wallis. 1968. Noah, Joseph, and operational hydrology. *Water Resources Research* 4(5): 909–918.

Meerschaert, M.M., D.A. Benson and B. Bäumer. 1999. Multidimensional advection and fractional dispersion. *Physical Review* E59(5): 5026.

Metzler, R. and J. Klafter. 2000. The random walk's guide to anomalous diffusion: a fractional dynamics approach. *Physics Reports* 339(1): 1–77.

Moroni, M. and J.H. Cushman. 2001. Statistical mechanics with three-dimensional particle tracking velocimetry experiments in the study of anomalous dispersion. II. Experiments. *Physics of Fluids* 13(1): 81–91.

McQuarrie, D.A. 1976. Statistical Mechanics. Harper and Row.

Neuman, S.P. 1993. Eulerian-lagrangian theory of transport in space-time nonstationary velocity fields: Exact nonlocal formalism by conditional moments and weak approximation. *Water Resources Research* 29(3): 633–645.

Nolan, J.P. 1997. Numerical calculation of stable densities and distribution functions. Communications in statistics. *Stochastic Models* 13(4): 759–774.

O'Malley, D. and J.H. Cushman. 2010. Fractional Brownian motion run with a nonlinear clock. *Physical Review* E82(3): 032102.

O'Malley, D., J.H. Cushman and G. Johnson. 2011. Scaling laws for fractional Brownian motion with power-law clock. *Journal of Statistical Mechanics: Theory and Experiment* 2011(01): L01001.

O'Malley, D. and J.H. Cushman. 2012a. A renormalization group classification of nonstationary and/or infinite second moment diffusive processes. *Journal of Statistical Physics* 146(5): 989-1000.

O'Malley, D. and J.H. Cushman. 2012b. Two-scale renormalization-group classification of diffusive processes. *Physical Review* E86(1): 011126.

O'Malley, D., V.V. Vesselinov and J.H. Cushman. 2014. A method for identifying diffusive trajectories with stochastic models. *Journal of Statistical Physics* 156(5): 896–907.

O'Malley, D. and V.V. Vesselinov. 2014. Analytical solutions for anomalous dispersion transport. *Advances in Water Resources* 68: 13–23.

Panja, D. 2011. Probabilistic phase space trajectory description for anomalous polymer dynamics. *Journal of Physics: Condensed Matter* 23(10): 105103.

Park, M., N. Kleinfelter and J.H. Cushman. 2005a. Scaling Laws and Fokker-Planck Equations for 3-Dimensional Porous Media with Fractal Mesoscale. *Multiscale Modeling and Simulation* 4(4): 1233–1244.

Park, M., N. Kleinfelter and J.H. Cushman. 2005b. Scaling laws and dispersion equations for Lévy particles in one-dimensional fractal porous media. *Physical Review* E72(5): 056305.

Park, M. and J.H. Cushman. 2006. On upscaling operator-stable Lévy motions in fractal porous media. *Journal of Computational Physics* 217(1): 159–165.

Park, M. 2013. Upscaling Lévy motions in porous media with long range correlations. *Journal of Mathematical Physics* 54(8): 083302.

Regner, B.M., D. Vučinić, C. Domnisoru, T.M. Bartol, M.W. Hetzer, D.M. Tartakovsky and T.J. Sejnowski. 2013. Anomalous diffusion of single particles in cytoplasm. *Biophysical Journal* 104(8): 1652–1660.

Samoradnitsky, G. and M.S. Taqqu. 1994. Stable non-Gaussian random processes: Stochastic models with infinite variance (Vol. 1). CRC Press.

CHAPTER
7

Multifractals and Geostatistics

A.M. Tarquis[1,2,3*], Juan J. Martín-Sotoca[1], M.C. Morató[2], M.T. Castellanos[2], J. Borondo[3], N.R. Bird[1] and A. Saa-Requejo[1,4]

[1]Research Centre for the Management of Agricultural and Environmental Risk (C.E.I.G.R.A.M.), E.T.S.I.A.A.B. UPM, Madrid, Spain
[2]Dpto. Matemática Aplicada. Universidad Politécnica de Madrid (UPM), Avda. Complutense s/n. 28040, Madrid, Spain
[3]Grupo de SistemasComplejos. UPM, Avda. Complutense s/n. 28040, Madrid, Spain
[4]Dpto de Producción Agraria. Universidad Politécnica de Madrid (UPM), Avda. Complutense s/n. 28040, Madrid, Spain

1. Introduction

The idea of describing natural phenomena using statistical scaling laws is not new. In the literature there exist several studies on such a subject (see seminal works of Bachelier 1900, Kolmogorov 1941, Hurst 1951, De Wijs 1951, Renyi 1955). Recently, statistical scaling techniques have been recalled again as a great number of physical systems tend to present similar behaviours at different scales of observation.

In the 1960s, the mathematician Benoît Mandelbrot introduced the term "fractal" to describe objects whose complex geometry cannot be characterized by an integral dimension. Fractal geometry was introduced to characterize irregular or fragmented shape of natural features as well as other complex phenomena (Mandelbrot 1963, 1967). The notion of fractal sets was then used to quantify the degree of regularity of a hierarchical structure. This phenomenon is often expressed by spatial or time-domain statistical scaling laws and is mainly characterized by the power-law behaviour of real-world physical systems. This concept can be found in many different fields, such as geophysics, biology or fluid mechanics.

*Corresponding author: anamaria.tarquis@upm.es

Fractal structures may be synthetically generated by applying an exact rule (exact self-similarity or affinity). Such structures are known as deterministic (or, mathematical) fractals and characterized by the same fractal dimension in all scales (scale-invariant characteristic): a structure is composed of objects whose smaller scales replicate exactly their larger ones, up to infinity. There are a lot of such structures, for instance, Cantor sets, Koch's curves, Sierpinski gasket and carpet, etc. (Mandelbrot 1983, Evertsz and Mandelbrot 1992). Instead, a variety of objects, structures, and phenomena in nature may exhibit self-similarity (or self-affinity) in a statistical (randomized) way, meaning that the reproduced feature may not be *exactly* the same as the original generator. These structures are known as random fractals. Furthermore, natural fractals are not self-similar over all scales. There are both upper and lower size limits, beyond which a structure is no longer fractal.

Multifractals could be seen as an extension of fractals and considered as a superposition of homogeneous monofractal structures. This type of analysis initially appeared with multiplicative cascade models to study energy dissipation in the context of the fully developed turbulence and then was applied on the measurement of the turbulent flow velocity in the 1980s (Hentschel and Procaccia 1983, Halsey et al. 1986, Chhabra and Jensen 1989). The purpose of multifractal analysis is to quantify the singular structure of a measure, and to provide models of different scaling power laws when dimension changes (Falconer 1996).

Multifractals techniques are mainly used to characterize the scaling behaviour of a system. They investigate the arrangement of quantities such as population or biomass densities (Saravia et al. 2012). Since scaling laws are an emergent general feature of ecological systems and reflect constraints in their organization that can provide tracks about the underlying mechanisms (Solé and Bascompte 2006), in the next two sections we provide some examples of how multifractals analysis and geostatistics are related.

2. Structure Function and Semivariogram

Soil properties vary spatially and exhibit strong fluctuations even over short distances. This variability is due to the combined action of physical, chemical and biological processes that operate with different intensities and at different scales. The description and quantification of the spatial variability of soil properties are important for modeling soil processes (Burrough et al. 1994). This variability is composed of "functional" (defined) variations and random fluctuations or noise (Goovaerts 1997, 1998). Geostatistical methods and, more recently, multifractal/wavelet techniques have been used to characterize the scaling and heterogeneity of soil properties along with other methods originating from complexity science (de Bartolo et al. 2011).

In many soil studies, researchers have characterized the spatial dependency of a variable measured along a transect as a mass distribution on a spatial domain (Zeleke and Si 2004, 2006). For this characterization, the transect is divided into a number of self-similar segments. The differences among the subsets are identified using fractal dimension $D(q)$ and a multifractal spectrum (Folorunso et al. 1994, Caniego et al. 2005, Morató et al. 2016). For example, several authors (Siqueira et al. 2013, López de Herrera et al. 2016) have applied multifractal analysis to profiles of soil penetrometer resistance data sets and found that these modern methods added complementary information to describe the spatial arrangement to more common geostatistical methods such as kriging and co-kriging.

Some recent works on agricultural soils have studied the application of these modern methods to cases in which a measure along a transect is observed as a random signal. For instance, Pozdnyakova et al. (2005) evaluated the spatial variability of cranberry yield by applying a generalized structure function (GSF) and proved the influence of multiscale factors (nonlinear structure functions). Kravchenko (2008) approached the spatial features of environmental and agronomic variables using multifractal characteristics in a stochastic simulation. García-Moreno et al. (2010) assessed the variability of soil surface roughness using the generalized structure function of transects to compare soil types and tillage tools, which yielded promising results. We will apply the generalized structure function to data on soil properties along a transect of arable fields, to compare and evaluate the results obtained for characterizing their structure and variability.

General Structure Function

The structure function (SF) analysis basically consists of studying the scaling behaviour of the non-overlapping fluctuations of a variable for different scale increments. The statistical moments of these fluctuations are estimated, which depend only on the scale increment (Monin and Yaglom 1975).

For non-stationary processes GSF of order q is defined as the qth moment of the increments of initial values $\mu(i)$. The equation is:

$$M_q(\Delta i) \equiv \left\langle |\mu(i+\Delta i) - \mu(i)|^q \right\rangle \tag{1}$$

where i denotes the ith data point, and $\langle\ \rangle$ represents the ensemble average.

GSF are generalized correlation functions. This is particularly evident from Eq. (1) for the case of $q = 2$ giving the variogram frequently used in geostatistics:

$$M_q(\Delta i) \equiv \left\langle \left|\mu(i+\Delta i) - \mu(i)\right|^2 \right\rangle \tag{2}$$

In general, q may be any real number not just integers, and can even be negative. However, there are divergence problems inherent to the negative-order exponent so that computations are best restricted to positive real numbers (Davis et al. 1994). If the process $\mu(i)$ is scale-invariant and self-similar or self-affine over some range of space lags $\Delta i_{\min} \leq \Delta i \leq \Delta i_{\max}$, then the qth-order structure function is expected to scale as

$$M_q(\Delta i) \equiv C_q \Delta i^{\zeta(q)} \approx \Delta i^{\zeta(q)} \tag{3}$$

where C_q can be a function of Δi which varies more slowly than any power of Δi, and $\zeta(q)$ is the exponent of the structure function. $\zeta(q)$ is a monotonically non-decreasing function of q if $\mu(i)$ has absolute bounds (Marshak et al. 1994, Frisch 1995). From Eq. (2) we can see that the statistics of the fluctuations over space lags Δi has two components; first because it depends on the fluctuations at low Δi values, and second because of the scaling relation between the fluctuations and Δi. From $\zeta(q)$ the first moment is related to the degree of non-conservation of a given field, $H = \zeta(q = 1)$.

The behaviour described by Eq. (1) and Eq. (3) is called "multiscaling" because each statistical moment scales with a different exponent. Therefore, a hierarchy of exponents can be defined using $\zeta(q)$ as

$$H(q) = \frac{\zeta(q)}{q} \tag{4}$$

where $H(q)$ is the generalized Hurst exponent or self-similarity scaling exponent (Davis et al. 1994). Calculation of $H(q)$ allows the straightforward identification of persistence, or long-space correlation, as well as the stationary/non-stationary and monofractal/multifractal nature of the data (Lovejoy et al. 2001).

Stationary processes have scale-independent increments and $\zeta(q) = H(q) \equiv 0$, due to the invariance under translation. Processes with a linear $\zeta(q)$ (or a constant $H(q)$) are non-stationary and monofractal, otherwise they are non-stationary and multifractal.

Application to Soil Transect

The data used in this chapter were collected in a survey on a common 1024 m transect across arable fields at Silsoe in Bedfordshire, east-central England. The data have previously been analyzed by Lark et al. (2004). The first sample point on the transect was at UK Ordnance Survey (OS) co-ordinates 508570, 235605, and the soil was sampled at 256 locations at 4 m intervals on a line running on a bearing of 188 degrees relative to UK OS grid north.

More specifically, the data selected for analysis in this chapter were porosity and nitrous oxide flux. The values of all these variables are shown in Fig. 1 (left side). For each of the variables of this study, the first four statistical moments i.e., average, variance, kurtosis and asymmetry (skewness) were calculated to study their similitude with a Gaussian distribution. The same calculations were performed on each variable after differentiating the series at several non-overlapping lags, from 1 to 64. In this way, we could study the statistical moments of the frequency distribution of the values obtained in each lag.

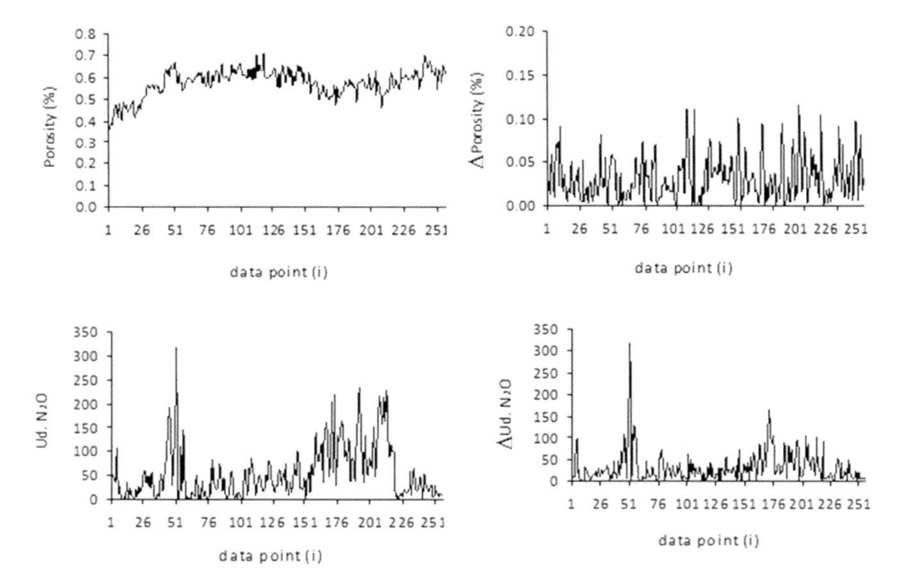

Figure 1: Original data of the soil variables: Porosity (%) and N_2O flux on the left column. On the right, the absolute differences obtained with lag 1 of the corresponding variables. Adapted from Morató et al. (2016).

The first four statistical moments were calculated for the two variables (Table 1). N_2O flux has greater variance compared to porosity, as expected (see values given in Fig. 1, left side, showing the original series of μ_i). With respect to higher-order moments, either asymmetry or kurtosis is closer to values corresponding to a normal distribution. Both variables have a positive kurtosis indicating a "peaked" distribution. The calculated asymmetry for porosity is negative showing slightly skewed left, while N_2O flux has a larger positive value indicating that the distribution is skewed right. The differentiation with lag 1 of the absolute values is given in Fig. 1 for each variable (right side, showing the $|\mu_i - \mu_{i+1}|$ series).

Focusing our attention on the differentiate values at different lags (Table 2), we can observe that lags greater than 8 do not have enough data points to yield a good estimation in the statistical moments of the frequencies distribution and therefore we will concentrate on lags between 1 and 8. For both variables the average values for different lags have a value close to zero; porosity slightly less than zero and N_2O slightly greater (Table 2). As the lag increases, the average value of porosity decreases from –0.0009 to –0.0149, meanwhile in the N_2O case increases from 0.0938 to 1.5 (see Table 2). The values of kurtosis for N_2O, higher than Gaussian distribution, decrease as the lag increases. However, porosity has kurtosis values very close to the Gaussian distribution. In the case of asymmetry, all the variables and lags from 1 to 8 show values close to zero (see Table 2).

Table 1: Descriptive statistics using the first four moments (average, variance, asymmetry and kurtosis) of soil porosity (Porosity) and N_2O flux values (N_2O). Adapted from Morató et al. (2016)

Statistics	*Porosity*	*N_2O*
Average	0.5736	54.42
Variance	0.0040	2970.74
Asymmetry	-0.8559	1.59
Kurtosis	0.9440	2.81

Table 2: Descriptive statistics using the first four moments (average, variance, kurtosis and asymmetry) of the differences in value of soil porosity (Porosity) and N_2O flux values (N_2O) at different lags. Adapted from Morató et al. (2016)

Statistics			*Porosity(%)*		
Lag	1	2	4	8	16
Average	-0.0009	-0.0019	-0.0037	-0.0075	-0.0149
Variance	0.0017	0.0021	0.0024	0.0038	0.0086
Kurtosis	0.1756	-0.0304	-0.0144	-0.5959	-0.8971
Asymmetry	0.1317	0.3275	-0.4688	-0.143	0.487
Data points	256	128	64	32	16
Statistics			*N_2O (%)*		
Lag	1	2	4	8	16
Average	0.0938	0.1875	0.375	0.75	1.5
Variance	2314.4225	3698.5315	3521.7937	1887.6129	3261.4667
Kurtosis	8.2063	7.4699	1.3533	0.7432	4.0086
Asymmetry	0.9069	0.8156	-0.6926	-0.0886	1.7007
Data points	256	128	64	32	16

The GSFs obtained from the variables studied here are presented in Fig. 2, where the maximum increment chosen was 128 data points ($L = \Delta i_{max}$) equivalent to 512 m. As can be observed, the bilog plot of $M_q(\Delta i)$

does not show linear behaviour in every interval. For this reason, a range of $\frac{\Delta i}{\Delta i_{max}}$ intervals was selected for each variable, as indicated by the arrows, giving a minimum R^2 of 0.97 for all q values used.

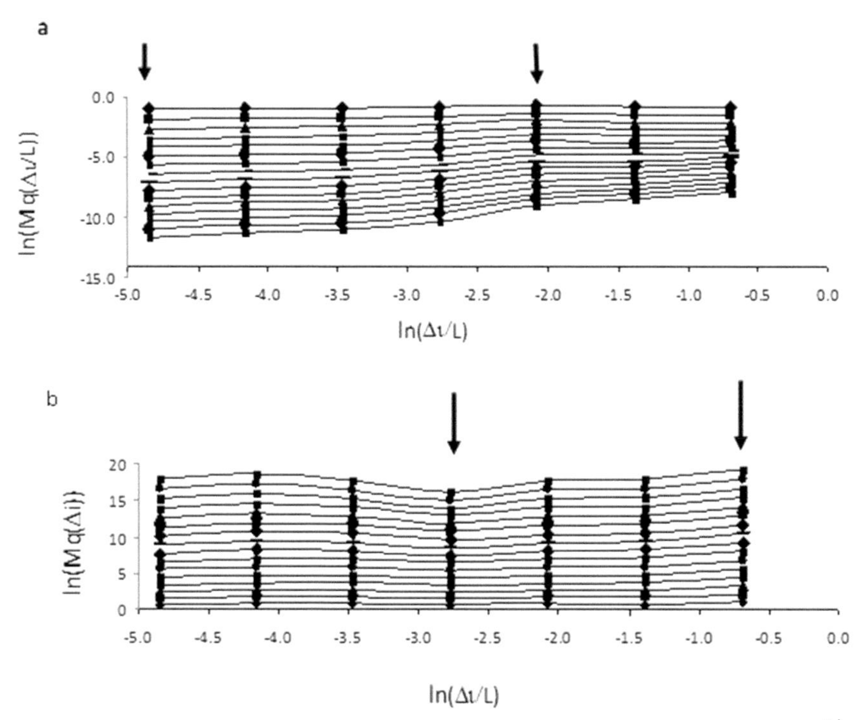

Figure 2: Log-log plot of the generalized structure function (M_q (Δi)) versus $\frac{\Delta i}{\Delta i_{max}}$ with $L = \Delta i_{max}$ for: (a) Porosity and (b) N_2O flux. The arrows point out the range of scale selected. Different symbols correspond to various values of the exponent q (from top $q = 0.25$ to 4). Adapted from Morató et al. (2016).

The results obtained from the $\zeta(q)$ curves, shown in Fig. 3, strongly corroborate the multifractal nature of N_2O flux, but weakly the multifractal nature of porosity, which looks almost like a straight line in this plot. For comparison, we also plotted the line $\zeta(q) = \frac{q}{2}$, corresponding to the Brownian motion. In both cases, the $\zeta(q)$ function is significantly different from the line corresponding to the Brownian motion, demonstrating the correlation of the increments among several scales.

The generalized Hurst exponents derived from the $\zeta(q)$ function based on GSF are shown in Fig. 4 in which the straight line represents the case of pure noise or uncorrelated noise (i.e., $H(q) = 0.5$). At lower q values

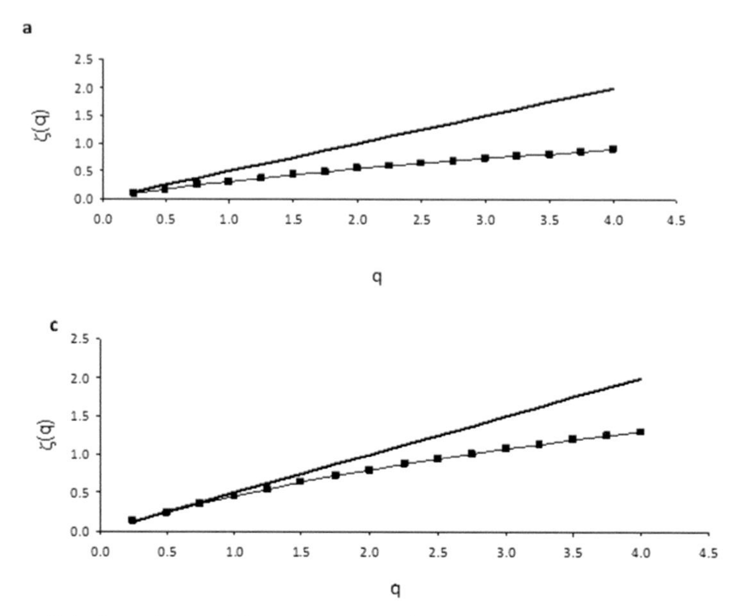

Figure 3: $\zeta(q)$ plots obtained from the generalized structure function (GSF) for: (a) Porosity, and (b) N_2O flux. Continuous solid lines represent straight lines with the slope of 0.5 denoting a non-correlated noise. Adapted from Morató et al. (2016).

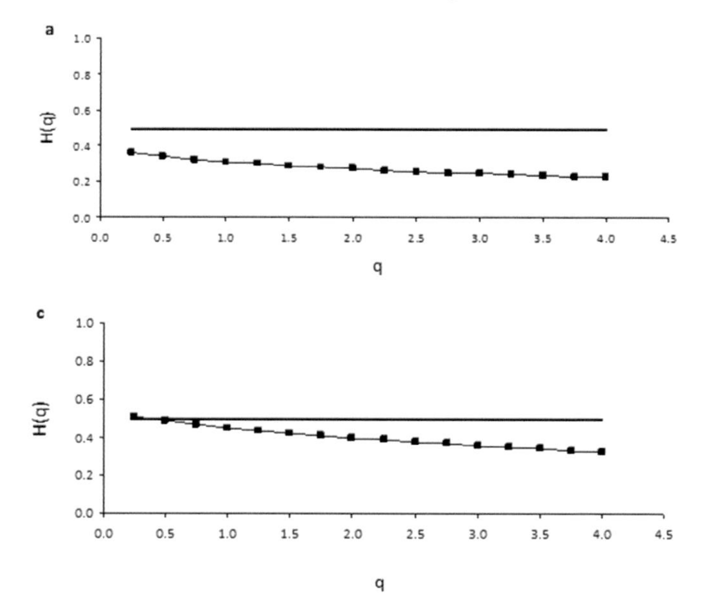

Figure 4: Generalized Hurst exponent ($H(q)$) for: (a) Porosity, and (b) N_2O flux. Continuous solid lines represent straight lines with the slope of 0 corresponding to non-correlated noise with Hurst value of 0.5. Adapted from Mortató et al. (2016).

the value of $H(q)$ is closer to 0.5; however, $H(q)$ decreases for the higher moments, attaining an anti-persistent character. The highest amplitude of $H(q)$ corresponds to N_2O flux, which showed a more pronounced curvature in $\zeta(q)$ (see Fig. 3).

At this point, we should point out that the $\zeta(q)$ function is based on the behaviour of the increments of the variables at different lags. However, these increments are in absolute value, implying a lack of ability to interpret the statistics of the probability density function of the increments as other scaling models do (Riva et al. 2015).

3. Hölder Exponents and Spatial Statistics

Multifractal (or cascade) models are related to specific probability distributions and geostatistics. For example, Cheng (1999) showed the links between multifractal parameters and semivariogram and autocorrelation. Using statistics based on second-order moments such as variograms, we focus our attention to characterize our spatial variables only around the mean and variance value. In applications where a variable shows high heterogeneity and its extreme values should be taken into account, higher-order statistical moments and multifractal analysis should be used (Cheng and Agterberg 1996).

Mapping of singularities in multifractal analysis has been proved to be a very effective tool to delineate areas with anomalies in measure distributions. Cheng (1999) elaborated on a local singularity analysis based on multifractal modelling that provides a powerful tool to characterize the local structural properties of spatial patterns. Under the assumption of local singularity, the values of the variable around a singular location often display a scale invariant property (Cheng 1999). This can be described as a power-law relation between the size of the area centred in the singular location and the average of the variable values in that area (Xie et al. 2007). The exponent of such a power-law local model is the Hölder exponent and characterizes the degree of singularity. This method has been successfully applied to detect anomalies in the concentration of a given mineral (Cheng 2001, 2006, 2007, Xie et al. 2007), meaning that these areas are the target of exploiting new deposits and texture analysis of remote-sensing images (Cheng 1999).

In section 3.3, Normalized Difference Vegetation Index (NDVI) is considered as a spatial measure to be studied. NDVI map has a multifractal character as shown by Poveda and Salazar (2004), Alonso et al. (2005, 2007, 2008) and Lovejoy et al. (2008). Thus, it would be interesting to check if in this case multifractal character holds, estimate singularity maps of NDVI images and extrapolate the conclusions obtained by Cheng for mineral concentration maps.

3.1 Multifractal Spectrum

The aim of the multifractal analysis (MFA) is to study how a normalized probability distribution of a variable (μ_i) varies with scale. In this sense, the density levels of these probabilities are evaluated through the scaling behaviour, r being the size of the scale, of a range of statistical moments (q) of the partition function ($\chi(q, r)$). We assume that the spatial distribution of NDVI corresponds to a variable or mass distribution μ on $\mathbb{R}^2$. Let's consider a grid of cells of size r covering an NDVI image, with total length of size L. The variable value of the ith cell is defined as $M_i(r)$. We now perform a weighted summation over all cells giving

$$\mu_1(q, r) = \frac{M_i^q(r)}{\sum_{i=1}^{N(\delta)} M_i^q(r)} \tag{5}$$

For a multifractal measure, the partition function $\chi(q, r)$ scales with r as follows (Evertsz and Mandelbrot, 1992),

$$\chi(q, r) \sim r^{r(q)} \tag{6}$$

in which,

$$\chi(q, r) = \sum_{j=1}^{N(r)} \mu_j^q(r) \tag{7}$$

where $\tau(q)$ is a nonlinear function of q called the "mass exponent function" (Feder 1989). For each q, $\tau(q)$ may be obtained as the slope of a log-log plot of $\chi(q, r)$ versus r, which is known as the method of moments (Halsey et al. 1986).

The singularity index or Lipschitz-Hölder exponents (α) can be determined using the Legendre transformation of the $\tau(q)$ curve as:

$$\alpha(q) = \frac{d\tau(q)}{dq} \tag{8}$$

The number of cells of size r with the same α, $N_\alpha(r)$, is related to the cell size as $N_\alpha(r) \propto r^{-f(\alpha)}$ in which $f(\alpha)$ is a scaling exponent of the cells with the same value α. Parameter $f(\alpha)$ can be calculated as:

$$f(\alpha) = q\alpha(q) - \tau(q) \tag{9}$$

Properties of functions $\alpha(q)$, $\tau(q)$ and $f(\alpha)$ have been discussed by several authors (see e.g., Feder 1989, Schertzer and Lovejoy 1991, Cheng and Agterberg 1996).

Multifractal spectrum (MFS), i.e. a graph of α vs. $f(\alpha)$, quantitatively characterizes variability of the measure studied with asymmetry to the right and left indicating scaling domination of small and large values, respectively. The width of the MF spectrum indicates the overall variability (Tarquis et al. 2001).

3.2 Singularity Maps

Singularity map is defined as the locus of the points x that have the same Lipschitz-Hölder exponent

$$\alpha = T(x) \tag{10}$$

where $T(.)$ is the function that gives us the Lipschitz-Hölder exponent for each point x. To calculate the singularity map or Lipschitz-Hölder exponents we follow Cheng (2001) and Falconer (2003). In multifractal measures the values of the local mass concentration $\mu(B(x, r))$, calculated for various cell sizes r centred at x, obeys a power law in terms of r as follows:

$$\mu(B(x, r)) \sim r^{\alpha(x)} \tag{11}$$

This power law is fulfilled in a certain range of r, $[r_{min}, r_{max}]$. Locations $\alpha(x) \neq E$, where E is the topological dimension of the support, called singular locations. Moreover, we can differentiate positive singularities $\alpha(x) < E$ and negative singularities $\alpha(x) > E$. Lipschitz-Hölder exponents $\alpha(x)$ can be calculated by the expression

$$\alpha(x) = \lim_{r \to 0} \frac{\ln \mu(B(x,r))}{\ln r} \tag{12}$$

Cheng (2001) applied these concepts to a concentration map of a certain mineral. He stated that average concentration $Z(x)$, calculated for various cell sizes centred at x, obeys a power law in terms of r as

$$Z(x) \sim r^{\alpha(x) - E} \tag{13}$$

where $E = 2$ for two-dimensional problems and $E = 1$ for one-dimensional. Positive singularities with $\alpha(x) < 2$ correspond to high values of concentration in a geochemical map, while negative singularities with $\alpha(x) > 2$ correspond to low concentration values. Therefore, calculating the singularity map for a geochemical concentration map may be used to characterize concentration patterns which provide useful information for interpreting anomalies related to local mineralization processes.

In the next section, instead of concentration of a chemical element, we discuss concentration of photosynthetic activity. We are particularly interested in places where $\alpha(x) \cong 2$, since average photosynthetic concentration in these areas would be relatively uniform (equal average concentrations independent of location). Given that the semivariogram only depends on the distance between locations (Cheng and Agterberg 1996), these areas meet the requirements to be smooth areas.

The region around $\alpha = 2$, $[\alpha_{a\_min}, \alpha_{a\_max}]$, incorporates α values within a homogeneous zone. To calculate these lower and upper values of α we apply the concentration-area method (C-A method) (Cheng et al. 1994, Liu et al. 2005). This method distinguishes anomalous α values so that

$$A(\alpha > C) \propto C^{\beta} \tag{14}$$

where A is the accumulated area whose singularity values α are greater than a given cutoff C and β is the fractal dimension of the C-A method. On the log-log graph, some α value at which slope changes represents α_{a_min}, above which anomalous values start. To calculate the upper cut-off α_{a_max} the following expression is analogously used

$$A(\alpha \leq C) \propto C^{\gamma} \tag{15}$$

In this study, all singularity map calculations were performed in MATLAB. "Polyfit" function was used to calculate Lipschitz-Hölder exponents, which uses the least-square method to optimize the power-law model parameters. We set $r_{min} = 1$ as a minimum size and $r_{max} = 10$ as a maximum size of r. Therefore, a frame of nine pixels thick is not analyzed in the original data.

3.3 Application to NDVI Maps

In this study, we have images of 300 × 280 pixels corresponding to an area of 150 × 140 km that includes the Community of Madrid, Spain. For each pixel we have calculated NDVI values for each season (Figs 5 and 6; left columns).

Let $A(a_{j,k})$ be the matrix of 300 rows × 280 columns with NDVI values. Firstly, we normalize the matrix, so that the probability value of the ith cell of side r would be

$$\mu_i = \frac{\sum_{j,k\in i-cell} a_{j,k}}{\sum_{j=1}^{256}\sum_{k=1}^{256} a_{j,k}} \tag{16}$$

In all cases, we found a linear relationship on log-log scale of the partition function ($\chi(q, r)$) versus the length of the scale (r) with R^2 values greater than 0.98. From the slope determined for each mass exponent (q), a nonlinear $\tau(q)$ function was obtained reflecting a hierarchical structure from one scale to another with $\tau(q = 1) = 0$ indicating the conservative character of the variable. We estimated the Hölder exponents and the multifractal spectrum (MFS) in an interval of $q = \pm 10$ with an increment of 1 (Fig. 7).

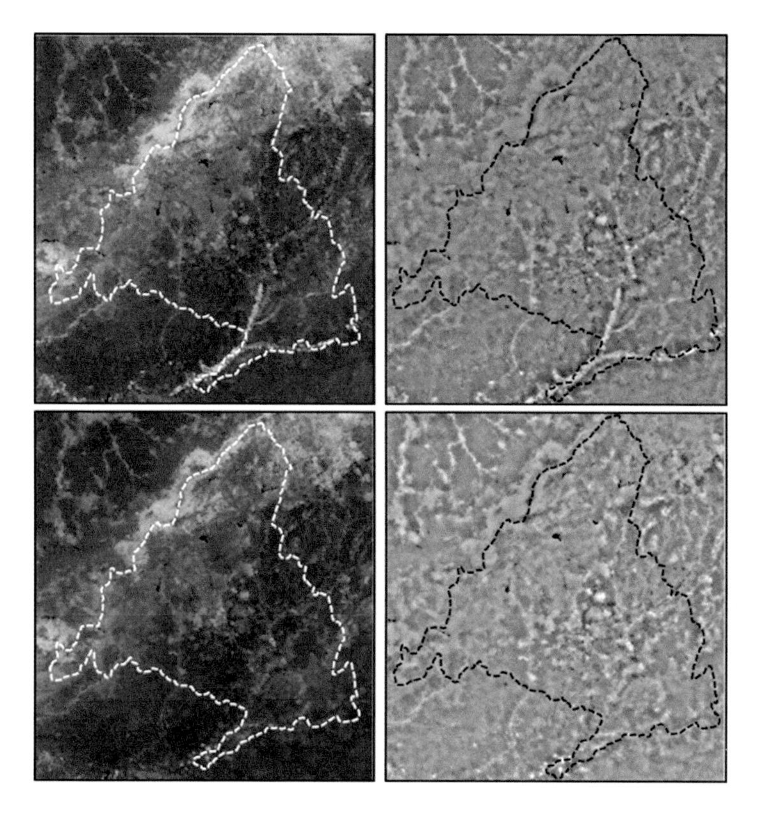

Figure 5: Study area images representing summer (upper row) and autumn (lower row) seasons. From left to right column: NDVI map, singularity map. Highest value in white and lowest value in black.

All the MFS showed concave down parabolic curves (Fig. 7B) with a variable symmetry depending on the studied NDVI season-dependent maps. Recall that in a fractal system the MFS is shown as a single point; therefore, our results support the hypothesis of a multifractal behaviour, rather than fractal, for NDVI profiles. As can be seen from all the MFS plots there are three differentiated groups on the left hand of the spectrum (positive q values) i.e., summer-autumn, winter and then spring, while on the right part of the MFS (negative q values) there are only two groups i.e., spring and the rest of the seasons.

MFS showed remarkable differences in amplitudes and symmetries, both can be quantified with the difference of the extreme singularities ($\Delta\alpha = \alpha_{max} - \alpha_{min}$) and the difference of their respective $f(\alpha)$ values ($\Delta f = f(\alpha_{max}) - f(\alpha_{min})$). In this way, the greater $\Delta\alpha$ the higher is the complexity of the structure studied in the NDVI map; the greater Δf the higher is the asymmetry presented in the MFS (left handed if $\Delta f < 0$ and right handed if $\Delta f > 0$).

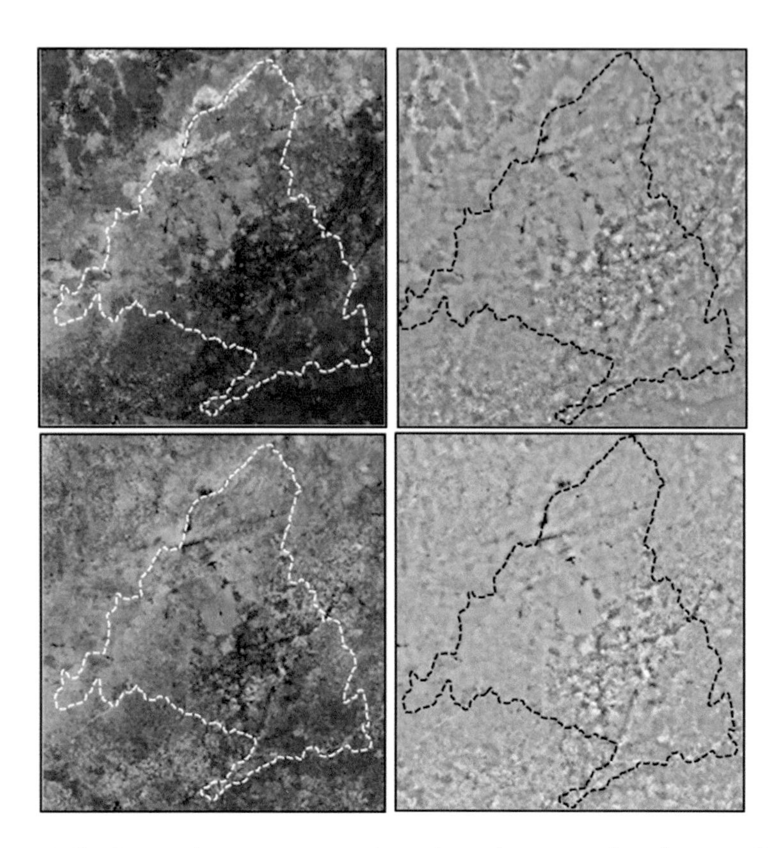

Figure 6: Study area images representing winter (upper row) and spring (lower row) seasons. From left to right column: NDVI map, singularity map. Highest value in white and lowest value in black.

By investigating the evolution of $\Delta\alpha$ along the year, we can see that during spring it presents the smallest $\Delta\alpha$ and then during summer and autumn it increases its value reaching a maximum. Finally, at winter season $\Delta\alpha$ decreases. This can be easily observed in Fig. 7A. This points out that during the dry season (i.e., summer) and beginning of the wet season (i.e., autumn) the NDVI map shows a higher hierarchical spatial structure among scales.

With respect to Δf, variation through the seasons follows a similar pattern as $\Delta\alpha$. During summer and autumn we found the highest value of Δf showing a clear asymmetry and a left handed shape, then winter with a right handed shape and next spring with the most negative value of Δf. This implies that during summer and autumn the higher values of NDVI have a stronger influence in the scaling behaviour as the left part of the multifractal spectrum corresponds to values using positive mass exponents (q) as shown in Equations 8 and 9. Through winter and

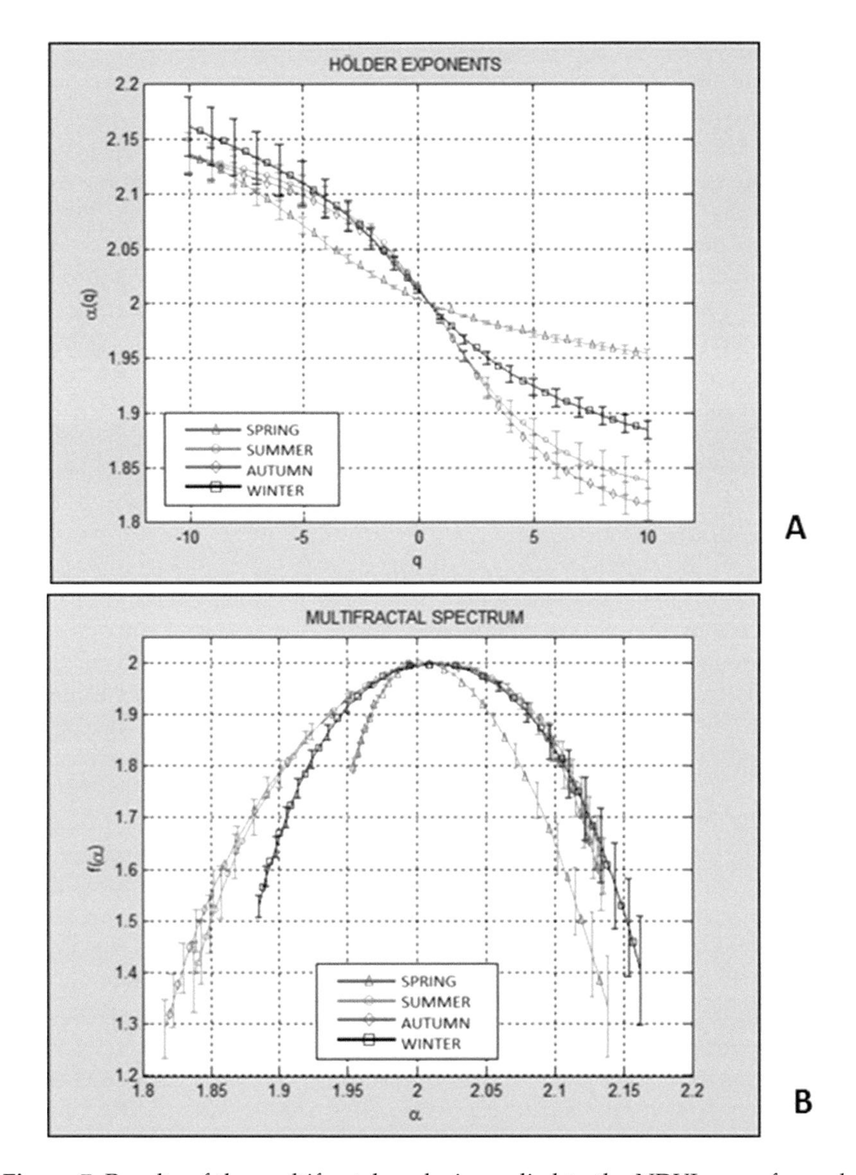

Figure 7: Results of the multifractal analysis applied to the NDVI maps for each season: A. Hölder exponents; B. Multifractal Spectrum. Segments represent the standard errors of the estimation based in the s.e. of the slope.

spring this disappears being the lower NDVI values which influence more the scaling as the right part of the multifractal spectrum corresponds to negative mass exponents.

Regarding singularity maps, for each date we have two images: (1) NDVI in gray scale (colour white represents $NDVI = NDVI_{max} = 1$) and

(2) singularity map in gray scale (colour white represents $\alpha = \alpha_{min}$). The results are presented in Figs 5 and 6; right columns.

Singularity maps clearly show areas of anomalies in the NDVI concentration. Positive singularities with $\alpha << 2$ (near-white areas on the singularity map) are highly correlated with riverbanks in the dry season (i.e., summer) as shown in Fig. 5. Negative singularities with $\alpha >> 2$ appear mostly on the city of Madrid and outlying cities.

For the calculation of non-singular areas, we first calculated the upper and lower cut-offs around the value $\alpha = 2$. Using the C-A method, we determined $[\alpha_{a\_min}, \alpha_{a\_max}]$ for each of the dates analyzed. In Fig. 8, log-log graphs are shown with different slopes at various scales showing how the lower and upper α values were calculated.

In Fig. 9, non-singular areas (white pixels) are shown for each analyzed date. Those pixels are the points for which $\alpha_{a\_min} \leq \alpha(x) \leq \alpha_{a\_max}$. For spring (Fig. 9; top left) large connected homogeneous zones appear, some of them identifiable as the natural park "Monte del Pardo". For summer (Fig. 9; top right) the extension of non-singular areas gets smaller, mainly due to extremely low values reached by NDVI. During this season, singularity areas along riverbanks are remarkable localized indicating the areas that retain a high NDVI value because the vegetation is in an area of higher humidity. In autumn (Fig. 9; bottom left) the extension and number of non-singular areas continues to decrease since spatial variability of the index increases, which indicates that for several reasons there are certain parts that achieve a high NDVI value faster than others. Indirectly, it is displaying the variability in soil moisture content and water availability for vegetation. Finally, during winter (Fig. 9; bottom right) large non-singular areas, that were observed during spring, begin to appear again pointing out a higher homogeneity in the space. Therefore, singularity maps are given a spatial and temporal study of the NDVI heterogeneity.

In summary, multifractal analysis (MFA) can be interpreted as one- or two-dimensional spatial statistics, as each example has illustrated, based on several order of moments, not only the second order moment as, for example, kriging does. In this way, MFA characterizes the scaling behaviour of several statistical moments of the variable being of high importance when its extreme values need to be studied, as it is in some cases of geology and soil science studies.

Acknowledgements

Our gratitude to the editors for their invitation to participate in this book, and their suggestions that have improved this work. This research has been partially supported by funding from MINECO under contract No. MTM2015-63914-P and CICYT PCIN-2014-080.

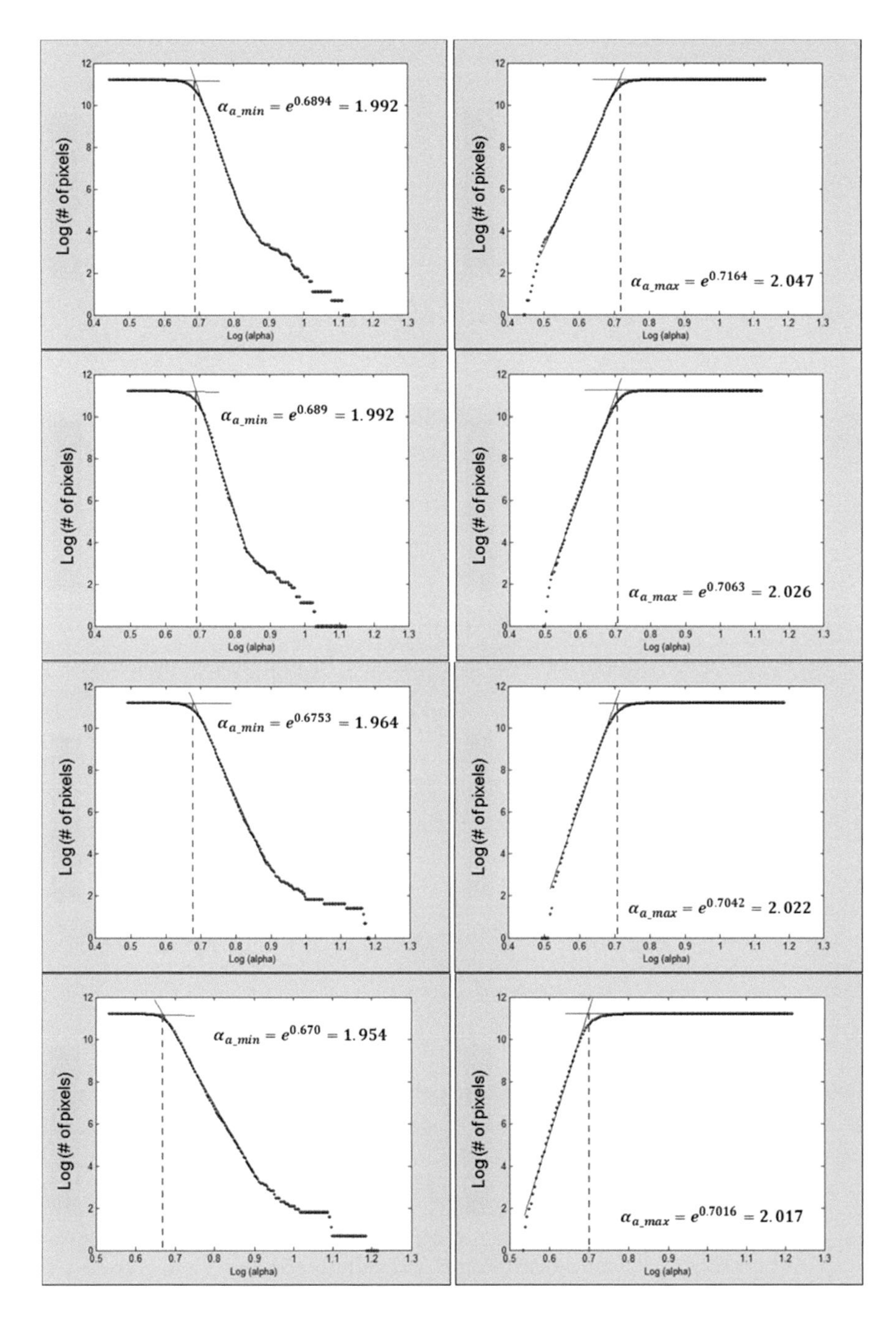

Figure 8: Singularity values from log-log graphs for summer, autumn, winter and spring seasons.

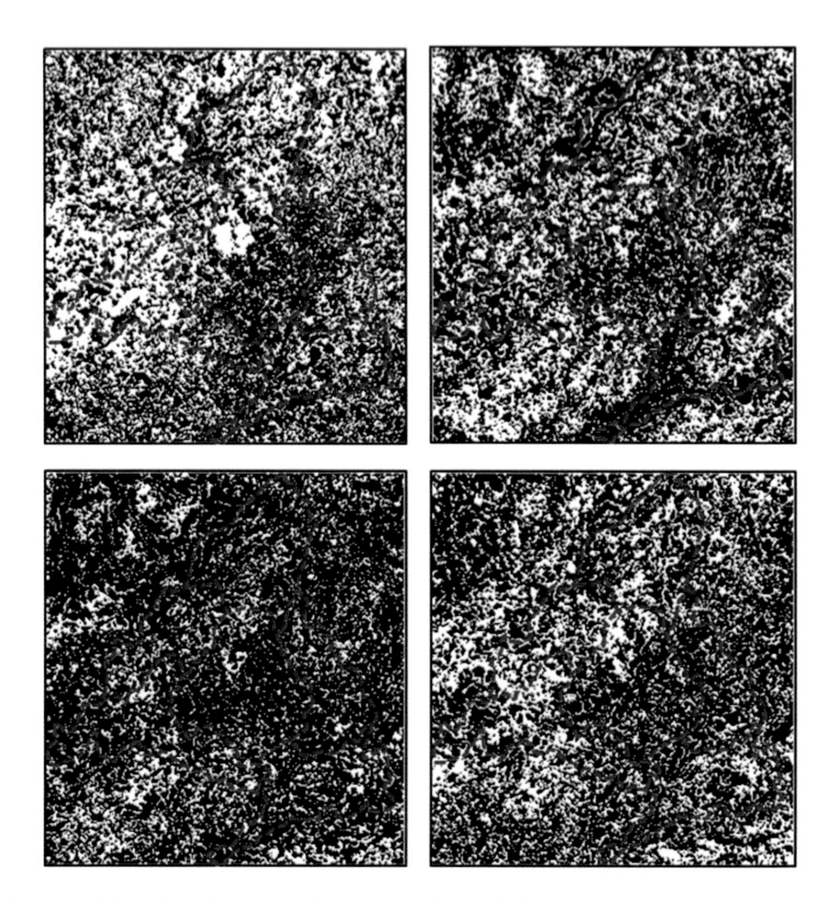

Figure 9: Non-singular areas shown as white pixels, where $\alpha_{a\_min} \leq \alpha(x) \leq \alpha_{a\_max}$, dated on (from left to right and from top to bottom): (a) spring, (b) summer, (c) autumn and (d) winter. Dashed red lines are the limits of agricultural regions.

REFERENCES

Alonso, C., A.M. Tarquis and R.M. Benito. 2005. Multifractal Characterization of Multispectrum Satellite Images. *Geophysical Research Abstracts* 7: 05413.

Alonso, C., A.M. Tarquis, R.M. Benito and I. Zúñiga. 2008. Influence of spatial and radiometric resolution of satellite images in scaling/multiscaling behavior. *Geophysical Research Abstracts* 10: 06184.

Alonso, C., A.M. Tarquis, R.M. Benito and I. Zúñiga. 2007. Scaling properties of vegetation and soil moisture indices: Multifractal and joint multifractal analysis. *Geophysical Research Abstracts* 9: 11643.

Bachelier, L. 1900. Théorie de la speculation. *Annales scientifiques de l'École Normale Supérieure* 17(3): 21–86.

Bartolo, S. de., W. Otten, Q. Cheng and A.M. Tarquis. 2011. Modeling soil system: Complexity under your feet. *Biogeosciences* 8: 3139–3142.

Burrough, P.A., J. Bouma and S.R. Yates. 1994. The state of the art in pedometrics. *Geoderma* 62: 311–326.

Caniego, F.J., R. Espejo, M.A. Martín and F. San José. 2005. Multifractal scaling of soil spatial variability. *Ecol. Model.* 182: 291–303.

Cheng, Q. and F.P. Agterberg. 1996. Multifractal modeling and spatial statistics. *Math. Geol.* 28: 1–16.

Cheng, Q., F.P. Agterberg and S.B. Ballantyne. 1994. The separation of geochemical anomalies from background by fractal methods. *J. Geochem. Explor.* 51: 109–130.

Cheng, Q. 1999. Multifractality and spatial statistics. *Computers & Geosciences* 25: 949–961.

Cheng, Q. 2006. GIS based fractal/multifractal anomaly analysis for modeling and prediction of mineralization and mineral deposits. In: GIS Application in the Earth Sciences – GAC Special Paper, Geological Association of Canada Special Book, edited by Harris, J.R., 285–296.

Cheng, Q. 2007. Mapping singularities with stream sediment geochemical data for prediction of undiscovered mineral deposits in Gejiu, Yunnan Province, China. *Ore Geol. Rev.* 32: 314–324.

Cheng, Q. 2001. Singularity analysis for image processing and anomaly enhancement. In: Proceedings IAMG'01, International Association for Mathematical Geology, Cancun, Mexcico, 6–12 September, CD-ROM.

Chhabra, A.B. and R.V. Jensen. 1989. Direct Determination of the $f(\alpha)$ singularity spectrum. *Phys. Rev. Lett.* 62, 1327.

Davis, A., A. Marshak, W. Wiscombe and R. Cahalan. 1994. Multifractal characterizations of nonstationarity and intermittency in geophysical fields: Observed, retrieved, or simulated. *J. Geophys. Res.* 99: 8055–8072.

De Wijs, H.H. 1951. Statistics of ore distribution. *I. Geol Mijnbouw* 13: 365–375.

Evertsz, C. and B. Mandelbrot. 1992. "Multifractal Measures", Appendix B. In: H. Peitgen, H. Jurgens, P. Andrews (Eds.), Andrews Chaos and fractal new frontiers of science. Springer-Verlag, New York. 922–953.

Falconer, K. 1996. Techniques in Fractal Geometry. John Wiley & Sons Ltd.

Falconer, K. 2003. Fractal Geometry. Mathematical foundations and applications (2nd edition). West Sussex: John Wiley & Sons.

Feder, J. 1989. Fractals. Plenum Press. New York.

Folorunso, O.A., C.E. Puente, D.E. Rolston and J.E. Pinzón. 1994. Statistical and fractal evaluation of the spatial characteristics of soil surface strength. *Soil Sci. Soc. Am. J.* 58: 284–294.

Frisch, U. 1995. Turbulence: The Legacy of A. Kolmogorov. Cambridge University Press, Cambridge, UK.

Garcia Moreno, R., M.C. Diaz, A.R. Saa, J.L. Valencia and A.M. Tarquis. 2010. Multiscaling analysis of soil roughness variability. *Geoderma* 160(1, 30): 22–30.

Goovaerts, P. 1997. Geostatistics for natural resources evaluation. Oxford Univ. Press, New York.

Goovaerts, P. 1998. Geostatistical tools for characterizing the spatial variability of microbiological and physico-chemical soil properties. *Biol. Fertil. Soils* 27: 315–334.

Halsey, T., M. Jensen, L. Kadanoff, I. Procaccia and B.I. Shraiman. 1986. Fractal measures and their singularities: The characterization of strange sets. *Physical Review* A33(2): 1141–1151.

Hentschel, H.G.E. and I. Procaccia. 1983. The infinite number of generalized dimensions of fractals and strange attractors. *Physica* D.8: 435–444.

Hurst, H.E. 1951. Trans. Amer. Soc. Civ. Eng. 116–770.

Kolmogorov, A. 1941. The local structure of turbulence in incompressible viscous fluid for very large Reynolds number. *Comptes Rendus de l'Académie des sciences* 30: 9–13.

Kravchenko, A.N. 2008. Stochastic simulations of spatial variability based on multifractal characteristics. *Vadose Zone J.* 7(2): 521–524.

Lark, R.M., A.E. Milne, T.M.A. Addiscott, K.W.T. Goulding, C.P. Webster and S. O'Flaherty. 2004. Scale and location-dependent correlation of nitrous oxide emissions with soil properties: An analysis using wavelets. *Eur. J. Soil Sci.* 55: 611–627.

Liu, Y., Q. Xia, Q. Cheng and X. Wang. 2005. Application of singularity theory and logistic regression model for tungsten polymetallic potential mapping. *Nonlin. Processes Geophys.* 20: 445–453.

Lopez de Herrera, J., T. Herrero, A. Saa and A.M. Tarquis. 2016. Effects of tillage on variability in soil penetration resistance in an olive orchard. *Soil Res.* 54(2): 134–143.

Lovejoy, S., D. Schertzer and J.D. Stanway. 2001. Fractal behaviour of ozone, wind and temperature in the lower stratosphere. *Phys. Rev. Lett.* 86: 5200–5203.

Lovejoy, S., A.M. Tarquis, H. Gaonac'h and D. Schertzer. 2008. Single and Multiscale remote sensing techniques, multifractals and MODIS derived vegetation and soil moisture. *Vadose Zone Journal* 7(2): 533–546.

Mandelbrot, B. 1963. The variation of certain speculative prices. *J. of Business* 36: 394–419.

Mandelbrot, B. 1967. How long is the coast of Britain? Statistical self-similarity and fractional dimension. *Science* 156: 636–638.

Mandelbrot, B. 1983. The Fractal Geometry of Nature. WH Freeman, Oxford.

Marshak, A., A. Davis, R.F. Cahalan and W.J. Wiscombe. 1994. Bounded Cascade Models as non-stationary multifractals. *Phys. Rev. E.* 49: 55–69.

Monin, A.S. and A.M. Yaglom. 1975. Statistical Fluid Mechanics: Mechanics of Turbulence. MIT Press, Boston.

Morató, M.C., M.T. Castellanos, N.R. Bird and A.M. Tarquis. 2016. Multifractal analysis in soil properties: Spatial signal versus mass distribution. *Geoderma*. http://dx.doi.org/10.1016/j.geoderma.08.004.

Poveda, G. and L.F. Salazar. 2004. Annual and interannual (ENSO) variability of spatial scaling properties of a vegetation index (NDVI) in Amazonia. *Remote Sensing of Environment* 93: 391–401.

Pozdnyakova, L., D. Giménez and P.V. Oudemans. 2005. Spatial Analysis of Cranberry Yield at Three Scales. *Agron. J.* 97(1): 49–57.

Renyi, A. 1955. On a new axiomatic theory of probability. *Acta Mathematica Hungarica* 6(3–4): 285–335.

Riva, M., S.P. Neuman and A. Guadagnini. 2015. New scaling model for variables and increments with heavy-tailed distributions. *Water Resour. Res.* 51: 4623–4634. doi:10.1002/2015WR016998.

Saravia, L.A., A. Giorgi and F. Momo. 2012. Multifractal Spatial Patterns and Diversity in an Ecological Succession. *PLoS ONE*, 7(3): e34096. doi:10.1371/journal.pone.0034096.

Schertzer, D. and S. Lovejoy (Eds.). 1991. Nonlinear Variability in Geophysics. Kluwer Academic Publishers. Dordrecht, The Netherlands. 318 pp.

Siqueira, G.M., E.F.F. Silva, A.A.A. Motenegro, E. Vidal Vázquez and J. Paz-Ferreiro. 2013. Multifractal analysis of vertical profiles of soil penetration resistance at the field scale. *Nonlinear Proc. Geoph.* 20: 529–541.

Solé, R.V. and J. Bascompte. 2006. Self-Organization in Complex Ecosystems. Princeton University Press.

Tarquis, A.M., J.C. Losada, R. Benito and F. Borondo. 2001. Multifractal analysis of the Tori destruction in a molecular Hamiltonian System. *Phys. Rev. E.* 65: 0126213(9).

Xie, S., Q. Cheng, G. Chen, Z. Chen and Z. Bao. 2007. Application of local singularity in prospecting potential oil/gas targets. *Nonlin. Processes Geophys.* 14: 285–292.

Zeleke, T.B. and B.C. Si. 2004. Scaling properties of topographic indices and crop yield: Multifractal and joint multifractal approaches. *Agron. J.* 96: 1082–1090.

Zeleke, T.B. and B.C. Si. 2006. Characterizing scale-dependent spatial relationships between soil properties using multifractal techniques. *Geoderma* 134: 440–452.

CHAPTER

8

Why the Warming Can't be Natural: The Nonlinear Geophysics of Climate Closure

Shaun Lovejoy
Physics Department, McGill University, 3600 University St., Montreal, Que. H3A 2T8, Canada
lovejoy@physics.mcgill.ca

1. Introduction

The atmosphere is a turbulent fluid whose temperature, humidity and wind vary from submillimetric eddies visible in cigarette smoke to huge planetary sized weather systems. It has been changing ever since the earth was formed several billion years ago and it changes at millisecond scales. Sixty five million years ago, the temperature was five or even ten degrees warmer than today and dinosaurs roamed an ice-free south pole. As little as fourteen thousand years ago, the earth was still in the throes of an ice age with global temperatures 2-4 degrees cooler than today. Historical viniculture records show that in the middle ages, England was significantly warmer than today, yet only centuries later in the "little ice age", Europe had cooled enough so that sixteenth century Dutch skaters were immortalized in Breugel's famous paintings.

These facts are supported by several converging lines of evidence and while the quantitative amounts of the warming and cooling are debated, the basic events are undisputed. Indeed, as shown in Fig. 1a-e, proxy and instrumental records show that there is strong variability at all observed time scales and Fig. 2a quantitatively confirms this with a modern spectrum showing that contrary to conventional wisdom, the "background" spectrum varies by a factor of more than 10^{15} over the range from hours to hundreds of millions of years. In space – and indeed in space-time – Fig. 2b shows that the scaling of satellite infrared radiance

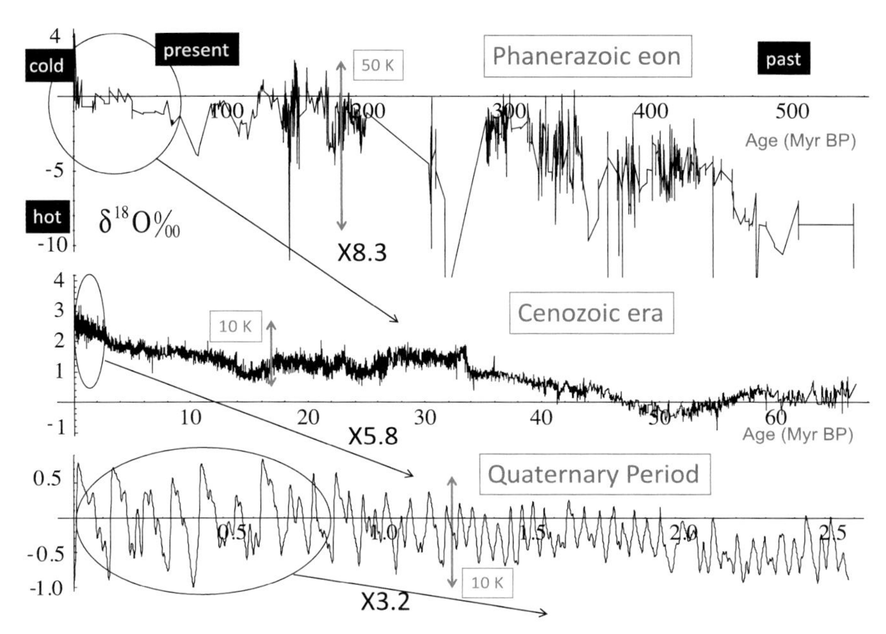

Figure 1a: δ^{18}O from assemblies of cores from ocean sediments of benthic organisms. Large values correspond to small temperatures and vice versa. The top series is an update of a global assemblage by Veizer et al. (1999), covering the Phanerozoic; the current geological eon during which abundant animal life has existed and goes back to the time when diverse hard-shelled animals first appeared: this figure goes as far back as this technique will allow. Although 2980 values were used, they are far from uniformly distributed; the figure shows a linear interpolation. The corresponding temperature range is indicated based on the "canonical" calibration of –4.5 K/δ^{18}O, and may be as much as a factor 3 too large. Note the negative sign in the calibration: large δ^{18}O corresponds to small temperatures and visa versa. The middle series is from a northern high latitude assemblage by Zachos et al. (2001) based on global deep-sea isotope records from data compiled from more than 40 Deep Sea Drilling Project and Ocean Drilling Project sites; it has 14828 values covering the period back 67 million years ago, again non uniformly distributed in time and considered to be globally representative. The bottom series is from Huybers (2007), it uses 2560 data points 12 benthic and 5 planktic δ^{18}O records over the Quaternary (the recent period during which there were glacials and interglacials; the rough oscillations that are visible, the series is mostly from high northern latitudes). For both of these series a roughly 50% larger calibration constant –6.5 K/δ^{18}O was used in order to take into account the larger high latitude variations. The ellipses, arrows and numbers indicate the parts of the time axis and zoom factor needed to go from one series to the next. This figure is continued in Fig. 1b. Reproduced from Lovejoy (2015b).

extends to planetary scales and up to about 10 days in time. It shows more: that over this range the (horizontal – not vertical) spatial and temporal statistics are related by an isotropic space-time scale invariance symmetry.

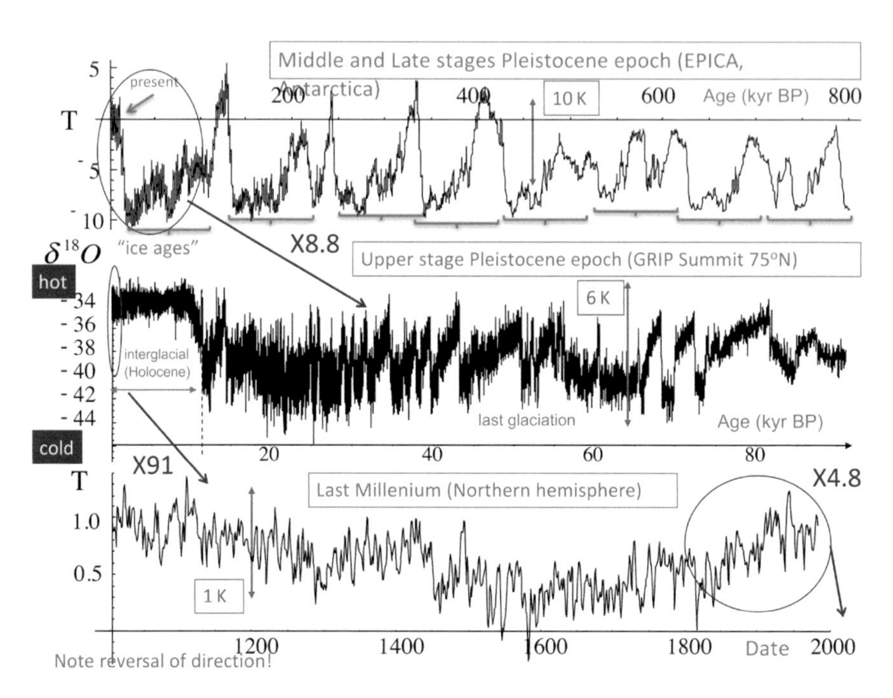

Figure 1b: The top series is the temperature anomaly from the Epica Antarctic core using a Deuterium based paleotemperature. The temperature anomalies are in degrees K (see (Lovejoy 2015b) for information on the data). We may note the loss in resolution (the apparent increase in smoothness) of the curve as we move into the past (to the right); it is an artefact of the compression of the ice column. Clearly, the neat classification of the series into 8 glacial and interglacial (parentheses) epochs is a somewhat subjective simplification of the true variability. Up to at least periods $\approx 10^5$ years, we see that the temperature seems to "wander" i.e. as we consider the change in temperature for increasingly long time periods, the temperature changes more and more. The series in the second row is over the time period indicated by the circle in top row, it is from a high resolution GRIP core (Summit Greenland). The current interglacial – the Holocene – is at the far left and an ellipse indicates the most recent 1000 year period. This last millennium is indicated in the bottom series which – conversely to the preceding – shows the present is on the right, the past on the left. This is a multiproxy temperature estimate from (Moberg et al. 2005), the ellipse (right) shows the industrial epoch of global warming; not all of this variability is natural in origin. The ellipses, arrows and numbers indicate the parts of the time axis and zoom factor needed to go from one series to the next. Reproduced from Lovejoy (2015b). This figure is continued in Fig. 1c.

Without human intervention, over sufficiently long periods, the temperature of the earth can clearly change by several degrees. But what about this: since the end of the 19th century instrumental records show that the earth has warmed by about one degree centigrade. The evidence

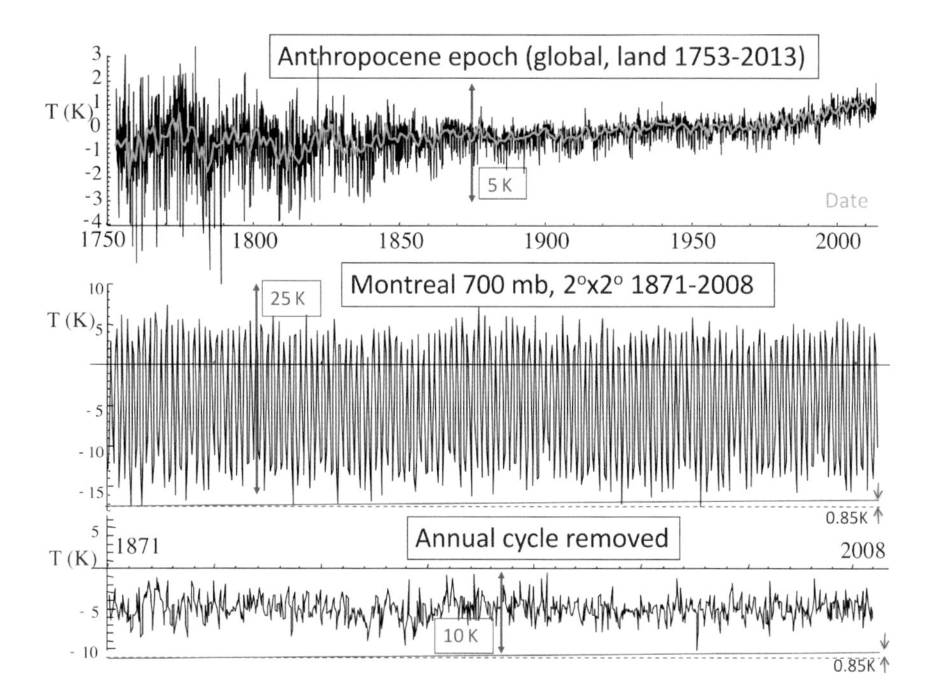

Figure 1c: The top series shows the longest available instrumentally based global temperature estimates (monthly, land only, 3129 values, 1753-2013 (Rohde et al. 2013), the grey line in the top plot is the annual averaged temperature. The data go back to 1753 but due to the very large uncertainties at the early dates (due to limited availability), the thickness of the zigzagging at the far left is large. This covers the epoch of the industrial revolution; the anthropocene, the geological period strongly influenced by humans. Starting in 1871, reanalysis data at 2°×2°, 6 hour resolution is available from the 20th C reanalysis (Compo et al. 2011); data at 700 mb are shown. There were over 200,000 values, we averaged so as to only display 720 points (the resolution displayed here is thus about 3 months). The middle series shows the raw data that includes the dominant annual cycle; the bottom series is the same but with this removed. We also show for reference an estimate the amplitude of the anthropogenic change (from Lovejoy (2014c) close to the IPCC AR4 estimate; for the global change since 1880, it is ≈0.85 K. For the land only (top series (Rohde et al. 2013)), the estimate is 1.5 K. Reproduced from Lovejoy (2015b).

is all around us: from the melting of polar sea ice – including the summer opening of the Northwest passage – to rising sea levels to deadly heat waves. But what is the cause? Is it simply another natural fluctuation, or is it something different, something artificial, something that only *we* could have done? More precisely, is a one degree warming of the whole planet *in only a single century* an ordinary – even common – event in the history of the earth, or is it so rare as to be demand a non natural explanation?

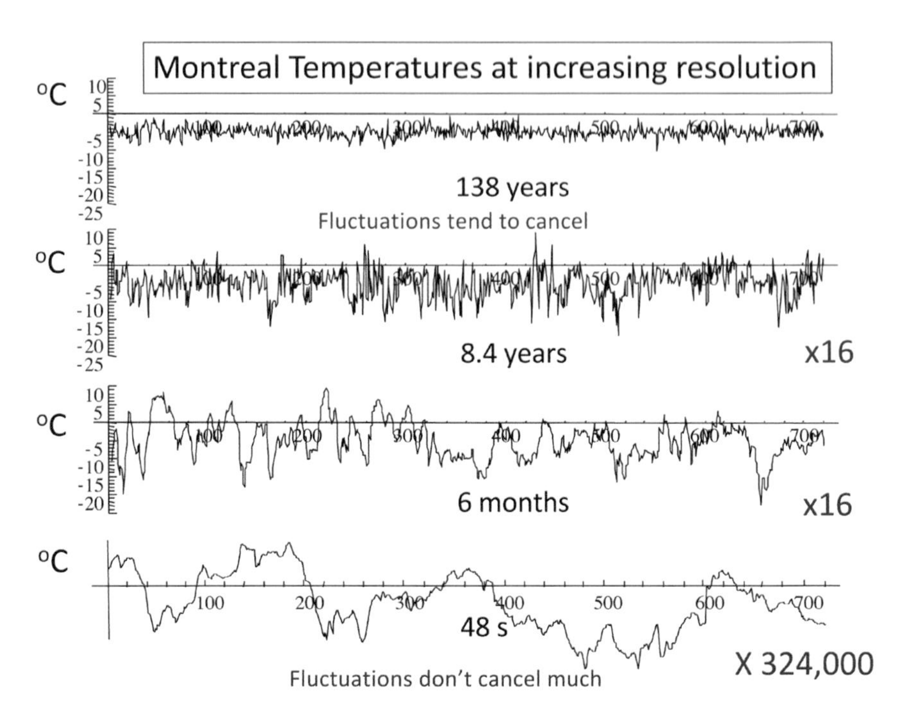

Figure 1d: The upper left is the same as the lower series in Fig. 1c. We successively take the left sixteenth of the series and blow it up by a factor of 16, retaining 720 points at each step until we get to the 6 hour resolution series (third from top), the total length of each series is indicated in each plot. The bottom series is also from Montreal, but from a millimetre sized thermistor on the roof of the McGill physics building at 0.067 s resolution. The temperature scale is the same for all the series except the bottom one. Higher resolution data would show that the variability continues for at least another 2 orders of magnitude to kHz scales. Starting at the lower left we see that – as for the Epica series (Fig. 1a, top row) – that the temperature appears to wander like a drunkard's walk with temperature differences $\Delta T = T(t+\Delta t) - T(t)$ tending to grow with time intervals Δt. This character is still apparent at the next (6 hour resolution, lower left) – at least for intervals as long as 10-20% of the series length (i.e. up to 10-20 days long). As we move upwards to longer and longer resolutions to the series indicated 8.5 years (which is at 4 day resolution), notice that the overall variation of the series doesn't change much (i.e. the rough range between the maximum and minimum is nearly independent of the resolution). Reproduced from Lovejoy (2015b).

The modern answer to this question emerged well before the warming itself was felt or even before human emissions had significantly changed the atmospheric composition. In 1896, in an attempt to understand the causes of the ice ages, Svante Arrhenius estimated that if the concentration of carbon dioxide (CO_2) in the atmosphere was doubled, that global

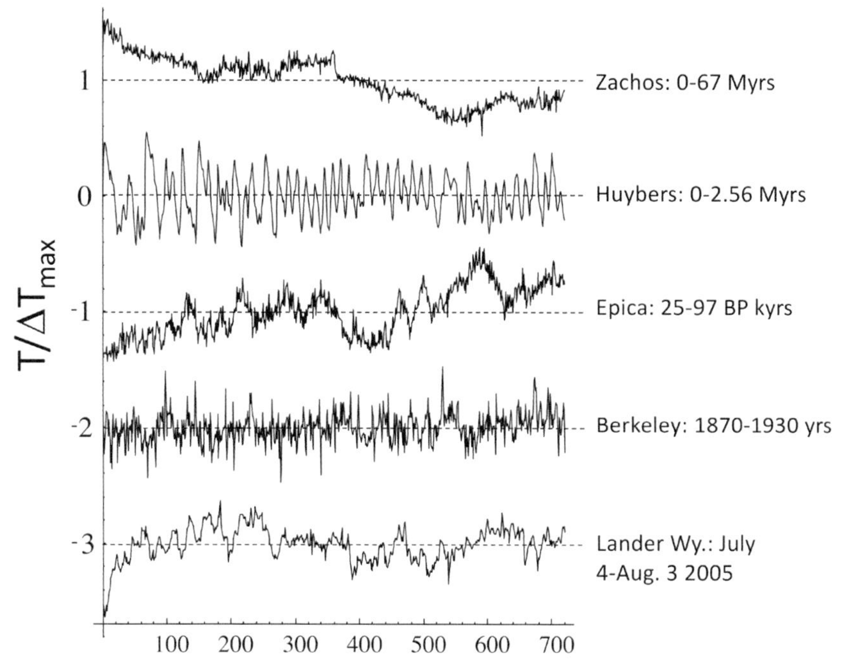

Figure 1e: Representative series from each of the five scaling regimes taken from Figs 1a-d with the addition of an hourly surface temperatures from Lander Wyoming, (bottom, detrended daily and annually). The Berkeley series was taken from a fairly well estimated period before significant anthropogenic effects and was annually detrended. The Veizer series was taken over a particularly data rich epoch, but there are still traces of the interpolation needed to produce a series at a uniform resolution. In order to fairly contrast their appearances, each series had the same number of points (180) and was normalized by its overall range (the maximum minus the minimum), and each series was offset by 1K in the vertical for clarity. The resolutions were adjusted so that as much as possible, the smallest scale was at the inner scale of the regime indicated. The series resolutions were 1 hour, 1 month, 400 years, 14 kyrs, 370 kyrs and 1.23 Myrs bottom to top respectively. In the macroclimate regime, the inner scale was a bit too small and the series length a bit too long. The resulting megaclimate regime influence on the low frequencies was therefore removed using a linear trend of 0.25 $\delta^{18}O$/Myr. The resolutions and time periods are indicated next to the curves. The black curves have $H > 0$, the grey, $H < 0$. From top to bottom the ranges used for normalizing are: 10.1, 4.59, 1.61 (Veizer, Zachos, Huybers respectively, all $\delta^{18}O$), 6.87 K, 2.50 K, 25 K (Epica, Berkeley, Lander). Reproduced from Lovejoy (2015b).

temperatures would rise by 5-6 °C, quite close to the modern value of 1.5-4.5 °C, (International Panel on Climate Change, fifth assessment report: IPCC AR5). From a scientific point of view, the basic result is straightforward:

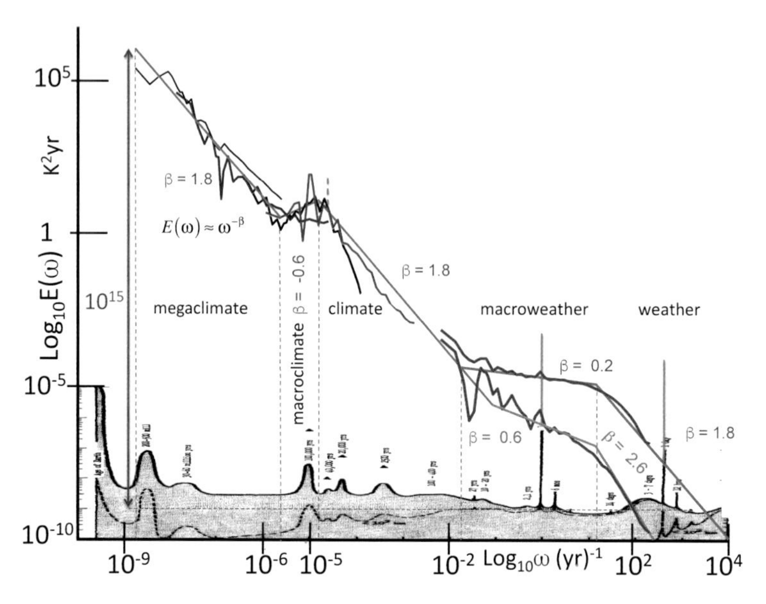

Figure 2a: A comparison of Mitchell's relative scale, "educated guess" of the spectrum (bottom, Mitchell 1976) with modern evidence from spectra of a selection of the series displayed in Fig. 1 (the plot is log-log). There are three sets of red lines; on the far right, the spectra from the 1871-2008 20CR (at daily resolution) quantifies the difference between the globally averaged temperature (bottom) and local averages (2° × 2°, top). Mitchell's figure has been faithfully reproduced many times. The upper left red curve is from the calibrated Epica Antarctic core (interpolated to 276 yrs resolution). All the spectra were averaged over logarithmically spaced frequency intervals (10 per order of magnitude), thus "smearing out" the daily and annual spectral "spikes". These spikes have been re-introduced without this averaging, and are indicated by green spikes above the red daily resolution curves. Using the daily resolution data, the annual cycle is a factor ≈ 1000 above the continuum, whereas using hourly resolution data (from the Lander series, Fig. 4a), the daily spike is a factor ≈3000 above the background. Also shown is the other striking narrow spectral spike at (41 kyrs)$^{-1}$ (obliquity; ≈ a factor 10 above the continuum), this is shown in dashed green since it is only apparent in the Huyber series over the period 0.8-2.56 Myr BP. At the upper left, the one brown curve and two black curves are δ^{18}O spectra from the benthic (i.e. ocean sediment) assemblages, the rightmost black is the Huybers series (at 10 kyr resolution), the middle (brown), is the Zachos series (interpolated to 18 kyrs), the leftmost (black) is Veizer series (interpolated to 185 kyrs). See (Lovejoy 2014b) for more details. The blue lines have slopes indicating the scaling behaviours ($E(\omega) \approx \omega^{-\beta}$) deduced from the real space Haar analyses (Fig. 7). The scaling exponents ξ are related to the slopes in Fig. 7 (ξ(2)/2) by $\beta = 1+\xi(2)$. The thin dashed green lines show the transition frequencies deduced from the spectra; these are at (20 days)$^{-1}$, (50 yrs)$^{-1}$, (80 kyrs)$^{-1}$, and (500 kyrs)$^{-1}$ close to those deduced in real space in Fig. 7. This adaptation is from Lovejoy (2015b).

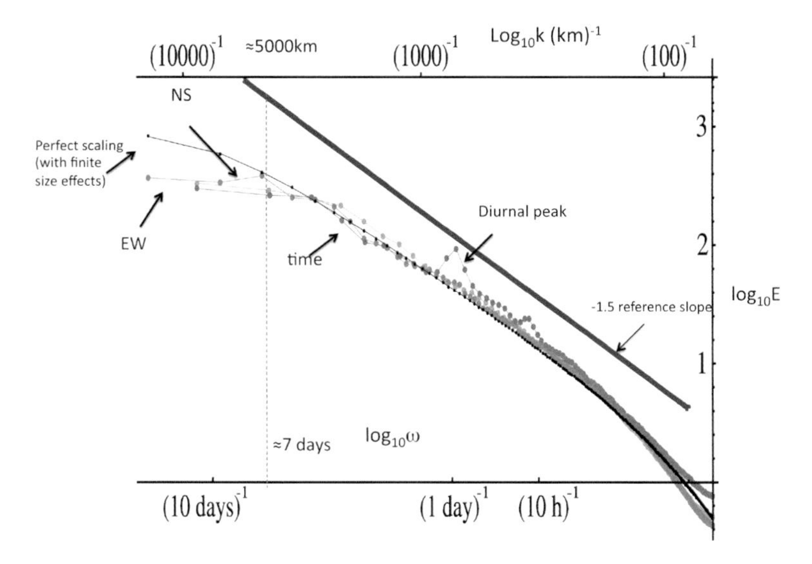

Figure 2b: 1D spectra of MTSAT thermal IR radiances; 1440 consecutive hourly images at 30 km were used and covered a region over 13,000 km across centered over the tropical Pacific. In black (curved, top): the theoretical spectrum using parameters estimated in (Pinel et al. 2014) and taking into account the finite space – time sampling volume. The spectra are $E_x(k_x) \approx k_x^{-\beta_x}$, $E_y(k_y) \approx k_y^{-\beta_y}$, $E_t(\omega) \approx \omega^{-\beta_t}$ with $\beta_x \approx \beta_y \approx \beta_t \approx 1.4 \pm 0.1$; $s \approx 3.4 \pm 0.1$. The straight line is a reference line with slope –1.5 (top). Curved "EW" is the zonal spectrum; Curved "NS" is the meridional spectrum; the "time" curve (with the diurnal spike and harmonic prominent) is the temporal spectrum. Reproduced from (Pinel et al. 2014). The full (horizontal space - time spectrum thus has the scaling symmetry $P(\lambda k_x, \lambda k_y, \lambda\omega) = \lambda^{-s} P(k_x, k_y, \omega)$ where P is the space-time spectral density.

CO_2 is a "Greenhouse Gas": it lets visible light from the sun through to the surface while absorbing part of the earth's outgoing heat radiation.

Arrhenius's theory signalled the beginning of modern attempts to prove the anthropogenic provenance of a warming that only became strongly apparent in the 1980's. From a purely scientific point of view, the main difficulty is that there are complicated feedbacks between CO_2, water vapour and clouds: these are the main effects responsible for the uncertainty. Arrhenius spent the best part of a year with pencil and paper grappling with these complications; today scientists use the world's most powerful supercomputers.

Is the warming mostly human-made, through the emission of CO_2 and other Greenhouse gases, or is it is mostly natural? Today, the theory of anthropogenic warming is entering a mature phase in which continued efforts to prove it more convincingly are starting to suffer from diminishing returns. Take for example the IPCC's Fifth Assessment Report (AR5 2013):

not withstanding massive improvements in computers and algorithms, it cited exactly the same range of temperature increase for a doubling of CO_2 as did the US National Academy of Science report in 1979: 1.5 to 4.5 °C. Whereas the fourth report (AR4, 2007) stated that it is "likely that human influence has been the dominant cause of the observed warming since the mid-20th century", six years later, the AR5 only upgraded this to "extremely likely".

In spite of the strong evidence in favour of the anthropogenic theory, it still faces a chorus of denial with entire organizations – such as Canada's "Friends of Science" – dedicated to the proposition that the "The sun is the main driver of climate change. Not CO_2. Not you" a slogan that adorned billboards across Canada in November 2014 and – at least in Quebec – prompted counter-billboards financed by the Association of Science Communicators. Solar, volcanic and internal climate system variability are all invoked by various proponents as plausible – or even proven – alternatives to the anthropogenic theory. So who is right?

In order to break through the impasse, to "close" the debate (Lovejoy 2015a) it is helpful to recall that science progresses not only from attempting to prove certain theories to be true, but also by rejecting theories that are false. In this, it benefits from a fundamental methodological asymmetry: while no theory can ever be proven true "beyond all doubt", even a single decisive experiment can disprove one that is otherwise highly seductive. For example in medical testing ineffective treatments are often rejected with high levels of confidence: the enormous complexity of the human body is irrelevant. Indeed, in their day-to-day work, scientists constantly reject ideas and theories that are incompatible with observations or with more powerful theories that are known to be true.

In this chapter, we review such a statistical disproof. Rather than exploiting large scale deterministic numerical models, we use past data combined with the new theoretical understanding of the natural variability that has been made possible by advances in nonlinear geophysics, in particular in scaling and multifractals. In section 2 we review elements of turbulence theory; the high level turbulent laws and explain how they can be generalized to take into account both strong nonclassical variability (intermittency, extremes), as well as strong scale dependent anisotropy (especially vertical stratification). In section 3 we discuss fluctuations and use them to give an objective definition of the climate. In section 4 we show how these elements can be combined first to make statistical tests of the giant natural fluctuation hypothesis, and second, by using concrete stochastic models, to accurately hindcast the recent "hiatus" or pause in the warming that is often invoked by skeptics as evidence that the warming is over. In section 5 we conclude.

2. Turbulence, Scaling, Multifractals, Emergent Laws

2.1 Fluctuations

The undisputed father of GCM's is Lewis F. Richardson. He was the first to write down the modern complete (closed) set of nonlinear, partial differential equations for the evolution of the atmosphere, publishing – at his own expense – the seminal "Weather prediction by numerical process" (Richardson 1922). In addition, he spent six weeks with pencil and paper numerically integrating the equations to estimate the pressure change at a single grid point (due to numerical "initialization" issues that were not resolved until the 1970's, he was off by a factor of 100, see the interesting history by Lynch (2006)). Realizing its importance, Richardson proposed the creation of a "forecasting factory" for numerical weather prediction involving tens of thousands of human computers.

Richardson was more than 30 years ahead of his time – the first numerical weather model was in 1956 – and is rightly revered by meteorologists as a pioneer of numerical weather prediction. However, he is also revered by the (quite different) turbulence community for proposing the first law of turbulence: the "Richardson 4/3 law" of turbulent diffusion (Richardson 1926) which is the precursor of Kolmolgorov's famous law (Kolmogorov 1941). More recently, he has been recognized as the grandfather of cascade models of turbulence. Indeed, hidden in the middle of his book, Richardson slyly inserted the now iconic poem "Big whorls have little whirls that feed on their velocity and little whorls have smaller whorls and so on to viscosity (in the molecular sense)".

Like the other classical turbulence theorists – Kolmogorov, Obhukhov, Monin, Corrsin, Bolgiano to name a few – Richardson believed that at high enough levels of nonlinearity (quantified by the Reynold's number, the ratio of the typical nonlinear to linear terms in the equations) that new laws would emerge. Although the laws of continuum mechanics are deterministic, the emergent higher level turbulence laws are statistical and govern the behaviour of huge numbers of eddies ("whorls", structures). The situation is analogous to the higher level laws of thermodynamics which "emerge" from the lower level (more basic) laws of statistical mechanics in the thermodynamic limit i.e. for large numbers of degrees of freedom. Note that from a mathematical point of view, if one starts with the simplest system of fluid equations - the incompressible Navier-Stokes equations – then the behaviour in the high Reynold's number limit – "fully developed turbulence" is an open mathematical problem so that – even seventy years after Kolomogorov's law was proposed – there is no mathematically rigorous derivation of any of the proposed laws of turbulence (there are however many, many physical arguments).

The classical turbulence laws can be expressed in the form:

$$\Delta I\,(\underline{\Delta r}) = \varphi\ |\underline{\Delta r}|^{H} \tag{1}$$

where $\Delta I\,(\underline{\Delta r})$ is the fluctuation in the quantity I over a vector displacement $\underline{\Delta r}$, (for the moment, take the fluctuation ΔI to be the difference of I over $\underline{\Delta r}$), φ is a driving, turbulent flux, the vertical lines represent the norm of the vector and H is the fluctuation exponent. The most famous example of Eq. (1) is the Kolmogorov law which is recovered by taking $I = v =$ a velocity component, $H = 1/3$ and $\varphi = \varepsilon^{1/3}$ where ε is the flux of energy (per mass) from large to small scales (strictly speaking it is a Fourier space flux). Notice that the form of Eq. (1) and the exponent H are scale invariant: they don't change under isotropic scale changes ("zooms") i.e. when $\underline{\Delta r} \rightarrow \lambda\underline{\Delta r}$ where λ is a scale ratio. ΔI changes in a power law way and is said to be "scaling".

Two key aspects of the classical laws make them poor approximations to the atmosphere. The first is that the flux φ that was originally considered to be fairly homogeneous (uniform), at most having statistics that were no more variable than Gaussian. This is unrealistic since even a cursory consideration of the weather indicates that most of the atmospheric fluxes occur in only small fractions of the available space, especially in storms, even in their centres: the turbulence is highly *intermittent*. When applying Eq. (1) to the atmosphere, the second limitation is implicit in the use of the vector norm in Eq. (1) to quantify the scale of $\underline{\Delta r}$. Since the norm is independent of orientation, it implies that the laws are isotropic, whereas the atmosphere is anisotropic, in particular it is highly stratified: gravity strongly imposes a preferred direction.

2.2 Intermittency

Starting in the 1960's, intermittency was explicitly modelled with the help of Richardson – inspired cascade processes. One first assumes that at large scales, the fluid is stirred in a quasi-steady manner (in the atmosphere by the solar gradient between equator and poles). Since the corresponding energy flux (energy per mass, per time from larges scales to small) is exactly conserved by the nonlinear terms, the latter act to break large eddies up into "daughter eddies", transferring their energy fluxes to smaller and smaller scales until eventually (in the atmosphere at scales of less than a millimetre), viscosity dissipates the energy as heat. Cascade models are phenomenological models of this process; the original versions were "toy" models (believed to capture the basic physics while being relatively simple to analyse) in which the parent eddies are large cubes and the daughters are subcubes, with half the parent diameter (see Fig. 3a for a schematic, Fig. 3b for an anisotropic extension and Figs 3c, d for corresponding multifractal simulations). For each daughter, one flips a coin to decide how the parent energy flux will be multiplicatively modulated over the daughter. In the simplest (fractal) "beta model" (Novikov and Stewart 1964, Mandelbrot 1974, Frisch et al. 1978) the daughters occasionally (with well defined probability) receive zero flux, they are "dead", the others have their fluxes multiplicatively boosted just enough to (on average) conserve the flux. If the process is slightly altered such that rather than receiving a boost or a decrease to zero flux, the alternative is instead a boost or a decrease to a finite positive flux (the "alpha model", Schertzer and Lovejoy 1985) then rather than just black or white (active/inactive, dead or alive), there will be intermediate levels of activity, each one concentrated on a different fractal set with a different fractal dimension: the result is a self-similar multifractal (see Figs 3c, d which are continuous in scale cascades).

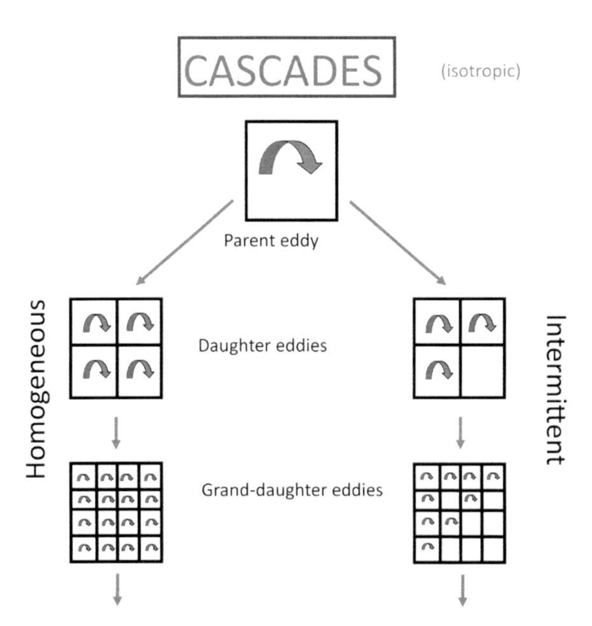

Figure 3a: A schematic diagram showing the first few steps in a (discrete in scale) cascade process. At each step, the parent eddy is broken up into "daughter" eddies, each reduced by a factor of 2 in scale, indicated as squares. The left shows a homogeneous cascade (corresponding to Kolmogorov's 1941 homogeneous turbulence) in which the energy flux is simply redistributed from large to small structures, while keeping its density constant. The right-hand side shows an improvement: "on/off" intermittency is modelled by an "alive/dead" alternative at each step (here only the bottom right sub-eddy becomes dead); the mean conservation of energy flux can be taken into account by boosting the density of the flux in the "active" eddies. For pedagogical reasons, the alternative displayed is purely deterministic, but could be easily randomized (see text). Adapted from Schertzer and Lovejoy (1987).

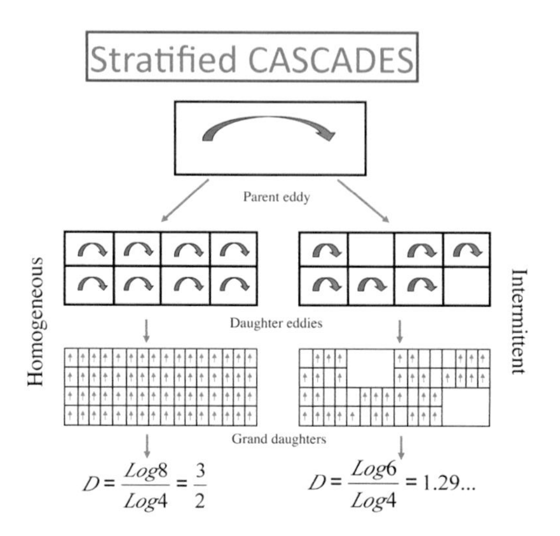

Figure 3b: Schematic of an anisotropic cascade; compare with its isotropic counterpart (Fig. 3a). The exponent governing the decrease in area (equivalently the increase in number) of the sub-eddies with each iteration is D_{el} = log8/log4 = 3/2. On the right-hand side we illustrate the inhomogeneous (intermittent) anisotropic cascade in which 6 of the 8 sub-eddies on average survive so the corresponding elliptical (anisotropic) dimension of the active grand daughters is D = log6/log4 = 1.29. Adapted from Schertzer and Lovejoy (1987).

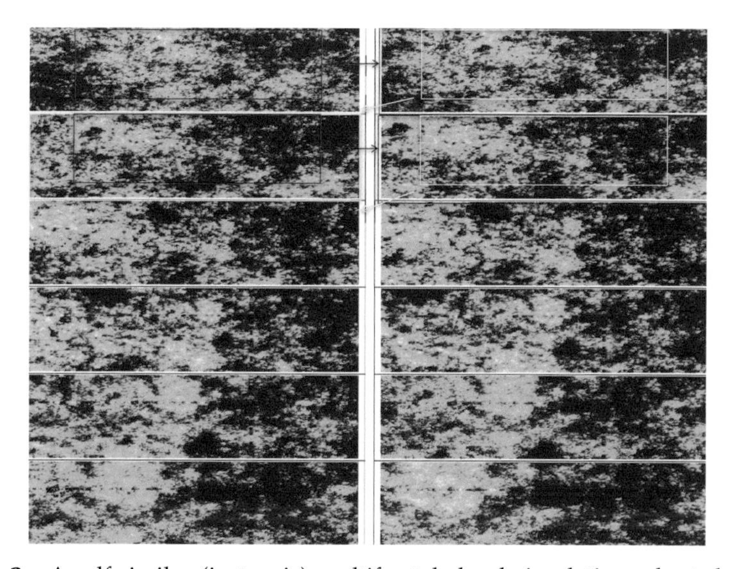

Figure 3c: A self-similar (isotropic) multifractal cloud simulation adapted from Lovejoy and Schertzer (2013). Each image is an enlargement by factor of 1.7 (the areas enlarged are shown in yellow and red rectangles for first few enlargements, top rows).

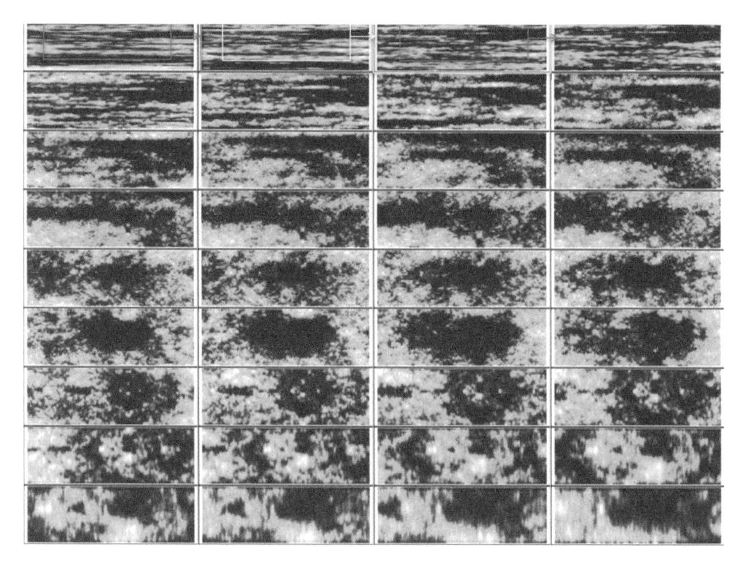

Figure 3d: A sequence "zooming" into a vertical cross section of an anisotropic multifractal cloud with $H_z = 5/9$. Starting at the upper left corner, moving from left to right, from top to bottom, we progressively zoom in by factors of 1.21 (total factor $\approx$ 1000). Notice that while at large scales, the clouds are strongly horizontally stratified, when viewed close up they show structures in the opposite direction (lower right). The sphero-scale is equal to the vertical scale in the left most simulation on the bottom row. The film version of this (and other anisotropic space-time multifractal simulations can be found at: http://www.physics.mcgill.ca/~gang/multifrac/index.htm). Adapted from Lovejoy and Schertzer (2013).

A technical point that is often missed is that there are two variants of the cascade conservation: canonical and microcanonical. The former was described above: at each step in the cascade, the only constraint is that the probabilities of boosts and decreases are such that on average they equal 1 so that on average the fluxes neither increase nor decrease as the cascade proceeds from large to small scales. The alternative, "microcanonical" conservation is much more strict, it requires that at each step of the cascade (i.e. everywhere in space, at every resolution, scale), the multipliers sum exactly to a constant (not only over a statistical average). Interestingly, the two state version of this was first discovered by de Wijs (1951) as a model for the distribution of ores in the earth's crust. It was rediscovered by Meneveau and Sreenivasan (1987) who baptised it the "p-model". Microcanonical models are still popular because they simplify the cascade mathematics: by construction, if spatially averaged over a microcanonical cascade step, then the small scales are completely averaged out, one is simply left with a low resolution cascade, one constructed with fewer steps. However, this is not true in the case of canonical conservation, indeed, it turns out that if we spatially average a canonical cascade taken to its small scale limit (i.e. after an infinite number of cascade steps), that the small scale activity is not only still present, but for moments exceeding a critical value q_D, that they dominate the statistics and lead to the phenomenon of divergence of statistical moments (Mandelbrot 1974, Schertzer and Lovejoy 1987). This means that the extreme tails of the probability distributions are power laws, a fact we exploit below for statistical testing of the natural warming hypothesis.

To make this more precise, consider a multifractal cascade developed over a range of scales from L to l (ratio λ), the statistics may be characterized by considering the various (q^{th} order) statistical moments:

$$\left\langle \varphi_\lambda^q \right\rangle = \lambda^{K(q)};\ \lambda = L/l \tag{2}$$

where "< >" indicates statistical (ensemble) averaging and $K(q)$ is a convex function of the order of moment q. Low order moments (small q) will be dominated by the numerous small fluctuations, high order moments (large q) will be dominated by rare large fluctuations; $K(q)$ generally gives a complete statistical characterization of the cascade at all scales and all intensities. Physically, L is the largest (outer) scale of the cascade, l is the smallest (inner, dissipation) scale. Since the mean flux ($q = 1$) is conserved from scale to scale, $K(1) = 0$ and in addition, for a nonzero cascade, $K(0) = 0$ also. Figure 4 gives an example of the temperature statistics relevant to both weather and climate showing that in space their fluxes obey Eq. (2). For the general canonical cascades discussed above, the high order statistical moments of the spatially integrated fluxes diverge: $K(q) \to \infty$ for $q > q_D$ where q_D is a critical order, typically about 5-7 for turbulence (see the review by Lovejoy and Schertzer (2013), Ch. 5). A divergence of moments of order $q \geq q_D$ is mathematically equivalent to a power law tail on the probability distribution, hence:

$$Pr(\Delta I > s) \approx s^{-q_D};\ s >> 1 \tag{3}$$

where "*Pr*" indicates "probability" and s is a threshold. If a random variable ΔI follows Eq. (3), then extreme values occur much more often than for classical (exponentially bounded) distributions such as the Gaussian.

In Taleb (2010), the author popularized the expression "Black Swans", originally to designate events that are not only unexpected because they are rare, but to events that are *epistemologically unexpected* in the sense that they are totally outside the ken of the reigning view, ideology or theory. Inspired by Mandelbrot's use of Levy distributions (which have power law tails but with Levy index $\alpha = q_D$ restricted to values below 2 – the result of additive, not multiplicative processes), Taleb goes on to refer to the corresponding extreme Levy events as "Grey Swans". He justifies this since extreme Levy events can be anticipated but only on the basis of unconventional (non Gaussian) theory. The cascade extension from Levy to more general power law tails (i.e. with $q_D > 2$) should therefore rightly also be referred to as "Grey Swans", but this term never stuck, hence we use the term "Black swans" more generally for any power law extremes (Eq. (3)).

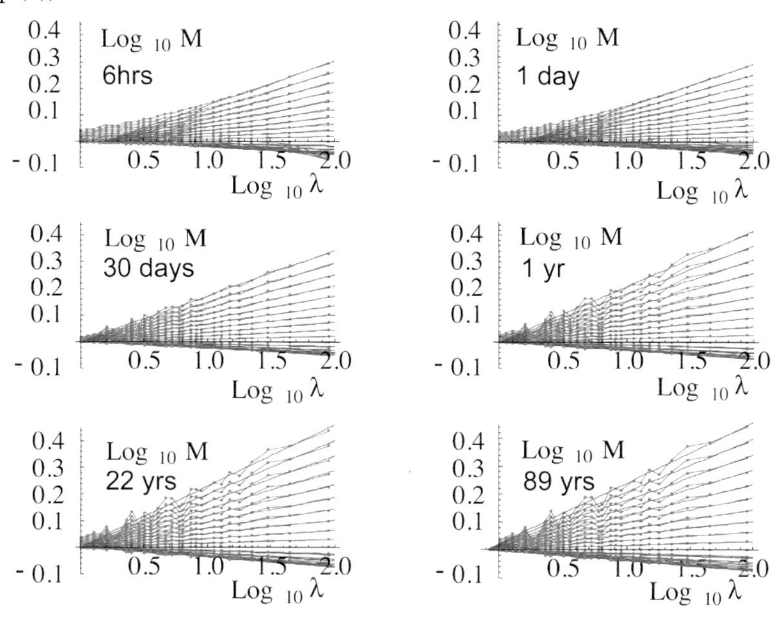

Figure 4: The verification of Eq. 2 on temperature data from the twentieth century reanalysis (20CR, Compo et al. (2011)) from zonal (east-west) transects at 45°N using data at 6 hour temporal and 2° spatial resolution, from 1871-2008. $M = \langle \varphi_\lambda^q \rangle$ (Eq. (2)) with moments q = 2, 1.8, 1.6,..., 0.2 corresponding to the points and regression lines; positive slopes for $q > 1$, negative slopes for $q < 1$. Each graph shows the spatial statistics of the fluxes φ_λ estimated at various temporal averaging scales ranging from 6 hours to 89 years by taking the absolute finite difference of the temperature temporally averaged over the indicated duration. Equation 2 predicts that on such log-log plots, that the lines for different moments q will converge to the "effective" outer scale of the process, i.e. the scale at which the process must start if it is to explain the observed variability at the smaller scales (to the right). In this analysis, the largest scale ($\lambda = 1$) corresponds to the largest distance along the 45°N latitude line i.e. 14100 km. It is approximately, but not exactly the outer scale. We can see that for all the different time scales, the spatial statistics well obey the predictions of multiplicative cascade theories. The large spatial intermittency – characterized near the mean by the codimension of the mean $C_1 = K'(1)$ slowly increases from 0.095 (6 hours, the weather regime) to 0.13 (89 years, the climate regime) implied by these plots is the statistical expression of the existence of various climate zones. This figure is reproduced from Lovejoy and Schertzer (2013).

We now illustrate such black swan extremes with an example from the climate that will be relevant later. First consider the probability distribution of daily temperature changes from a single station (Fig. 5a). The figure shows the cumulative distributions of the temperature changes accumulating from the largest, not smallest value (it is one minus the usual cumulative distribution function). We see that for both positive and negative temperature changes that the distribution has far more extreme events than would be expected from the classical Gaussian distribution, indeed, the extremes are 7 standard deviation events corresponding to Gaussian probabilities of less than 10^{-20}. On the other hand, the data closely follow a power law with exponent $q_D \approx 5$. Moving to longer times (Fig. 5b), we see the same type of behaviour, even in paleotemperatures. For the latter, modern data (far right) allow the tails to be examined more closely, yielding a more convincing result, again with $q_D \approx 5$. Note that taking differences over longer time scales shifts the tails by a constant factor corresponding to fluctuation exponent $H = 0.4$ (see below; the left two plots in Fig. 5b).

Finally, to evaluate the statistics of natural temperature changes – and to avoid biases due to anthropogenic effects – consider global scale pre-industrial (1500-1900) temperatures. In the pre-industrial period, global scale temperatures can be estimated using "multiproxy" reconstructions. As the name suggests, these statistically combine data from diverse sources ("proxies") typically including tree rings, ice cores, lake sediments in order to estimate the temperature in the absence of instruments, Fig. 5c shows the corresponding distributions for differences of 1, 2, 4, … 64 years. Figure 5c shows that Eq. (3) is a reasonable approximation to the tails of

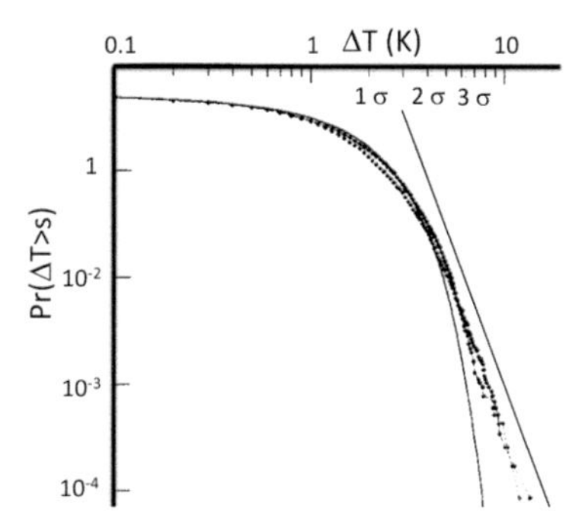

Figure 5a: The probability distribution of daily temperature differences in daily mean temperatures from Macon France for the period 1949-1979 (10,957 days). Positive and negative differences are shown as separate curves. A best fit Gaussian is shown for reference indicating that the extreme fluctuations correspond to more than 7 standard deviations, for a Gaussian this has a probability of 10^{-20}. The straight reference line (added) has a slope of $-q_D$ with $q_D = 5$. Adapted from Ladoy et al. (1991).

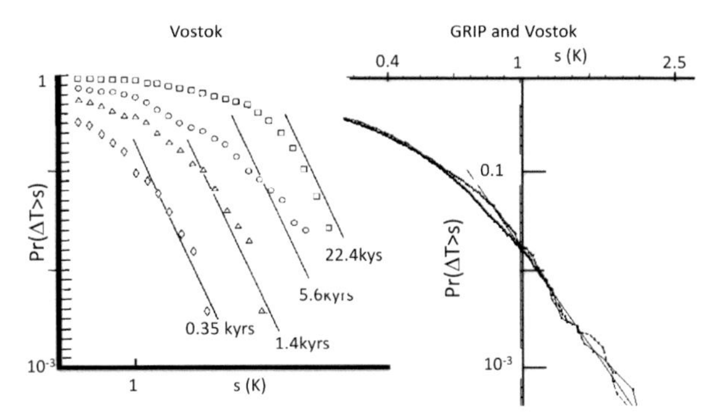

Figure 5b: Probability distributions of paleotemperature changes for Vostok (left), (Antarctica, paleo temperatures from ^{18}O proxies reproduced from Lovejoy and Schertzer (1986)) and a modern comparison of GRIP (Greenland) and Vostok right, reproduced from Lovejoy and Schertzer (2013). The graphs differ not only due to the much improved sampling density of the more modern data, but also, the rightmost graph is at constant depth intervals (0.55 m for GRIP and 1 m for Vostok); this avoids issues of uncertain chronologies. In all cases, the straight reference lines indicate extreme s^{-q_D} behaviour with $q_D = 5$ where s is a temperature change. The reference lines in the left graph are spaced $H\log_{10}4$ apart with scaling exponent $H = 0.4$.

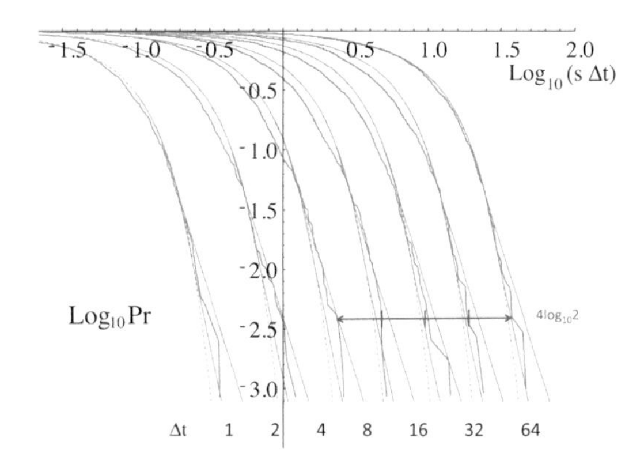

Figure 5c: The total probability of random absolute pre 1900 temperature differences exceeding a threshold s (in K), using three multiproxies to increase the sample size (the distribution are very similar in form for each of the multiproxies). To avoid excessive overlapping, the latter were compensated by multiplying by the lag Δt (in years, shifting the curves to the right successively by $\log_{10}2 \approx 0.3$), the data are the pooled annual resolution multiproxies from 1500-1900. The blue double headed arrow shows the displacement expected if the difference amplitudes were constant for 4 octaves in time scale (corresponding to $H = 0$ for differences, the standard deviations each octave is indicated by a vertical tick mark on the arrow). The (dashed) reference curves are Gaussians with the corresponding standard deviations and with (thin, straight) tails ($Pr \approx$ <3%) corresponding to bounding s^{-4} and s^{-6} behaviours. Reproduced from Lovejoy (2014b).

the (pre-industrial) distributions of temperature changes. This result will be used below for estimating the probability that industrial warming is no more than a giant natural fluctuation.

3. The Climate

3.1 GCM's: The Climate as a Boundary Value Problem

At first, General Circulation Models (GCM's; large numerical models of the atmosphere) were weather models designed to model the atmosphere over periods of days: the slowly varying ocean was taken as a fixed lower boundary condition. In order to extend GCM's for modelling the longer time scales associated with the climate, at the very least they had to be coupled to ocean models; modern GCM's are also coupled to cryosphere and carbon cycle models. Changing land use and atmospheric composition (CO_2, methane, aerosol concentrations), are taken into account as changes in boundary condition as are the external "forcings" consisting variable solar output, volcanic eruptions and changing land use.

So how do GCM's work? Due to their sensitive dependence on initial conditions (the "butterfly effect", deterministic chaos), errors grow quickly (due to the scaling, only algebraically, not exponentially fast (Schertzer and Lovejoy 2004)), so that for planetary sized structures there are deterministic predictability limits of the order of ten days. This means that even with all the correct couplings, that the exact state of the atmosphere cannot be well predicted beyond 10 days or so. Indeed, for climate modelling – and even though all the details are known to be wrong – GCM's are routinely integrated at subhourly time steps for decades or hundreds of simulated years. However, it is hoped that the resulting fields will have the correct type of statistical variability and that any low frequency trends imposed by slowly varying boundary conditions (such as changing CO_2 concentrations) will change the averages (e.g. of the temperature) in a more or less realistic manner. Mathematically, whereas below ten days, GCMs consider atmospheric forecasting as an *initial* value problem, beyond this, it is considered to be a *boundary* value problem. Notice the irony: that in this "climate" prediction mode, the world's largest supercomputers (some with $\approx 10^6$ coprocessors) are essentially used as random number generators: generating no more than random "weather" so that the climate can be deduced by averaging almost all of it out. We return to this below.

3.2 Fluctuations and the Fluctuation Exponent

Surely, if scale invariance is a basic symmetry respected by the atmosphere and its models, then shouldn't it be used to categorize the different regimes of atmospheric dynamics? Indeed, since the horizontal velocity field is scaling out to planetary scales (see Fig. 2b for the horizontal scaling of satellite radiances, for the wind, see Lovejoy et al. (2009), Pinel et al. (2012), or for a review, Ch. 2 of Lovejoy and Schertzer (2013)), this can be used (at least dimensionally) to convert from space to time so that we find that the temporal fluctuations of the wind, temperature and other fields are scaling in time out to scales corresponding to the lifetime of planetary structures: about 10 days (this can be determined from first principles using the energy input from the sun, the size of the planet and the Kolmogorov's law, Lovejoy and Schertzer (2010), Lovejoy et al. (2014)). Beyond this timescale, one is dealing with the statistics of structures over many lifetimes, new and quite different scaling regimes are established (see below).

To understand the different regimes, let us simplify Eq. (1) by considering only the mean fluctuations $\langle \Delta I \rangle$ over time intervals Δt:

$$\langle \Delta I \rangle = \langle \varphi \rangle \, \Delta t^{H} \tag{4}$$

Since $\langle \varphi \rangle \approx$ constant ($K(1) = 0$ in Eq. (2)), the mean behaviour of the fluctuations is determined by the exponent H. For example, when $H > 0$, fluctuations tend to grow with the time interval, I will fluctuate like a drunkard's walk – indeed, the usual Brownian motion is the case $H = 1/2$ and φ is a Gaussian (with $K(q) = 0$). When $H < 0$, on the contrary, fluctuations will tend to cancel each other out so that averaging over longer and longer intervals tends to converge. For this to be possible, the notion of fluctuation needs to be appropriately defined with the help of wavelets. When $1 > H > 0$ using differences $\Delta I = I(t+\Delta t) - I(t)$ is sufficient (the "poor man's wavelets"). However any process whose correlations decay with time Δt have average differences that increase with Δt so that when $H < 0$ they cannot be used to estimate fluctuations (see Eq. (4)). In this case, one could use "anomaly fluctuations" that are equal to the average over time Δt of the series after its overall mean has been removed. Anomaly fluctuations on the contrary, can only decrease with time scale and therefore cannot be used when $H > 0$. A simple definition that combines the advantages of both – and is valid over the range $-1 < H < 1$ – is simply to use the difference between the average over the first and second halves of the interval: "Haar fluctuations" (the coefficients of Haar wavelets).

Since Haar fluctuations are equal to the difference fluctuation of the anomalies (or equivalently the anomaly fluctuation of the differences), they are easy to apply, and – most importantly – easy to interpret. In contrast, the popular Detrended Fluctuation Analysis (Peng et al. 1994) defines fluctuations as the root mean square residual of a polynomial regression against the running sum of the process. Not only is this mathematically difficult to analyse (DFA fluctuations aren't wavelets) but – more importantly – the interpretation of the fluctuations is so opaque that in DFA papers the authors don't even both to indicate the units of the fluctuations on their scaling plots; they only use them to estimate scaling exponents! A final comment on the exponent H: it is denoted "H" in honour of Edwin Hurst but in general – unless the process is Gaussian – it is not the same as the Hurst (i.e. "R/S") exponent (e.g. for a standard random walk – Brownian motion – they both yield $H = ½$).

In order to understand the H exponent, consider the simple (essentially pedagogical) fractal construction shown in Figs 6a, b, that – for want of a better name – I call the "H model" (when $1 > H > 0$ it is close to the "pulse in pulse" model, and has divergence of statistical moments of order $>1/H$, i.e. $\langle \Delta T^q \rangle \to \infty$ for $q > 1/H$, see Lovejoy and Mandelbrot (1985)). To simulate a series with fluctuation exponent H over the unit interval, start with the basic fluctuation, the step function labelled "motif" in Fig. 6a (top); the dashed line indicates the horizontal axis so that the left half is negative, the right half is (symmetrically) positive. To obtain the 2nd generation of the construction, compress the motif by a factor two in the horizontal and 2^{-H} in the vertical and place the result in the left half of the interval, then multiply it by a random sign. Finally, repeat with another random sign and place the result in the right half interval. The figure shows the result for signs +, –; this defines the fluctuations at the corresponding reduced scale. Fig. 6b shows the result when this is iterated 8 times; the left column with $H > 0$, the right column, $H < 0$. The final fractal process is obtained by summing all the contributions. Notice that in the $H > 0$ process, the fluctuations decrease with decreasing scale so that the process is dominated by the larger scales, conversely for the $H < 0$ process. When $H < 1$ the process has mean fluctuations $\langle \Delta T(\Delta t) \rangle \propto \Delta t^H$.

3.3 What is the Climate?

When the wind, temperature, humidity, pressure and other atmospheric variables are considered – whether empirically or from GCM's – or from appropriately generalized turbulent cascade processes – it is found that the transition at 5-10 days is universally observed to be from high frequency, growing ($H > 0$) to low frequency, decreasing ($H < 0$) behaviour with successive fluctuations tending to cancel each other out. At first sight, this would appear to validate the dictum "the climate is what you expect, the weather is what you get", (Heinlein 1973) i.e. that the climate is a kind of average weather, which is obtained by averaging the weather over longer and longer time intervals. However, analysis of instrumental and paleo (i.e. proxy) data over decades, century and millennial scales shows that the $H < 0$ convergence of averages ends at about 30 years (in the industrial epoch) and after about 100 years (pre-industrial), and that at longer scales, the averages begin to change again (i.e. with $H > 0$). Figure 7 is a composite showing the variability over nearly 13 orders of magnitude based on instrumental and paleo data (Fig. 2a is the spectral equivalent using largely the same data). For example, the climate "normal" (defined as a thirty year average) itself tends to fluctuate becoming more and more

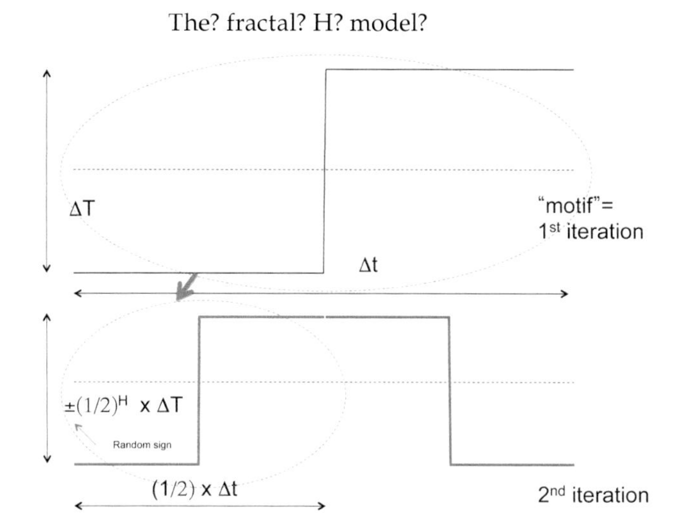

Figure 6a: The first two steps in the construction of the fractal H model. To obtain the second row, the motif (i.e. a basic "fluctuation", top row) is reduced by a factor 2 in the horizontal and by 2^{-H} in the vertical and then multiplied by a random sign, this is placed in the left hand half of the figure; the right hand half has the same shape but with another random sign. Reproduced from Lovejoy (2015b).

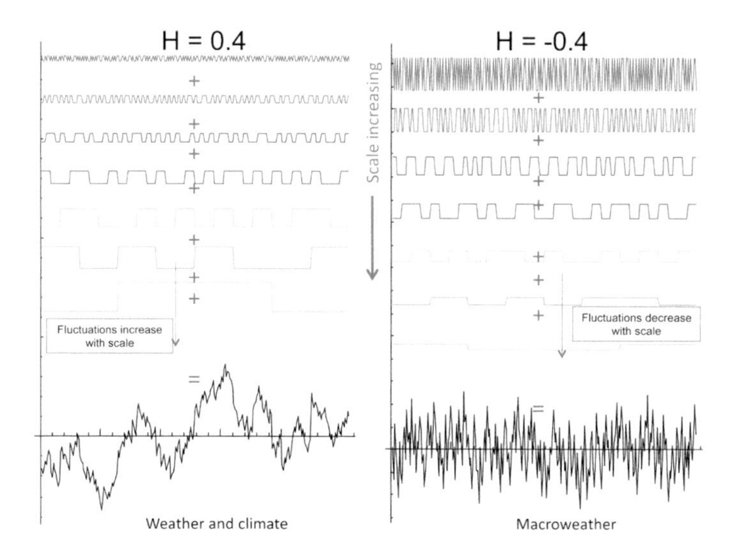

Figure 6b: The first eight steps in the construction of the fractal H model with the sum, bottom series. In the left hand column we show the result for $H > 0$, the right, $H < 0$. In the $H > 0$ case we see that the amplitude of the fluctuations decreases as we go to smaller scales whereas in the $H < 0$ case, they increase. Reproduced from Lovejoy (2015b).

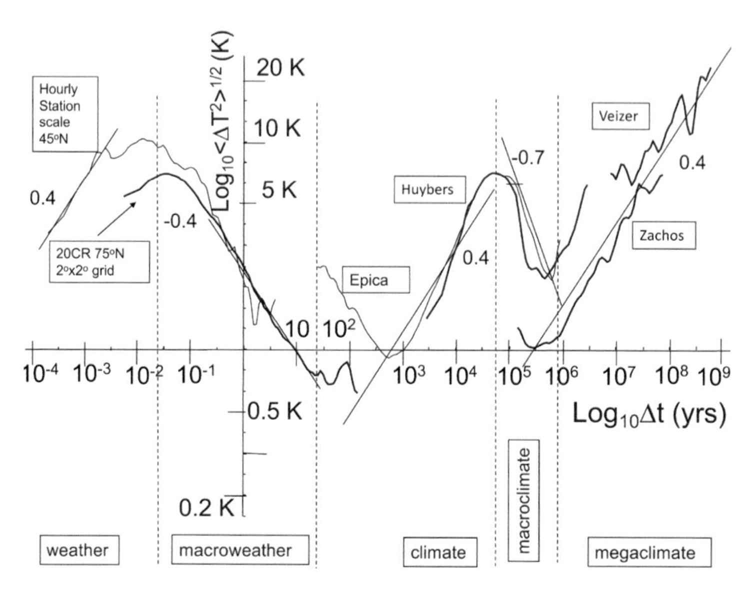

Figure 7: This is a wide scale range composite series showing atmospheric variability over the range from 1 hour to 553 million years. The 5 regimes: weather, macroweather, climate, macroclimate and megaclimate are also indicated. The curves to left are instrumental, the Epica curve is from Antarctica (ice core isotope proxy), the Huybers, Zachos and Veizer curves are from ocean core isotope proxies. Adapted from (Lovejoy 2015b).

variable right up to tens of thousands of years, the ice ages. Empirically – up to dozens of millennia at least – there are therefore three different regimes, not two, the intermediate regime – which is really a kind of "slow" weather and is called "macroweather" with the term "climate" being reserved for the longer period (unstable) variations up to ice-age scales of ≈ 100 kyrs (Lovejoy 2013). If one uses paleo data to extend such analyses to very long scales (up the limit of reliable proxies, 540 Myrs corresponding to the Phanerozoic eon), then in addition one finds a narrow "macroclimate" regime from ≈100 kyrs to ≈1 Myr (with $H \approx -0.8$) and then a "megaclimate" regime (with $H \approx 0.4$) from ≈1 Myr to (at least) 540 Myrs. Interestingly, this latter $H > 0$ regime is associated with "unstable" drunkard's walk like behaviour, it is incompatible with Lovelock's Gaia ("living earth") hypothesis (Lovelock 1995) that posits homeostasis (i.e. negative feedbacks that are strong enough to keep the temperature from wandering too far from optimal values for life; Lovejoy (2015b)).

The weather – macroweather – climate trichotomy is helpful for understanding anthropogenic warming: whereas in pre-industrial times, slowly acting natural processes eventually – at scales of a century or more – begin to dominate the (cancelling, diminishing) macroweather processes. In the industrial era, the anthropogenic effects are stronger than the natural climate processes and they begin to dominate macroweather after only about 30 years. This is important because it means that the (roughly century long) variation since 1880 is mostly due to anthropogenic not natural variability: it allows us to fairly accurately separate out the anthropogenic from the natural variations.

3.4 Scaling and Anthropogenic Warming

The weather, macroweather and then climate picture is based on the corresponding scaling regimes, it characterizes the natural variability. For the industrial period which for our purposes started roughly 125 years ago, the corresponding duration is still within (although perhaps not far from the limits) of the preindustrial macroweather regime. This means that the probability of these natural fluctuations can be safely extrapolated somewhat beyond the limits of the empirical probabilities in Fig. 5c. These probabilities can thus be used to estimate how long we may expect to wait for various temperature changes – an estimate of "return times", taking into account the scaling of the probabilities of temperature changes combined with the expected behaviour of the extremes (Fig. 5c, Eq. (3)) to estimate the probability of extreme temperature fluctuations. If it is found that the probability of the observed change since 1880 ($\approx 0.9°$ C, see below) is low enough – alternatively that the return times are long enough – then we can reject the hypothesis that the fluctuation was caused by natural variability.

Before discussing the statistical test, let us estimate the magnitude of the warming that has occurred. One way to do this is to attempt to separate the anthropogenic and natural variability (if the magnitude of the change is estimated in some other way, this step can be skipped, it is not essential to the conclusions). From fluid mechanical, turbulent and GCM viewpoints, the main anthropogenic forcings (Green House Gases (GHG), aerosols and land use changes) affect the boundary conditions, not the *type* of variability (see above). These arguments support the separation:

$$T_{globe}(t) = T_{anthro}(t) + T_{nat}(t) \tag{5}$$

where T_{globe} is the global temperature anomaly and T_{anthro} and T_{nat} are the contributions of anthropogenic (climate) and natural (macroweather) processes. The justification for this is that the anthropogenic forcings since 1880 are of the order of 2 W/m^2 which is less than 1% of the mean solar forcing of $\approx$240 W/m^2. The next step is to note that anthropogenic effects are tightly correlated with global economic activity, this justifies using the (relatively well measured) industrial epoch CO_2 forcing, as a linear surrogate for all the anthropogenic forcings (see Lovejoy (2014b)):

$$T_{anthro}(t) = \lambda_{2xCO_2,eff} \log_2\left(\rho_{CO_2}(t)/\rho_{CO_{2,pre}}\right) \tag{6}$$

where $\lambda_{2xCO_2.eff}$ is the "effective" sensitivity of the climate to a CO_2 doubling, ρ_{CO_2} is the global mean CO_2 concentration and $\rho_{CO_2,pre}$ is the pre-industrial value (277 ppm; the logarithmic form goes back to Arrhenius (1896)), it is a consequence of the saturation of the CO_2 absorption bands. Alternatively, we may use the "equivalent CO_2" concentration which results from the conversion of all the GHG and aerosols in CO_2 radiative equivalents. However, CO_2 and CO_{2eq} are very highly correlated (correlation >0.99) so that the residuals – the estimate of the natural variability is nearly the same in both cases. We could note that economic activity is more related to the emission rate rather than the CO_2 concentration (which depends on the cumulative emissions). However, since economic growth has been roughly exponential – and the integral of an exponential is again an exponential – the two are roughly proportional which is all that we require. More generally, one may assume that the anthropogenic contribution to the temperature is related to the forcing by a non instantaneous but still linear, transfer function. Due to the scaling, this is expected to be a power law – at least above some inner scale, but again, we find that the residuals have nearly the same variability, (work in progress, as macroweather).

Unlike approaches that attempt to separate internal variability from the responses to external natural and anthropogenic forcings, T_{nat} includes any temperature variations that are not anthropogenic in origin, i.e. it includes both "internal" variability and (implicitly) the responses to solar and volcanic forcings which are external but still natural. Similarly, T_{anthro} includes the warming due to the other GHG's as well as the (difficult to estimate) aerosol cooling: $\lambda_{2xCO_2.eff}$ is thus the "effective climate sensitivity". It is the sensitivity to the actual (historical) doubling of CO_2, it is thus conceptually distinct from the theoretical/model notions of "equilibrium" and "transient" sensitivity that have been empirically estimated elsewhere (although it is closer to the latter than to the former). It is only the effective climate sensitivity that permits one to estimate the natural variability during the industrial epoch (as a residual, as macroweather).

Figure 8 shows the results when Eqs. (8), (9) are applied using global, annual ρ_{CO_2} from 1880–2013 using the NASA GISS global temperature series (Hansen et al. 2010). Without sophisticated statistics, the linear trend is fairly convincing (correlation coefficient = 0.94), and the effective sensitivity (the slope) $\lambda_{2xCO_2.eff} = 2.33 \pm 0.36$ °C/CO_2 doubling which is close to the IPCC AR5 equilibrium sensitivity 1.5–4.5 °C/CO_2 (conceptually, it is closer to the somewhat smaller transient climate sensitivity). The total anthropogenic warming (the vertical range of the line in Fig. 8a) is 0.87 ± 0.1 °C which is close to the IPCC AR5 estimate 0.85 ± 0.20 °C (uncertainties are 90% confidence limits).

While Fig. 8a's linearity is impressive, are the residuals (Fig. 8b) really reasonable estimates of T_{nat}? One answer is to note that the amplitude

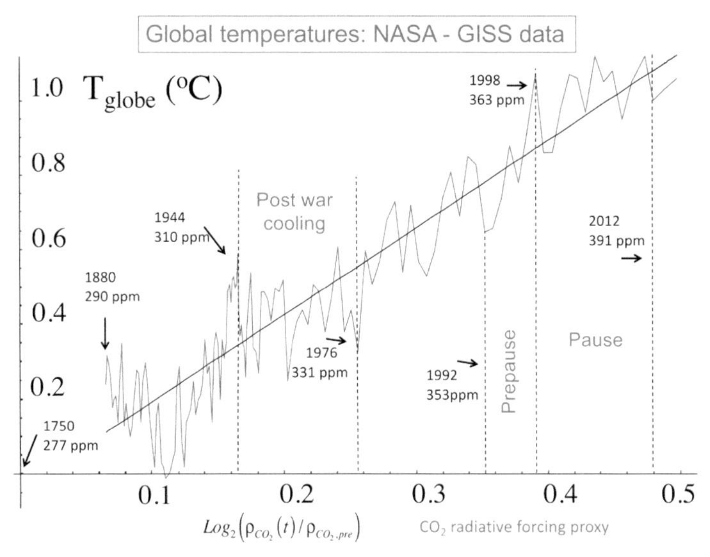

Figure 8a: Global temperature anomalies (NASA, 1880-2013) as functions of radiative forcing using the CO_2 forcing as a linear surrogate. The line has a slope of 2.33 °C per CO_2 doubling. Some of the dates and corresponding annually, globally averaged CO_2 concentrations are indicated for reference; the dashed vertical lines indicate the beginning and end of the events discussed in the text (1944, 1976, 1992, 1998). Adapted from Lovejoy (2014a).

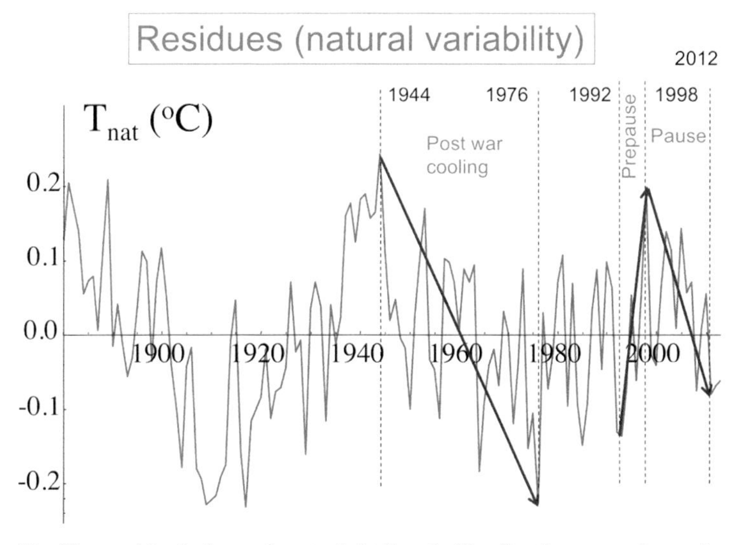

Figure 8b: The residuals from the straight line in Fig. 8a, these are the estimates of the macroweather (natural) variability. The vertical dashed lines are the same as in the previous. The arrows indicate the events discussed in the paper. Adapted from Fig. 1c (Lovejoy 2014a).

of the residuals is quite small: ±0.109 °C (this and the following are one standard deviations). Remarkably, it is virtually the same as the *errors* in GCM hindcasts of global temperatures at one year forecast horizons. For example, over the period 1983-2004, using different GCMs, and bias correction techniques, Smith et al. (2007) and Laepple et al. (2008), obtained RMS hindcast errors of ±0.105 °C and ±0.106 °C. (A hindcast is the use of models to make forecasts for historical time periods.) For example, we can use the data available at time t to make forecasts for time $t + \Delta t$. In a hindcast, $t + \Delta t$ is in the past, it is forecast using the data up to time t and verified with data from t to $t + \Delta t$). In other words, if all we knew was the global annual CO_2 concentration and the value of $\lambda_{2xCO_2.eff}$, then (on average) we could already *predict* the next year's global temperature to ±0.109 °C i.e. just as well as the GCM's! Clearly, this T_{nat} must be close to the true natural variability. Unsurprisingly, this *unconditional* prediction (i.e. using no information about the actual global temperature series) can be improved even further by exploiting the stochastic climate "memory" to make *conditional* predictions.

4. The Pause, Hiatus, Slowdown

4.1 Return Periods

We have used (Fig. 5c) the multiproxies to estimate the probabilities of natural macroweather temperature changes over various time intervals. For the theoretical reasons discussed above, over the empirically reliable range of time intervals (up to 64 years, Fig. 5c), these distributions turn out to be nearly scale invariant (their form is nearly independent of the interval), and we can extend the estimates to intervals of 125 years. As discussed earlier, the scaling is associated with power law probabilities. These were used to bound the extreme 3% of the distributions (Eq. (3)): this takes into account the fact that the extreme changes occur much more frequently than the Gaussian distribution would allow (the "black swans"). Finally, we can estimate the expected time interval between temperature changes of various magnitudes – their return periods (we ignore possible clustering of the extremes and estimate the return period as the inverse probabilities taken from Fig. 5c). The return periods are shown in Fig. 9 where we can see that the warming since 1880 is expected to occur naturally every 1000 to 20,000 years (if we use the traditional Gaussian – the red line – we find periods > 1 Myrs, the warming is nearly a five standard deviation event!). The possibility that 1880 just happened to near the beginning of such a natural warming can thus be dismissed at the 0.1% level. Similarly, the post war cooling – the largest event in the record since 1880 – should occur every 100-150 years. As expected, such an event does occurs in the record (it happens to start in 1944). We can also consider the natural cooling of about 0.3 °C since 1998 – needed to offset the anthropogenic warming over the period and account for the post 1998 flattening in Fig. 8a. This "pause" or "hiatus" is much touted by climate skeptics as proof that the warming has stopped and is therefore not anthropogenic. Although from Fig. 9 we see that such a cooling is expected to occur every 20-50 years, it is not so unusual. Yet it becomes quite probable when it is noticed that it immediately follows an equally strong warming "prepause" event from 1992-1992 (Figs 8a, b) so that the natural cooling since 1998 is no more than a return to the mean behaviour. Indeed, from Fig. 7 we see that it was only in 2012 that the temperature finally went below its long term (anthropogenic) trend. Recently Karl et

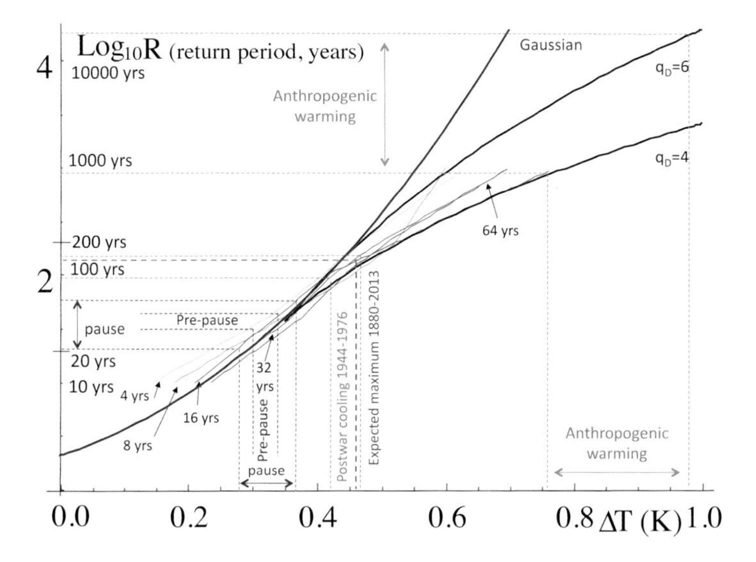

Figure 9: The typical amount of time one must wait to observe a global scale temperature fluctuation of the amplitude indicated on the horizontal axis, the curves are the scale invariant regressions to the empirical return times using the classical (Gaussian, red), and bounding hyperbolically tailed distributions (black, exponents q_D for the extreme tails 4, 6 as indicated). The pairs of vertical lines (one standard deviation bands about the mean) correspond to various events; the pairs of horizontal lines indicate the corresponding return periods from the more extreme (q_D = 4, bottom) or less extreme (q_D = 6, top) probabilities, respectively (see Eq. (3)). The events, from right to left are: global warming since 1880 (green range 0.76-0.98 °C), the largest event expected in the 134 years since 1880 (blue, 0.47 °C), the postwar cooling (green, 0.42-0.47 °C), the pre-pause 0.30-0.33 °C (1992-1998) and "pause" 0.28-0.37 °C (1998-2012). The horizontal lines indicate the corresponding return periods. Note that these curves should not be used for estimating the return times of temperature changes over periods much longer than a century (the rough duration of the macroweather regime). This figure is from Lovejoy (2014a).

al. (2015) produced a temperature series with new ocean and other bias corrections. In this warmer series, the amplitude of the corresponding natural cooling is 0.09 °C less than that shown in Fig. 8b (i.e. about 0.2 instead of 0.3 °C). Since the return period for this smaller natural cooling is only about 10 years (Fig. 9), decadal trends cannot (and did not) detect any statistically significant pause at all and the authors pronounced the pause nonexistent.

4.2 Stochastic Forecasting: The Stochastic Seasonal and Interannual Prediction System (StocSIPS) and the ScaLIg Macroweather Model (SLIMM)

In the previous subsection, we used the unconditional statistics – the return periods – to argue that the natural cooling that roughly offset the anthropogenic warming since 1998 was not so unusual, noting in particular that it happened to follow an even stronger pre-pause warming. While this argument is already fairly convincing, it would be better still to use the conditional statistics i.e. to quantitatively take into account the temperature variations that preceded the pause. In order to do this, we need a stochastic model of the series. In this subsection, we discuss a fairly accurate model, the ScaLIng Macroweather Model (SLIMM). This model is the core of the Stochastic Seasonal and Interannual Prediction System (StocSIPS: http://www.physics.mcgill.ca/StocSIPS/).

To understand the physics behind stochastic macroweather models, recall that when GCM's are used for macroweather forecasting, they are pushed far beyond their deterministic predictability limits: the weather they generate is just a "noise" forcing the lower frequencies. In this macroweather regime, control runs – with fixed climate forcings (boundary conditions) – converge "ultra slowly" (in a power law manner with small negative exponent) to the GCM climates (Lovejoy et al. 2013a). However, due to model "biases" (in both the means and in the annual cycle), neither the statistics of the driving noise (the weather), nor the model climates are fully realistic.

Following Hasselmann (1976), alternative stochastic models have been developed, the most sophisticated of which are the Linear Inverse Models (LIM), (Penland 1996, Penland and Sardeshmuhk 1995), (Newman et al. 2003), (Sardeshmukh et al. 2000), (Sardeshmukh and Sura 2009). In principle stochastic models have the advantage that their statistics can be made realistic and by exploiting empirical data (the system "memory") they can effectively be forced to converge to the real climate, so that for example a 20 component implementation of the LIM model (with >100 parameters) can already do somewhat better than GCM's for global annual temperature forecasts (Newman 2013).

The key question for stochastic macroweather models is thus how big is the system memory and how best to exploit it? The LIM approach is based on systems of coupled ordinary (integer ordered) differential equations whose solutions are essentially white noises and their integrals (Ornstein-Uhlenbeck processes). Their low frequency limits are (unpredictable) white noises so that for horizons beyond about 2 years, the errors rapidly increase (Newman 2013). However, we have seen that over the last decades, a scaling paradigm for atmospheric variability has evolved that implies the existence long-range – potentially huge – memories and these can be exploited for forecasting (Lovejoy and Schertzer 1986), (Pelletier 1998), (Koscielny-Bunde et al. 1998), (Franzke 2010, 2012), (Rypdal et al. 2013), (Rypdal and Rypdal 2014, Yuan et al. 2014), (Lovejoy 2015b), see the reviews: (Lovejoy and Schertzer 2010), (Lovejoy and Schertzer 2012), (Lovejoy and Schertzer 2013).

In a recent paper, we showed how to use the simplest relevant scaling model – fractional Gaussian noise – to exploit the system memory: SLIMM (Lovejoy et al. 2015). SLIMM was shown to make skillful hindcasts of natural variability from monthly to decadal scales. The key to overcoming the limitations of an earlier attempt to exploit the scaling (Baillie and Chung 2002) was to use the empirical effective climate sensitivity (λ_{2xCO_2}) to remove the anthropogenic effects (Lovejoy 2014b). For annually, globally averaged temperatures, the resulting two parameter SLIMM model (λ_{2xCO_2} and the scaling exponent H, see below) was already generally better than both initialized GCM's and LIM, although LIM (with hundreds of parameters) was marginally better for horizons up to about 2 years. The present results can be viewed as the conditional probability extensions of the unconditional (return period) results of the previous section.

In the weather regime, atmospheric dynamics are intermittent; however in the macroweather regime, the intermittency is much weaker so that as a first approximation, nonintermittent (quasi-Gaussian) models may be used (although not for the extreme 3%; Fig. 5c). The usual starting point (e.g. for LIM) is the ordinary differential equation:

$$\left(\frac{d}{dt}+\tau_w^{-1}\right)T=\sigma\gamma(t) \tag{7}$$

where τ_w is the weather/macroweather transition scale", σ is the amplitude of the forcing and $\gamma(t)$ is a Gaussian white noise forcing with mean $<\gamma(t)> = 0$. To understand this, operate on both sides by $\left(\frac{d}{dt}+\tau_w^{-1}\right)^{-1}$; this exponentially smooths the noise at scale τ_w (denoted by a subscript). At low frequencies, Eq. (7) is simply:

$$T(t) = \sigma\gamma_{\tau_w}(t) \tag{8}$$

so that at low frequencies $T(t)$ is a white noise.

The key to realistic modelling of frequencies lower than τ_w^{-1} (SLIMM) is therefore to use a (low frequency) fractional order generalization:

$$\frac{d^{1/2+H}T}{dt^{1/2+H}} = \sigma\gamma_{\tau_w}(t) \tag{9}$$

whose solution is obtained by (Riemann-Liouville) fractional integration of both sides of the equation by order H+1/2:

$$T(t) = \sigma G_{H,\tau_w}(t) \tag{10}$$

where $G_{H,\tau_w}(t)$ is a fractional Gaussian noise (fGn) process:

$$G_H(t) = K_H \int_{-\infty}^{t} (t-t')^{-(1/2-H)}\gamma(t')dt' \tag{11}$$

smoothed at scale τ_w (K_H is an appropriate normalization constant). From Eq. (11) we see that fGn has long range memory due to the slow fall-off in the weighting (the power law convolution kernel). When $H<0$, the above process is stationary; here it is in the range $-1/2<H<0$.

Equivalently, we obtain the same result by simply starting with:

$$\left(\frac{d}{dt} + \tau_w^{-1}\right)T = \sigma G_H(t) \tag{12}$$

i.e. by replacing the LIM white noise forcing by the SLIMM scaling noise forcing with long range statistical dependency (and hence long range memory). Since $G_{-1/2}(t)$ is a white noise, LIM is recovered with $H = -1/2$. The difference between LIM and SLIMM is (equivalently) either the order of the differential equation (c.f. Eqs. (8), (9)) or the scaling (long range dependency, memory) of the forcing (c.f. Eqs. (7), (12)). Note that – unless we are interested in the extremes – this macroweather model is nonintermittent, $K(q) = 0$, an approximation that turns out to be reasonable in the macroweather regime (to see this, in Eqs. (1), (2) take $\langle \Delta t \rangle = \underline{|\Delta r|} = L/\lambda$ to yield $\left\langle \Delta I(\Delta t)^q \right\rangle \propto \Delta t^{qH-K(q)}$ and since fGn has $\left\langle \Delta G_H(\Delta t)^q \right\rangle \propto \Delta t^{qH}$ we see that for fGn $K(q)$=0).

At small scales $G_H(t)$ has singular behaviour so that the temporal resolution τ of $T(t)$ is fundamentally important (as with a white noise, Eq. (11) should be understood in the sense of generalized functions, i.e. fGn defined this way is only strictly meaningful when integrated over a finite set). Although physically, the weather scales are responsible for the smoothing at τ_w in practice, we typically have macroweather data averaged at even lower resolutions: for example monthly or annually. The simplest procedure is to introduce the resolution as an averaging procedure yielding $T_\tau(t)$ so that the variance $\left\langle T_\tau^2 \right\rangle = \sigma_T^2 \tau^{2H} t$ diverges in the small resolution limit (recall $H < 0$), it follows that the spectrum of T is $E(\omega) \approx \omega^{-\beta}$, ω is the frequency and β = 1+2 H the integral of $T(t)$

is fractional Brownian motion process introduced by Kolmogorov (1940) and Mandelbrot and Van Ness (1968). In Lovejoy and de Lima (2015) it was shown how to extend this scalar SLIMM to spatially intermittent space-time SLIMM (accounting for different climatic regions).

To appreciate the huge difference between LIM with exponential correlations, (it is a continuous version of the discrete autoregressive processes) and SLIMM with power law correlations, we can calculate the fraction of the memory that resides in past event ("innovations, the white noise, $\gamma(t)$ in Eqs. (9) or (11)). Figure 10 compares the sizes of the LIM and SLIMM memories when forecasts are made for one (nondimensional) time step into the future. For example for LIM, the data more than 3 time steps in the past contains only $\approx$ 1% of the memory so that 99% of the basic "innovations" needed for the forecast is in the past one to three time steps. In comparison, the SLIMM memory contained in 4 and more past time steps for $H = -0.4, -0.3, -0.2$ and -0.1 respectively is $\approx$ 20%, 30%, 50% and 70% of the total memory. When $H = -0.1$ – a typical ocean value – thousands of time steps in the past are still being "felt" in the sense that they collectively contribute over 20% of the memory. As a

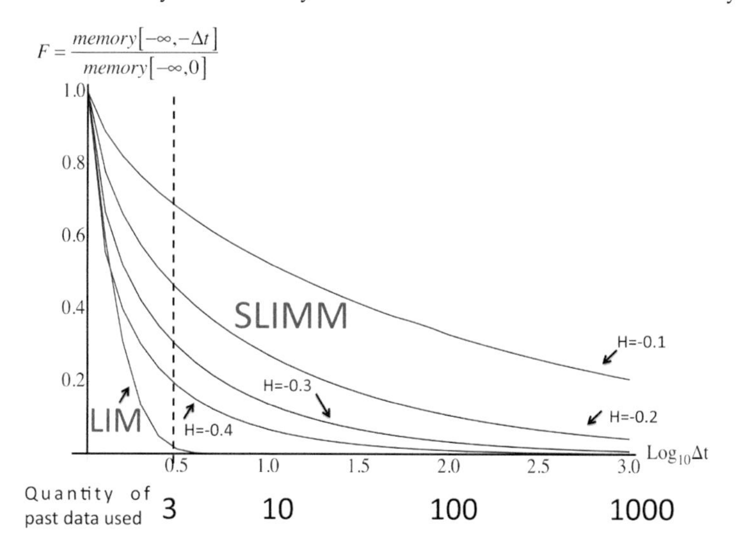

Figure 10: The large (power law) memory of SLIMM models (based on fractional Gaussian noise, the four rightmost curves) versus the short (exponential) memory of LIM processes (far left; time step is taken to be equal to the decorrelation time). F is the fraction of the memory that influences forecasts one time step into the future. The dashed line is roughly the limit of the LIM processes: if one neglects innovations from more than about 3 time steps in the past, only a few per cent of the information is lost. The comparable numbers for SLIMM are $\approx$70%, 50%, 30%, 20% for $H = -0.1, -0.2, -0.3, -0.4$ respectively (corresponding roughly to the oceans, the globe, land (temperatures) and (typical) precipitation respectively).

technical point, this does *not* mean that we need thousands of past time steps in order to make a good forecast. This is because the temperature values even 10 time steps in the past are also strongly influenced by the distant past innovations. Thanks to the solution of the (mathematical) fGn forecasting problem by Gripenberg and Norros (1996), Yaglom (1955), and the practical (finite difference) version by Hirchoren and Arantes (1998) we know that in practice 10-20 past time step temperature data give reasonable forecasts. Whereas in LIM processes, the weights of the past data decrease rapidly from present to past so that the most recent data are the most "influential", in fGn forecasts, the weight of the data furthest in the past is also singularly (strongly) weighted since they are the best available "witnesses" of the distance past.

4.3 Hindcasting the Pause with SLIMM

To illustrate the method, we used the NASA GISS annually globally averaged series from 1880 through 2013 (Fig. 8a). The first step was to estimate the natural variability as discussed in section 3.4. Unlike initialized GCM hindcasts that "optimistically" assume that future solar and volcanic forcings are known, our hindcasts (statistically) take these into account.

In Fig. 11 we compare the annual temperature residuals (T) and its running sum $S(t) = \sum_{t' \leq t} T(t')$ with their SLIMM hindcasts (red) and theoretical one standard deviation error bars (dashed). As expected, the actual temperatures are seen to lie almost entirely within the limits. The bottom row (right) shows a blow-up of the temperature hindcast and the actual temperature residuals, and Fig. 11 shows the difference (error); we see that over the entire period 1998-2013, the maximum forecast error is $\approx \pm 0.11$ °C (one standard deviation). However, the error for the hindcast "anomalies" is considerably smaller (i.e. the residues averaged over the hindcast horizon t: $(\hat{S}(t) - S(0))/t$): Fig. 11 lower left and Fig. 12 (blue). Beyond two years, the anomalies are within the theoretical one standard deviation limits, from 2002-2013, the anomaly errors are $\leq \pm 0.02$ °C i.e. below the estimated temperatures measurement errors (± 0.03 °C, Lovejoy et al. 2013b). (The term "anomaly" is a short hand for the anomaly fluctuation discussed earlier, for the residues this is simply the series at a specific resolution. For example, the RMS accuracy of annual GCM hindcasts are often quoted after averaging them over 4 or 5 years, this is the RMS of the 4 or 5 year anomaly). We could mention that other "mean reverting" processes such as LIM will also show qualitatively the same behaviour: the superiority of SLIMM was quantitatively established using over 100 hindcasts from 1900 to 2003.

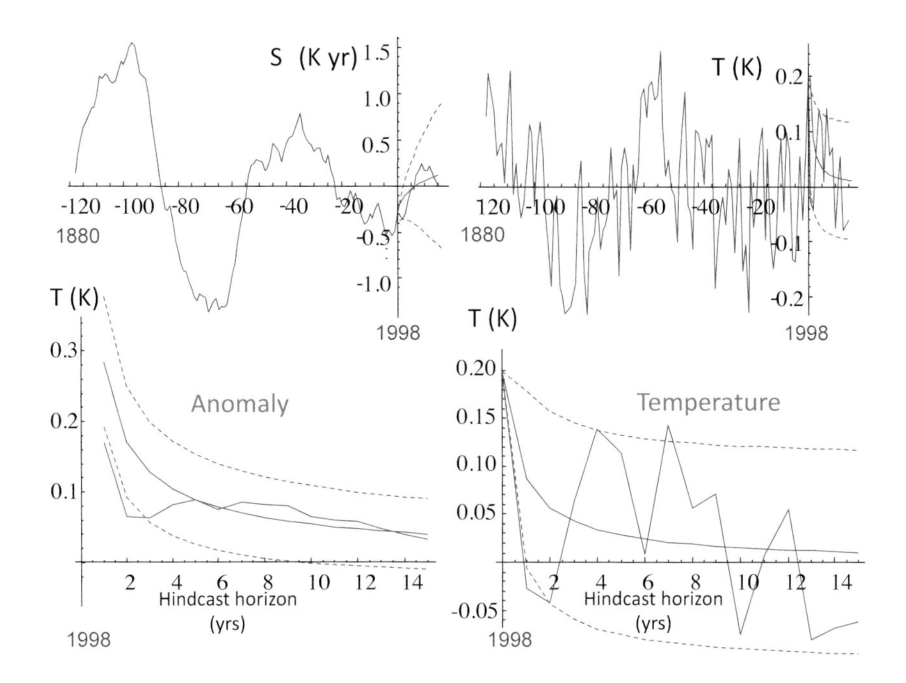

Figure 11: Upper left: The summed (natural) global, annual temperature $S(t)$ (blue) with the hindcast (red) from 1998 shown in red (here and elsewhere, the dashed red lines are one standard deviation error limits).
Upper right: The natural temperature (blue) with the hindcast from 1998 (red).
Lower left: The anomaly defined as the average natural temperature (i.e. residue) over the hindcast horizon (blue), red is the hindcast.
Lower right: The temperature since 1998 (blue) with hindcast (red), a blow-up of the hindcast part of the upper right plot.
A little more effort (using stochastic forecasting) shows that if the anthropogenic warming continues at its present rate, that there is reasonable chance (5%) that the pause will continue until 2019–2020. Alternatively, climate skeptics will have to wait another 5–6 years before using any continuing pause to reject the anthropogenic warming hypothesis at 95% levels. Reproduced from Lovejoy (2015c).

Since it can be so accurately hindcast, these results support the (unconditional) statistical analysis of the previous section that the pause must be due to natural variability. This is consistent with Steinman et al. (2015) who singled out particular high amplitude – but narrow scale range – low frequency natural processes including the Atlantic Multidecadal Oscillation (AMO) and argued that they explain the pause. However, due to the scaling (in space and in time) there is in fact a hierarchy of processes analogous to the AMO (Held et al. 2010): our hindcasts statistically account for the whole relevant scale range and show that they are all important.

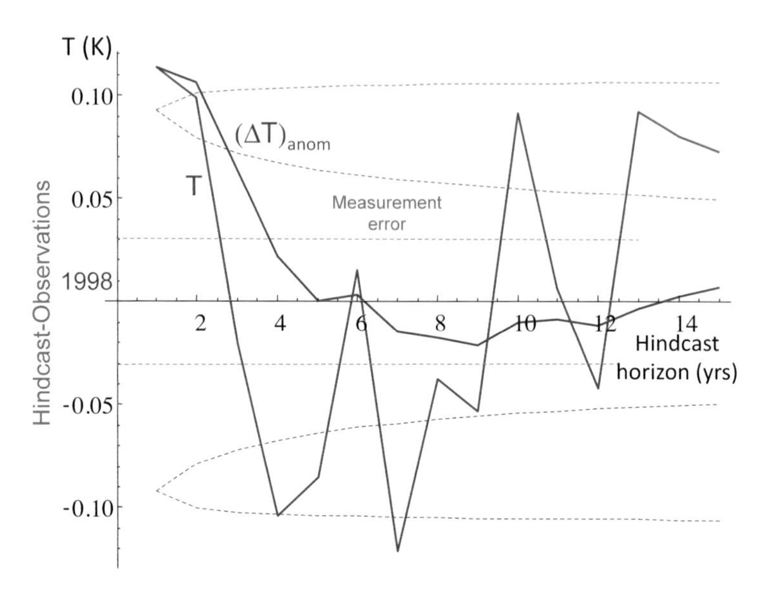

Figure 12: The hindcast errors: anomalies (blue) and temperatures (red), obtained from Fig. 11 lower left and right, respectively. The one standard deviation error limits are shown as dashed. The largest absolute errors are ±0.11 K. The estimate of the accuracy of the observations (dashed green), about ±0.03 K, is also shown. Reproduced from Lovejoy (2015c).

5. Conclusions

Due to strong dynamical nonlinearities, the atmosphere is highly variable from planetary down to millimetric dissipation scales, in time from the age of the earth to milliseconds (10, 20 orders of magnitude respectively). The typical ratio of nonlinear to linear terms is about 10^{12} and it is believed that at such high degrees of nonlinearity that new statistical turbulent laws emerge from the deterministic laws of continuum mechanics. These laws are based on scale invariance symmetries, but to be realistic, the scaling must be anisotropic: different in the horizontal and vertical directions. In addition, the intermittency (variability) is so high that they obey nonclassical, multifractal statistics with extreme power law tails. Global atmospheric data and models (GCM's) are now of high enough quality – they span wide enough ranges of scale – that these ideas can be tested, the scaling is indeed accurately obeyed by both models and data and is of the theoretically predicted type including extremes (Eq. (3)). It was argued that these statistical laws are the high level (emergent) consequences of the (lower level) deterministic laws of continuum mechanics in the limit of high nonlinearity.

This new understanding of natural variability is based on the intermittent anisotropic generalisations of classical turbulence theory described above, and can be used to statistically test the hypothesis that the industrial epoch warming is simply a giant natural fluctuation (GNF). By using temperature proxy data over the pre-industrial period 1500-1900 and the theoretically predicted form of the probabilities, we can estimate how long we must wait for fluctuations of various amplitudes over time scales up to $\approx$ 125 years. The only additional step is to estimate the magnitude of the industrial epoch warming (about 0.9 °C for the global average). When this was done, it was found that the probability was so low ($\approx$ 0.1%) that the GNF hypothesis could be dismissed.

Following 1998, the warming apparently slowed down – and due to the lack of a convincing model based explanation – the IPCC AR5 resorted to the vague: "Due to natural variability, trends based on short records are very sensitive to the beginning and end dates and do not in general reflect long-term climate trends" (see Hawkins et al. 2014). Our understanding of the natural variability allows us to quantitatively explain the slowdown as a cooling fluctuation that masked an ongoing anthropogenic warming trend. A first explanation was based on the *unconditional* statistics the natural cooling (of about 0.3 °C) had a fairly short return period (20-50 years), it is not so unusual. However, we noted that the natural cooling since 1998 immediately followed an even larger pre-pause warming (1992-1998) so that it was plausible that the cooling was simply a return to the long term trend.

In order to confirm this – i.e. to work out the *conditional* statistics, we exploited the scaling to make stochastic forecasts based on the fluctuation exponent H and the global temperatures that preceded the "pause". The model we used for the hindcast was fractional Gaussian noise which was a reasonable model for the time series since the latter had low intermittency ($C_1 \approx 0.01$ to 0.02). The resulting conditional hindcast was accurate to within 0.1 °C for the entire period following the 1998 peak and the 3-4 year anomalies were hindcast to with 0.03 °C. This was compared to the multimodel mean of CMIP 3 hindcasts that were about 0.2 °C too high. The problem was with the models, not the theory of anthropogenic warming itself which accurately predicted the pause as a necessary consequence of the pre-pause warming: without the "slowdown", the warming would have been too strong.

Using fGn as a model for the temporal development of macroweather fields (which have low intermittency) was convenient since the mathematical problem of forecasting fGn has recently been solved. Regional (spatial) extensions of this – with realistic strongly intermittent spatial variability – ScaLIng Macroweather Models (SLIMM) are currently under development since they lead to accurate monthly, seasonal, annual and decadal (i.e. macroweather) forecasts that are currently competitive with conventional methods since they avoid the model "drift" and poor seasonality that plague GCM's while skillfully forecasting the natural variability.

Although scientific opinion has for many years been virtually consensual about the theory of anthropogenic warming, the implications for humanity are so large that professional "skeptics" have continued to denigrate the models and tout the theory that the warming is no more than a giant natural fluctuation (see Lovejoy et al. 2016). It is therefore important to close the debate by eliminating this last source of rational criticism. Now that this has been done, any remaining skeptics are no more than deniers. Scientists can move on to understanding (and predicting) space-time climate variability (including regional forecasts) and the rest of the world can move on to dealing with the warming and its potentially catastrophic consequences.

REFERENCES

Arrhenius, S. 1896. On the influence of carbonic acid in the air upon the temperature on the ground. *The Philosophical Magazine* 41: 237–276.

Baillie, R.T. and S.-K. Chung. 2002. Modeling and forecasting from trend-stationary long memory models with applications to climatology. *International Journal of Forecasting* 18: 215–226.

Compo, G.P. et al. 2011. The Twentieth Century Reanalysis Project. *Quarterly J. Roy. Meteorol. Soc.* 137: 1-28. doi: 10.1002/qj.776.

de Wijs, H.J. 1951. Statistics of ore distribution. Part I. *Geologie en Mijnbouw* 13: 365–375.

Franzke, C. 2010. Long-range dependence and climate noise characteristics of Antarctica temperature data. *J. of Climate* 23: 6074–6081 doi: 10.1175/2010JCL13654.1.

Franzke, C. 2012. Nonlinear trends, long-range dependence and climate noise properties of temperature. *J. of Climate* 25: 4172–4183. doi: 10.1175/JCLI-D-11-00293.1.

Frisch, U., P.L. Sulem and M. Nelkin. 1978. A simple dynamical model of intermittency in fully develop turbulence. *Journal of Fluid Mechanics* 87: 719–724.

Gripenberg, G. and I. Norros. 1996. On the Prediction of Fractional Brownian Motion. *J. Appl. Prob.* 33: 400–410.

Hansen, J., R. Ruedy, M. Sato and K. Lo. 2010. Global surface temperature change. *Rev. Geophys.* 48: RG4004 doi: doi:10.1029/2010RG000345.

Hasselmann, K. 1976. Stochastic Climate models. Part I: Theory. *Tellus* 28: 473–485.

Hawkins, E., T. Edwards and D. McNeall. 2014. Pause for thought. *Nature Clim. Change* 4: 154–156.

Heinlein, R.A. 1973. Time Enough for Love. 605 pp. G.P. Putnam's Sons, New York.

Held, I.M., M. Winton, K. Takahashi, T. Delworth, F. Zeng and G.K. Vallis. 2010. Probing the Fast and Slow Components of Global Warming by Returning Abruptly to Preindustrial Forcing. *J. Climate* 23: 2418–2427.

Hirchoren, G.A. and D.S. Arantes. 1998. Predictors For The Discrete Time Fractional Gaussian Processes. *In:* Telecommunications Symposium, 1998. ITS '98 Proceedings. SBT/IEEE International, edited, pp. 49–53, IEEE, Sao Paulo.

Huybers, P. 2007. Glacial variability over the last two million years: An extended depth-derived agemodel, continuous obliquity pacing, and the Pleistocene progression. *Quaternary Science Reviews* 26(1-2): 37–55.

Kolmogorov, A.N. 1940. Wienershe spiralen und einige andere interessante kurven in Hilbertschen Raum. *Doklady Academii Nauk S.S.S.R.* 26: 115–118.

Kolmogorov, A.N. 1941. Local structure of turbulence in an incompressible liquid for very large Reynolds numbers. (English translation: Proc. Roy. Soc. A434, 9–17, 1991), *Proc. Acad. Sci. URSS., Geochem. Sect.* 30: 299–303.

Koscielny-Bunde, E., A. Bunde, S. Havlin, H.E. Roman, Y. Goldreich and H.J. Schellnhuber. 1998. Indication of a universal persistence law governing atmospheric variability. *Phys. Rev. Lett.* 81, 729–732.

Ladoy, P., S. Lovejoy and D. Schertzer. 1991. Extreme Variability of climatological data: Scaling and Intermittency. *In:* Non-linear variability in geophysics: Scaling and Fractals, edited by D. Schertzer and S. Lovejoy, pp. 241–250, Kluwer.

Laepple, T., S. Jewson and K. Coughlin. 2008. Interannual temperature predictions using the CMIP3 multi-model ensemble mean. *Geophys. Res. Lett.* 35. doi: L10701, doi:10.1029/2008GL033576, 2008.

Lovejoy, S. 2013. What is climate? *EOS* 94(1): 1–2, 1 January.

Lovejoy, S. 2014a. Return periods of global climate fluctuations and the pause. *Geophys. Res. Lett.* 41: 4704–4710. doi: 10.1002/2014GL060478.

Lovejoy, S. 2014b. Scaling fluctuation analysis and statistical hypothesis testing of anthropogenic warming. *Climate Dynamics* 42: 2339–2351. doi: 10.1007/s00382-014-2128-2.

Lovejoy, S. 2015a. Climate Closure. *EOS* 96. 10.1029/2015EO037499.

Lovejoy, S. 2015b. A voyage through scales, a missing quadrillion and why the climate is not what you expect. *Climate Dyn.* 44: 3187–3210. doi: 10.1007/s00382-014-2324-0.

Lovejoy, S. 2015c. Using scaling for macroweather forecasting including the pause. *Geophys. Res. Lett.* 42: 7148–7155, doi:DOI: 10.1002/2015GL065665.

Lovejoy, S., L. del Rio Amador, R. Hebert and I. de Lima. 2016. Giant natural fluctuation models and anthropogenic warming. *Geophys. Res. Lett.* 43: doi:10.1002/2016GL070428.

Lovejoy, S. and B.B. Mandelbrot. 1985. Fractal properties of rain and a fractal model. *Tellus* 37A: 209.

Lovejoy, S. and D. Schertzer. 1986. Scale invariance in climatological temperatures and the local spectral plateau. *Annales Geophysicae* 4B: 401–410.

Lovejoy, S. and D. Schertzer. 2010. Towards a new synthesis for atmospheric dynamics: Space-time cascades. *Atmos. Res.* 96: 1–52. doi: 10.1016/j.atmosres. 2010.01.004.

Lovejoy, S. and D. Schertzer. 2012. Low frequency weather and the emergence of the Climate. *In:* Extreme Events and Natural Hazards: The Complexity Perspective, edited by A.S. Sharma, A. Bunde, D.N. Baker and V.P. Dimri, pp. 231-254, AGU monographs, Washington D.C.

Lovejoy, S. and D. Schertzer. 2013. The Weather and Climate: Emergent Laws and Multifractal Cascades, 496 pp. Cambridge University Press, Cambridge.

Lovejoy, S. and M.I.P. de Lima, 2015. The joint space-time statistics of macroweather precipitation, space-time statistical factorization and macroweather models. *Chaos* 25. 075410 doi: doi: 10.1063/1.4927223.

Lovejoy, S., D. Schertzer and D. Varon. 2013a. Do GCM's predict the climate… or macroweather? *Earth Syst. Dynam.* 4: 1–16. doi: 10.5194/esd-4-1-2013.

Lovejoy, S., D. Scherter and D. Varon. 2013b. How scaling fluctuation analyses change our view of the climate and its models (Reply to R. Pielke sr.: Interactive comment on "Do GCM's predict the climate... or macroweather?" by S. Lovejoy et al.), *Earth Syst. Dynam. Discuss.* 3: C1–C12.

Lovejoy, S., J.P. Muller and J.P. Boisvert. 2014. On Mars too, expect macroweather. *Geophys. Res. Lett.* 41: 7694–7700. doi: 10.1002/2014GL061861.

Lovejoy, S., L. del Rio Amador and R. Hébert. 2015. The Scaling Linear Macroweather model (SLIM): Using scaling to forecast global scale macroweather from months to decades. *Earth System Dyn. Disc.* 6: 489–545. doi: 10.5194/esdd-6-489–2015.

Lovejoy, S., A.F. Tuck, D. Schertzer and S.J. Hovde. 2009. Reinterpreting aircraft measurements in anisotropic scaling turbulence. *Atmos. Chem. and Phys.* 9: 1–19.

Lovelock, J.E. 1995. Ages of Gaia, Oxford University Press.

Lynch, P. 2006. The emergence of numerical weather prediction: Richardson's Dream. 279 pp. Cambridge University Press, Cambridge.

Mandelbrot, B.B. 1974. Intermittent turbulence in self-similar cascades: Divergence of high moments and dimension of the carrier. *Journal of Fluid Mechanics* 62: 331–350.

Mandelbrot, B.B. and J.W. Van Ness. 1968. Fractional Brownian motions, fractional noises and applications. *SIAM Review* 10: 422–450.

Meneveau, C. and K.R. Sreenivasan. 1987. Simple multifractal cascade model for fully developped turbulence. *Physical Review Letter* 59(13): 1424–1427.

Mitchell, J.M. 1976. An overview of climatic variability and its causal mechanisms. *Quaternary Res.* 6: 481–493.

Moberg, A., D.M. Sonnechkin, K. Holmgren, N.M. Datsenko and W. Karlén. 2005. Highly variable Northern Hemisphere temperatures reconstructed from low- and high-resolution proxy data. *Nature* 433(7026): 613–617.

Newman, M. 2013. An Empirical Benchmark for Decadal Forecasts of Global Surface Temperature Anomalies. *J. of Clim.* 26: 5260–5269. doi: 10.1175/ JCLI-D-12-00590.1.

Newman, M.P., P.D. Sardeshmukh and J.S. Whitaker. 2003. A study of subseasonal predictability. *Mon. Wea. Rev.* 131: 1715–1732.

Novikov, E.A. and R. Stewart. 1964. Intermittency of turbulence and spectrum of fluctuations in energy-disspation. *Izv. Akad. Nauk. SSSR. Ser. Geofiz.* 3: 408–412.

Pelletier, J.D. 1998. The power spectral density of atmospheric temperature from scales of 10**-2 to 10**6 yr. *EPSL* 158: 157-164.

Peng, C.-K., S.V. Buldyrev, S. Havlin, M. Simons, H.E. Stanley and A.L. Goldberger. 1994. Mosaic organisation of DNA nucleotides. *Phys. Rev. E* 49: 1685–1689.

Penland, C. 1996. A stochastic model of IndoPacific sea surface temperature anomalies. *Physica D* 98: 534–558.

Penland, C. and P.D. Sardeshmuhk. 1995. The optimal growth of tropical sea surface temperature anomalies. *J. Climate* 8: 1999–2024.

Pinel, J., S. Lovejoy and D. Schertzer. 2014. The horizontal space-time scaling and cascade structure of the atmosphere and satellite radiances. *Atmos. Resear.* 140–141: 95–114. doi: doi.org/10.1016/j.atmosres.2013.11.022.

Pinel, J., S. Lovejoy, D. Schertzer and A.F. Tuck. 2012. Joint horizontal – vertical anisotropic scaling, isobaric and isoheight wind statistics from aircraft data. *Geophys. Res. Lett.* 39: L11803 doi: 10.1029/2012GL051698.

Richardson, L.F. 1922. Weather prediction by numerical process. Cambridge University Press republished by Dover, 1965.

Richardson, L.F. 1926. Atmospheric diffusion shown on a distance-neighbour graph. *Proc. Roy. Soc.* A110: 709–737.

Rohde, R., R.A. Muller, R. Jacobsen, E. Muller, S. Perlmutter, A. Rosenfeld, J. Wurtele, D. Groom and C. Wickham. 2013. A New Estimate of the Average Earth Surface Land Temperature Spanning 1753 to 2011. *Geoinfor Geostat: An Overview* 1:1 doi: doi: http://dx.doi.org/10.4172/2327-4581.1000101.

Rypdal, K., L. Østvand and M. Rypdal. 2013. Long-range memory in Earth's surface temperature on time scales from months to centuries. *JGR, Atmos.* 118: 7046–7062 doi: doi:10.1002/jgrd.50399.

Rypdal, M. and K. Rypdal. 2014. Long-memory effects in linear response models of Earth's temperature and implications for future global warming. *J. Climate* 27(14): 5240–5258. doi: 10.1175/JCLI-D-13-00296.1.

Sardeshmukh, P., G.P. Compo and C. Penland. 2000. Changes in probability assoicated with El Nino. *J. Climate* 13: 4268–4286.

Sardeshmukh, P.D. and P. Sura. 2009. Reconciling non-gaussian climate statistics with linear dynamics. *J. of Climate* 22: 1193–1207.

Schertzer, D. and S. Lovejoy. 1985. The dimension and intermittency of atmospheric dynamics. *In:* Turbulent Shear Flow, edited by L.J.S.B. et al., pp. 7–33, Springer-Verlag.

Schertzer, D. and S. Lovejoy. 1987. Physical modeling and Analysis of Rain and Clouds by Anisotropic Scaling of Multiplicative Processes. *Journal of Geophysical Research* 92: 9693–9714.

Schertzer, D. and S. Lovejoy. 2004. Uncertainty and Predictability in Geophysics: Chaos and Multifractal Insights. *In:* State of the Planet, Frontiers and Challenges in Geophysics, edited by R.S.J. Sparks and C.J. Hawkesworth, pp. 317–334, American Geophysical Union, Washington.

Smith, D.M., S. Cusack, A.W. Colman, C.K. Folland, G.R. Harris and J.M. Murphy. 2007. Improved Surface Temperature Prediction for the Coming Decade from a Global Climate Model. *Science* 317: 796–799.

Steinman, B.A., M.E. Mann and S.K. Miller. 2015. Atlantic and Pacific multidecadal oscillations and Northern Hemisphere temperatures. *Science* 347: 988–991. doi: 10.1126/science.1257856.

Taleb, N.N. 2010. The Black Swan: The Impact of the Highly Improbable. 437 pp. Random House, New York.

Veizer, J. et al. 1999. 87Sr/86Sr, d18O and d13C Evolution of Phanerozoic Seawater. *Chemical Geology* 161: 59–88.

Yaglom, A.M. 1955. Correlation theory of processes with random stationary nth increments (Russian). English translation. *Amer. Math. Soc. Trans. Ser.* 8: 87–141, *Mat. Sb. N.S.* 37: 141–196.

Yuan, N., Z. Fu and S. Liu. 2014. Extracting climate memory using Fractional Integrated Statistical Model: A new perspective on climate prediction. *Nature Scientific Reports* 4: Article number: 6577 doi:10.1038/srep06577.

Zachos, J., M. Pagani, L. Sloan, E. Thomas and K. Billups. 2001. Trends, Rhythms, and Aberrations in Global Climate 65 Ma to Present. *Science* 292(5517): 686–693. Doi: 10.1126/science.1059412.

CHAPTER

9

Fractals and Multifractals in Geophysical Time Series

Mikhail I Bogachev[1]*, Naiming Yuan[2] and Armin Bunde[3]

[1]Radio Systems Department, St. Petersburg Electrotechnical University, 197376 St. Petersburg, Russia

[2]CAS key Laboratory of Regional Climate-Environment for Temperate East Asia, Institute of Atmospheric Physics, Chinese Academy of Sciences, Beijing, 100029, China

[3]Institut für Theoretische Physik, Justus-Liebig Universität, D-35392 Giessen, Germany

1. Introduction

In recent years, there is growing evidence that hydroclimate data like river flows (Hurst 1951, Mandelbrot and Wallis 1968, Tessier et al. 1996, Montanari et al. 2000, Montanari 2003, Koutsoyiannis 2003, Koutsoyiannis 2006, Kantelhardt et al. 2006, Koscielny-Bunde et al. 2006, Mudelsee 2007, Livina et al. 2003), atmospheric and sea surface temperatures (Mandelbrot 2002, Bloomfield and Nychka 1992, Koscielny-Bunde et al. 1996, Pelletier and Turcotte 1997, Koscielny-Bunde et al. 1998, Malamud and Turcotte 1999, Talkner and Weber 2000, Weber and Talkner 2001, Monetti et al. 2003, Eichner et al. 2003, Fraedrich and Blender 2003, Blender and Fraedrich 2003, Gil-Alana 2005, Cohn and Lins 2005, Király et al. 2006, Rybski et al. 2006, Rybski et al. 2008, Zorita et al. 2008, Giese et al. 2007, Rybski and Bunde 2009, Halley 2009, Lennartz and Bunde 2009b, Fatichi et al. 2009, Franzke 2010, Lennartz and Bunde 2011, Franzke 2012, Lovejoy and Schertzer 2013, Franzke 2013, Bunde et al. 2014, Yuan et al. 2014, Ludescher et al. 2015, Yuan et al. 2015), sea level heights (Beretta et al. 2005, Dangendorf et al. 2014, Becker et al. 2014), or wind fields (Santhanam and Kantz 2005) and midlatitude cyclones (Blender et al. 2015) exhibit long-

*Corresponding author: Mikhail.Bogachev@physik.uni-giessen.de

term persistency. In previous reports of the Intergovernmental Panel on Climate Change (IPCC) (see, e.g. (Stocker et al. (eds) 2013) it had been anticipated that only short-term persistence occurs. In long-term persistent data sets the autocorrelation function (ACF) decays algebraically, without a characteristic time scale, while in short-term persistent data sets, in contrast, the autocorrelation function decays exponentially and there is a characteristic time scale above which the data can be considered as uncorrelated.

The way how the data are correlated is of enormous interest since it strongly influences the occurrence of extremes as well as their predictability. It also influences the estimation of the magnitude of possible external trends (like anthropogenic warming) in the data. When the data show some trend, one likes to know to which extent the trend can be considered as natural or not. This chapter deals with these questions.

First we show how monofractal (linearly) long-term correlated records can be generated numerically and how they can also be detected in the presence of external trends. We give examples for long-term correlated hydroclimate records. After that, we discuss how the formalism to detect monofractal (linear) correlations can be generalized to also detect non-linear long-term correlations (multifractality). We discuss the consequences of both linear and nonlinear correlations for the temporal arrangements of extreme events. Finally, we show how external trends can be detected in data with long-term memory.

2. Long-term Correlations and How to Detect Them

2.1 Long-term Correlations

Consider a record W_i of discrete numbers, where the index i runs from 1 to L. W_i may be daily, monthly or annual temperatures, precipitation data or river flows, or any other set of data consisting of L successive data points. We are interested in the fluctuations $y_i = W_i - \overline{W_i}$ of the data around their (sometimes seasonal) average value $\overline{W_i}$ (see Fig. 1).

Without loss of generality, we assume that the mean of the data is zero and its variance equals one. We call the data long-term correlated, when the corresponding autocorrelation function $C(s) = \langle y_i y_{i+s} \rangle \equiv \frac{1}{L-s} \sum_{i=1}^{L-s} y_i y_{i+s}$ decays, in the limit of $L \to \infty$ as a power law (see (Mandelbrot and Wallis 1968)),

$$C(s) \sim (1-\gamma)s^{-\gamma} \quad (1)$$

where γ denotes the correlation exponent, $0 < \gamma < 1$, L is the data size and s is the time lag. Such correlations are named "long-term" since the mean correlation time $s_x = \int_0^\infty C(s)ds$ diverges in the limit of infinitely long

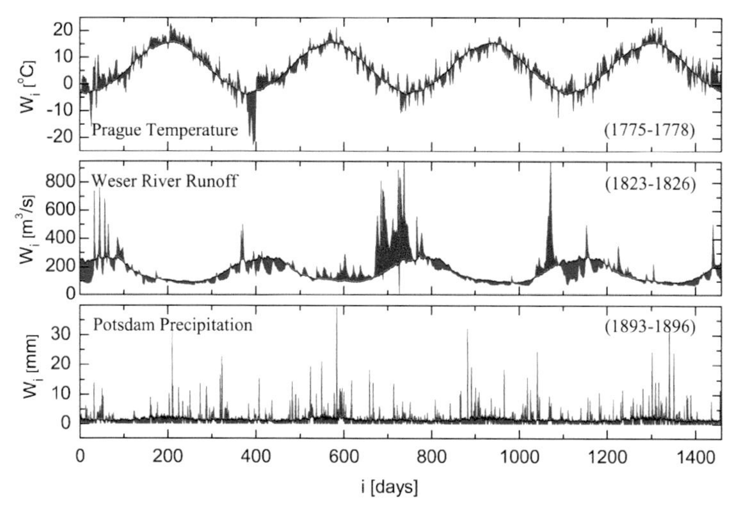

Figure 1: Daily temperature (Prague), river flow (Weser river) and precipitation (Potsdam) data. Data below (above) the seasonal average are in blue (red). One can see by eye that for the temperature and river flow data, blue and red patches occur which are an indication of persistence. While the patches are pronounced in the river flow data, they are very small in the precipitation data. The deviations y_i from the seasonal average represent the temperature, river flow and precipitation anomalies that we consider in this chapter.

series where $L \to \infty$. For data sets with *finite* length L, the autocorrelation function decays as (Lennartz and Bunde 2009a)

$$C(s) = \frac{1}{1-L^{-\gamma}}\left(\frac{(1-\gamma)(2-\gamma)}{2}s^{-\gamma} - L^{-\gamma}\right) \tag{2}$$

Accordingly, there are large finite-size effects in $C(s)$ that makes it difficult to estimate the correlation exponent γ for L of the order of the length of typical climate records.

If y_i are uncorrelated, $C(s) = 0$ for $s > 0$. More generally, if persistence (characterized by positive correlations) exists up to a certain correlation time $s_\times$, then $C(s) > 0$ for $s < s_\times$ and $C(s) = 0$ for $s > s_\times$. If y_i are short-term correlated and described iteratively by a first-order autoregressive process (AR1) where $y_{i+1} = c_1 y_i + \eta_i$, $i = 1, 2, \ldots, L-1$ and η_i is white noise, $C(s)$ decays exponentially, $C(s) = c_1^s$, i.e., c_1 is identical to the lag-1 autocorrelation $C(1)$. For $c_1 > 0$, $C(s)$ can be written as $C(s) = \exp(-s/s_\times)$ where $s_\times = 1/|\ln c_1|$ denotes the persistence time.

2.2 Power Spectrum and Generation of Long-term Correlated Records

Long-term correlated records can also be characterized by their power spectral density (PSD) $P(f) = |y(f)|^2$, where $y(f)$ is the Fourier transform of $\{y_i\}$. By definition, $P(f) = P(-f)$. With increasing frequency f, $f = 1/L$, $2/L, \ldots, 1/2 - 1/L, 1/2$, $P(f)$ decays by a power law

$$P(f) \sim f^{-\beta} \tag{3}$$

where $\beta = 1 - \gamma$ characterizes the long-term memory (Turcotte 1997). For uncorrelated data $\beta = 0$.

Equation (3) can be used to generate synthetic Gaussian distributed long-term correlated records, see, e.g., (Turcotte, 1997). The procedure is described in Fig. 2 and consists of 4 steps. First one generates uncorrelated Gaussian data (Fig. 2a) and transforms them to the Fourier domain (Fig. 2b). Then one multiplies the result by $f^{-\beta/2}$ (Fig. 2c) and transforms it back to the time domain (Fig. 2d). The final result in Fig. 2d is long-term correlated with $\gamma = 1 - \beta$. By definition, the resulting $P(f)$ follows Eq. (3).

Figure 3 compares parts of an uncorrelated record (a) with 2 long-term correlated records (b, c), with $\gamma = 0.5$, and 0.25; all series have been generated by the algorithm described above. The grey line is the moving average over 30 data points. For the uncorrelated data, the moving average is close to zero, while for the long-term correlated data sets, the moving average can have large deviations from the mean, forming some kind of mountain valley structure. This structure, which is best seen for $\gamma = 0.25$,

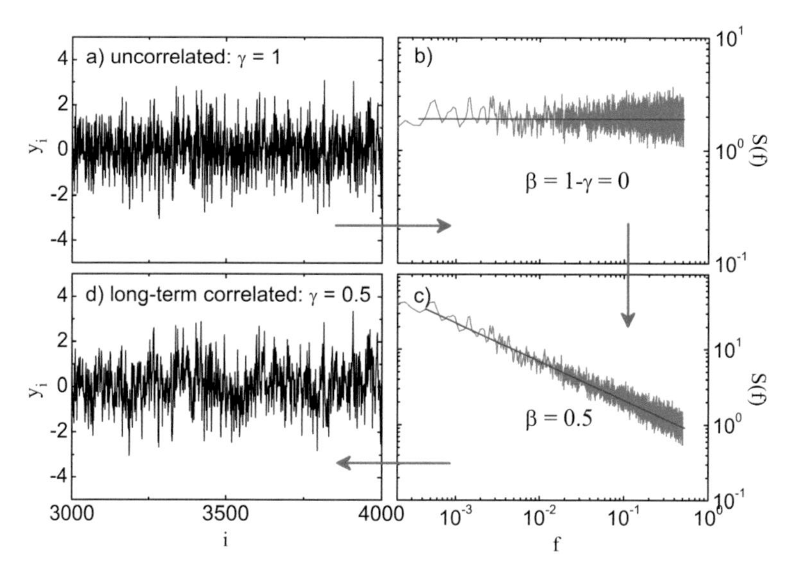

Figure 2: Sketch of the Fourier-Filtering method. (a) Gaussian distributed uncorrelated random numbers y_i are generated. (b) The power spectrum $S(f) = |y(f)^2|$ shows a plateau. (c) The Fourier transform $y(f)$ is multiplied by $f^{-\beta/2}$, here $\beta = 0.5$. Accordingly, the power spectrum is replaced by a power spectrum that decays as $f^{-\beta}$. (d) Backtransformation leads to Gaussian distributed long-term correlated data y_i with $\gamma = 1 - \beta$, here $\gamma = 0.5$.

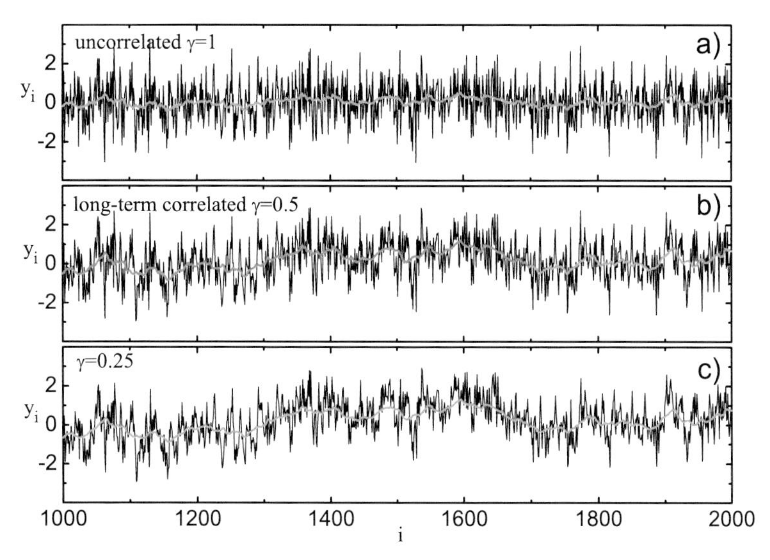

Figure 3: Comparison of an uncorrelated record with two long-term correlated records with $\gamma = 0.5$ and 0.25. The grey curve is the moving average over 30 data points.

is a consequence of the long-term power-law persistence. The mountains and valleys in Fig. 3 look as if they had been generated by external trends, and one might be inclined to draw a trend-line and to extrapolate the line into the near future for some kind of prognosis. But since the data are trend-free, only a short-term prognosis utilizing the persistence can be made, and not a longer-term prognosis which often is the aim of such a regression analysis.

2.3 Self-affinity of the Cumulated Record and the Hurst Exponent

Since trends resemble long-term correlations and vice versa, there is a general problem to distinguish between trends and long-term persistence and to determine the correlation exponent γ in the presence of external trends. To further characterize long-term correlated records and develop techniques that eventually will allow an accurate determination of the correlation exponent γ also in the presence of additive external trends, we now consider the cumulated sum of the record, $Y(i) \equiv \Sigma_{k=1}^{i} y_k$, $i = 1, \ldots, L$, which is also called 'profile' or 'landscape'. As we will see, this landscape forms a self-affine structure: When Y is considered on two time scales s and bs, then, on the average,

$$Y(bs) = b^H Y(s), \tag{4}$$

where H is called the Hurst exponent. To obtain H, we study how the fluctuations of the landscape $Y(s)$, in a given time window of size s, increase with s. We can consider $Y(s)$ as the position of a random walker on a linear chain after s steps. The random walker starts at the origin and performs, in the ith step, a jump of length $|y_i|$. The jump is to the right if y_i is positive, and to the left if y_i is negative. To find how the squared fluctuations of the profile scale with s, we first divide each record of L elements into $N_s = integer\,[N/s]$ non-overlapping segments of size s starting from the beginning and N_s non-overlapping segments of size s starting from the end of the considered record. Then we determine the fluctuations in each segment v. Different methods are distinguished by the way the fluctuations are determined.

In the standard fluctuation analysis (FA), we obtain the fluctuations simply from the values of the profile at both endpoints of each segment v,

$$F_v^2(s) = [Y(vs) - Y((v-1)s)]^2, \tag{5}$$

and average $F_v^2(s)$ over all $K_s = 2N_s$ subsequences to obtain the mean fluctuation $F_2(s)$,

$$F_2(s) \equiv \left[\frac{1}{K_s}\sum_{v=1}^{K_s} F_v^2(s)\right]^{1/2}. \tag{6}$$

By definition, $F_2(s)$ can be viewed as the root-mean square displacement of the random walker on the chain from the origin after s steps. For uncorrelated y_i values, we know from Fick's diffusion law that $F_2(s) \sim s^{1/2}$, providing $H = 1/2$ for uncorrelated records.

For the relevant case of long-term correlations, where $C(s)$ follows Eq. (1), $F_2(s)$ increases by a power law (Mandelbrot and Wallis 1968)

$$F_2(s) \sim s^H \tag{7}$$

where H is related to the correlation exponent γ and the power-spectrum exponent β by

$$H = 1 - \gamma/2 = (1 + \beta)/2. \tag{8}$$

For power law correlations decaying faster than $1/s$, we have $H = 1/2$ for large s values, like for uncorrelated data. It is worth mentioning that the standard fluctuation analysis is somewhat similar to the rescaled range analysis introduced by Hurst (for a review see Feder 1988), except that it focuses on the second moment $F_2(s)$ while Hurst considered the first moment $F_1(s)$. The drawback of this simple fluctuation analysis method is that it only works for stationary data where the standard deviation around the mean is finite. If we analyse a random walk with FA, we would obtain the false result $H = 1$ (instead of the correct result $H = 3/2$). Other drawbacks are the presence of large finite-size effects and the fact that external trends may lead to overestimations of H (Kantelhardt et al. 2001).

2.4 Detrended Fluctuation Analysis (DFA)

In recent years, several methods have been developed, mostly based on the hierarchical detrended fluctuation analysis (DFAn) or the Haar wavelet technique (WTn) where long-term correlations in the presence of smooth polynomial trends of order $n - 1$ can be detected (Peng et al. 1993, Bunde et al. 2000, Kantelhardt et al. 2001). The DFAn procedure is a generalization of the standard fluctuation analysis (Kantelhardt et al. 2001). As in FA, we consider the profile and divide the record into K_s windows of length s. Then we calculate the local trend for each of the K_s segments by fitting (least squares fit) a polynomial of order n to the data and determine the variance

$$F_v^2(s) \equiv \frac{1}{s}\sum_{i=1}^{s}[Y((v-1)s+i)-p_v(i)]^2 \tag{9}$$

for each segment v. Here $p_v(i)$ is the fitting polynomial representing the local trend in the segment v. Constant trends (DFA0) as well as linear, quadratic, cubic and higher-order polynomials can be used in the fitting procedure. When linear polynomials are used, the fluctuation analysis is called DFA1, for quadratic polynomials we have DFA2, for cubic polynomials DFA3, etc. By definition, DFA2 removes quadratic trends in the profile $Y(i)$ and thus linear trends in the original series y_i.

In the last step, finally, one averages $F_v^2(s)$ (as in Eq. (6)) over all segments and takes the square root to obtain the mean fluctuation function $F_2(s)$. One can show that for long-term correlated trend-free data the resulting DFA fluctuation functions $F_2(s)$ scale with the window size s as in Eq. (7), $F_2(s) \sim s^H$, with H from Eq. (8), irrespective of the order of the detrending polynomial. For short-term correlated records, the exponent is 1/2 for s well above the correlation time $s_\times$. Trends of order $k - 1$ in the original data are eliminated by DFAk, while they contribute to DFA($k - 1$) and DFA($k - 2$) etc., and this allows to determine the correlation exponent γ in the presence of trends. For example, in the case of a linear trend, DFA0 and DFA1 are affected by the trend and will overestimate H, while DFA2, DFA3 etc. are not affected by the trend and will show, in a double logarithmic plot, the same slope H, which then immediately gives the correlation exponent γ. The DFA method has comparatively small finite size effects and is valid for s up to $L/4$. It is easy to verify that for DFAn, $F_2(s) \equiv 0$ for $s \leq n + 1$. For synthetic long-term correlated records, the asymptotic power law behaviour of the DFA2 fluctuation function which is most commonly considered in geoscience, is reached at $s = 8$. Accordingly, the Hurst exponent can be determined by a regression analysis between $s = 8$ and $L/4$ in a double logarithmic presentation of $F_2(s)$.

2.5 Haar Wavelet Technique (WT)

In the Haar wavelet technique, ones usually considers the original data y_i and divides, as in FA and DFAn, the record in K_s segments of length s. Then one determines, in each segment v, the mean value $\overline{y}_v$ of the data. In WT0, one considers the quantity $G_v^2(s) = (\overline{y}_v)^2$, in WT1 $G_v^2(s) = (\overline{y}_v - \overline{y}_{v-1})^2$, and in WT2 $G_v^2(s) = (\overline{y}_v - 2\overline{y}_{v-1} + \overline{y}_{v-2})^2$. Then one averages $G_v^2(s)$ over all segments of length s and takes the square root to obtain the wavelet fluctuation function

$$G_2(s) \equiv \left[\frac{1}{K_s} \sum_{v=1}^{K_s} G_v^2(s) \right]^{1/2}. \quad (10)$$

One can show that for purely long-term correlated data, for not too large s values (typically for s below $L/100$),

$$G_2(s) = G_2(1)s^{H-1}. \quad (11)$$

Regarding trend elimination, DFAn corresponds to WTn. In particular, in WT2 linear external trends are eliminated. The advantage of the WT technique is that it, unlike the DFA, does not exhibit the initial crossover from the "perfect fitting" regime, and follows the proper asymptotic functional form also at very small scales, starting with $s = 1$ for synthetic data. Accordingly, in order to reveal the fluctuation properties for both short and long time scales, DFA and WT are complementary and can be used together.

3. Long-term Correlations in Climate Records

Figure 4 shows representative results of the DFAn and WTn analysis ($n = 1$ and 2), for temperature, precipitation and river runoff data. For continental temperatures, the exponent H is around 0.65, while for island stations and sea surface temperatures the exponent is considerably higher. There is no crossover towards uncorrelated behaviour at larger time scales. For the precipitation data, the exponent is between 0.51 and 0.55, not being significantly larger than for uncorrelated records.

Figure 5 shows a summary of the Hurst exponents H for a large number of climate records. It is interesting to note that, while the distribution of H-values is quite broad for river flows, sea-surface temperatures, and precipitation data, the distribution is quite narrow for continental atmospheric temperature records, located around $H = 0.65$. For the island records, the exponent is larger. The quite universal exponent $H = 0.65$ for continental stations can be used as an efficient testbed for climate models (Govindan et al. 2002, Vjushin et al. 2004, Rybski et al. 2008) and paleo reconstructions (Bunde et al. 2013).

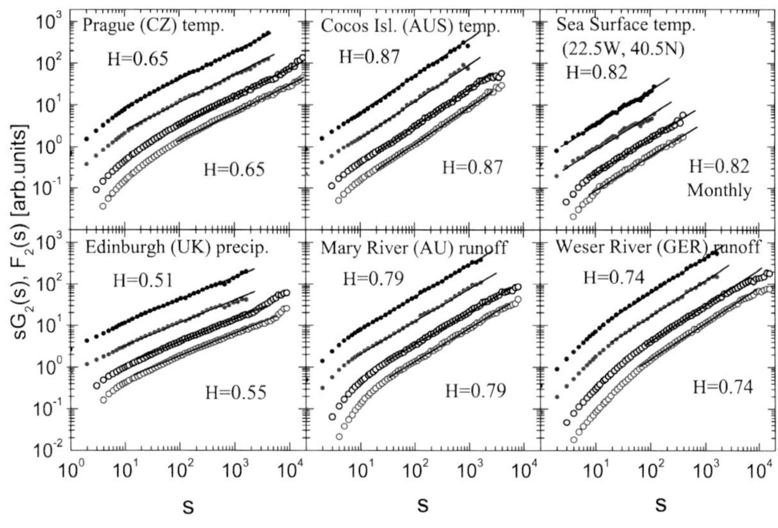

Figure 4: DFAn and WTn (n = 1, 2) analysis of 3 temperature records, one precipitation record and 2 runoff records. In each panel, the two upper curves represent $sG_2(s)$ for WT1 (black full circles) and WT2 (red full circles), respectively, while the two lower curves represent $F_2(s)$ for DFA1 (black open circles) and DFA2 (red open circles), respectively. The numbers denote the asymptotic slopes of $sG_2(s)$ and $F_2(s)$.

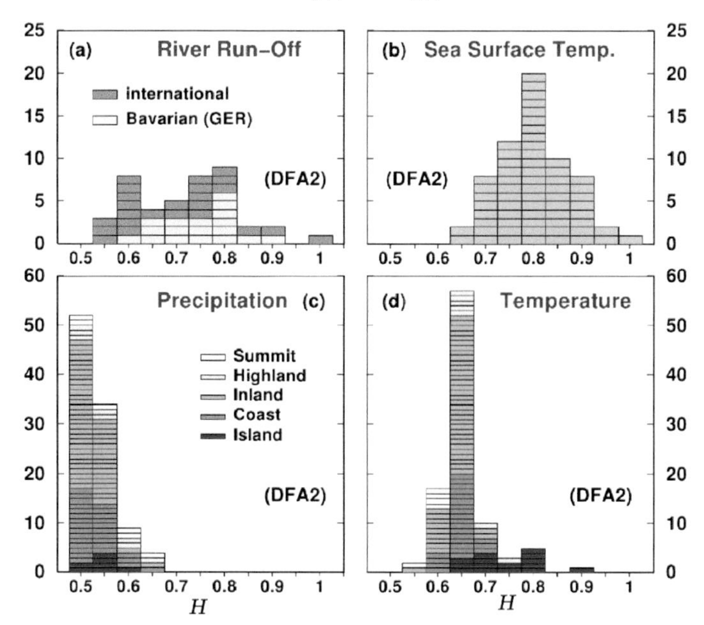

Figure 5: Histogramms of the fluctuation exponents H for several kinds of climate records (Eichner et al. 2003, Monetti et al. 2003, Kantelhardt et al. 2006).

The maximum time window accessible by DFAn is typically 1/4 of the length of the record. For instrumental records, the time window is thus restricted to about 50 years. For extending this limit, one can either take reconstructed records or model data, which roughly range up to 2000y. Both have, of course, large uncertainties, but it is remarkable that exactly the same kind of long-term correlations can be found in model data, thus extending the time scale where long-term memory exists to at least 500y (Rybski et al. 2006, Rybski et al. 2008). Paleo reconstructions show stronger long-term correlations where the exponent H is considerably exaggerated (Bunde et al. 2013, Franzke 2013).

4. Non-linear Correlations: Multifractality

In some climate records, not all subsets of the series can be described by Eq. (4) with the same scaling exponent H. Different subsets of the series have different scaling behaviour, and a single scaling exponent H may not be sufficient to characterize them. To quantify this situation, one traditionally considers the partition function

$$Z_q(s) \equiv \sum_{v=1}^{Ks} | Y(vs) - Y((v-1)s) |^q \sim s^{\tau(q)}, \tag{12}$$

where the variable q can take any real value except 0 and $\tau(q)$ is the Renyi scaling exponent. A record is called *monofractal*, when $\tau(q)$ is linear in q; otherwise it is called *multifractal*. To relate the partition function to the fluctuation analysis considered above, we extend Eq. (6) by (Barabasi and Viscek 1991)

$$F_q(s) \equiv \left[\frac{1}{K_s} \sum_{v=1}^{K_s} [[Y(vs) - Y((v-1)s)]^2]^{q/2} \right]^{1/q}. \tag{13}$$

For $q = 2$, the standard FA is retrieved. Since $K_s \propto 1/s$, it is easy to verify that $F_q(s) = [Z_q(s)/K_s]^{1/q} \sim s^{\tau(q)/q+1/q}$. This implies

$$F_q(s) \sim s^{h(q)}, \tag{14}$$

where

$$h(q) = [\tau(q) + 1]/q. \tag{15}$$

Thus the generalized Hurst exponent $h(q)$ is directly related to the classical multifractal scaling exponent $\tau(q)$.

For monofractal time series, $h(q)$ is independent of q, since the scaling behaviour of the variances $F_v^2(s)$ is identical for all segments v. If, on the other hand, small and large fluctuations scale differently, there will be a significant dependence of $h(q)$ on q: If we consider large positive values of q, those segments v with large variance $F_v^2(s)$ (i.e., large deviations from the corresponding fit) will dominate the average $F_q(s)$. Thus, for large positive values of q, $h(q)$ describes the scaling behaviour of the segments with large fluctuations. Usually large fluctuations are characterized by a smaller scaling exponent $h(q)$. On the contrary, for large negative values of q, those segments v with small variance $F_v^2(s)$ will dominate the average $F_q(s)$. Hence, for large negative values of q, $h(q)$ describes the scaling behaviour of the segments with small fluctuations, which are usually characterized by larger scaling exponents.

The multifractal analysis described above is a straightforward generalization of the fluctuation analysis and therefore has the same problems: (i) monotonous trends in the record may lead to spurious results for the fluctuation exponent $h(q)$ which in turn leads to spurious results for the correlation exponent γ, and (ii) non-stationary behaviour characterized by exponents $h(q) \geq 1$ cannot be detected by the simple method since the method cannot distinguish between exponents above 1 yielding $F_2(s) \sim s$ in this case (see above).

To overcome these drawbacks the multifractal detrended fluctuation analysis (MF-DFA) has been introduced (Kantelhardt et al. 2002), see also (Koscielny-Bunde et al. 1998, Weber and Talkner 2001). According to (Kantelhardt et al. 2002), the method is equivalent to the wavelet transform modulus maxima (WTMM) method (Muzy 1991, Arneodo et al. 2002) but easier to implement on the computer. In MF-DFA, the variance $F_v^2(s)$ in Eq. (9) is replaced by its $q/2$th power and the square root in the average over all K_s segments is replaced by the $1/q$th power, where $q \neq 0$ is a real parameter,

$$F_q(s) \equiv \left[\frac{1}{K_s} \sum_{v=1}^{K_s} [F_v^2(s)]^{q/2} \right]^{1/q} \sim s^{h(q)}. \tag{16}$$

Temperature records can be considered as monofractals, while precipitation and river flow records usually show weak to modest multifractality (Tessier et al. 1996, Kantelhardt et al. 2006, Koscielny-Bunde et al. 2006, Lovejoy and Schertzer 2013). Pronounced multifractality occurs in turbulent wind fields (Lovejoy and Schertzer 2013) as well as in financial data sets (Lux and Ausloos 2002, Ludescher and Bunde 2014) and references therein).

Figure 6 shows that moderate multifractal effects can be observed in both precipitation and river runoff representative records. To evaluate long-term memory, we are interested solely in the asymptotic behaviour of $F_q(s)$, that can be clearly observed at scales above the crossover caused by the seasonal trend. In particular, in the precipitation record from Edinburgh $h(q)$ ranges between 0.45 and 0.63 for q between 10 and –10, respectively. For the Weser river runoff record, $h(q)$ changes between 0.7 and 1.17 for the same q range.

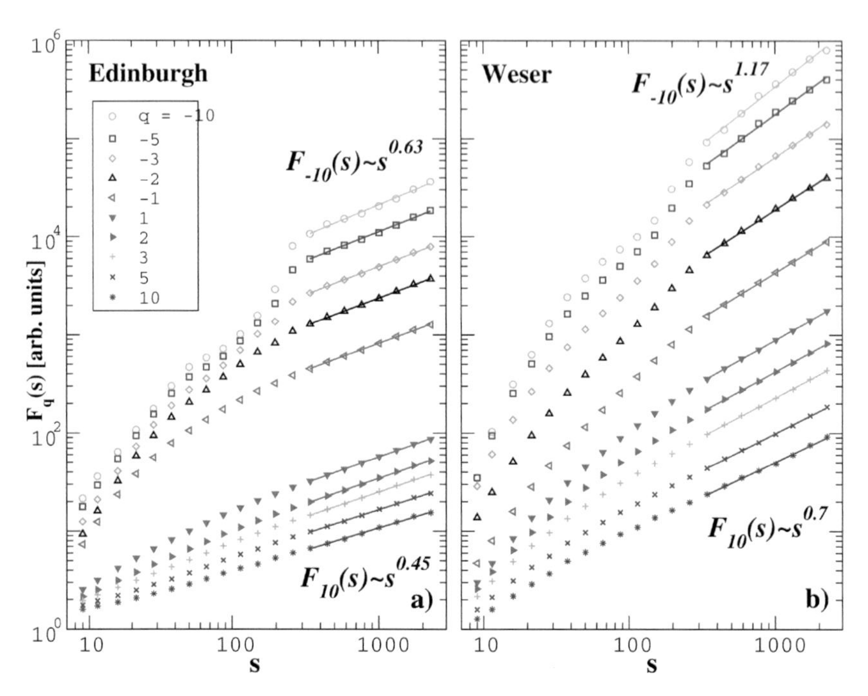

Figure 6: Generalized fluctuation functions obtained using MF-DFA for (a) precipitation in Edinburgh, Scotland and for the daily runoff data of the Weser river at Vlotho, Germany.

4.1 Multifractal Models

There are several models for generating multifractal fluctuations (Feder 1982, Peitgen et al. 1991, Bogachev et al. 2009, Ludescher and Bunde 2014). Here we focus on the phenomenological multiplicative random cascade (MRC) model where the data are obtained in an iterative way, such that the length of the record doubles in each iteration.

Figure 7 illustrates the MRC data generation algorithm. In the 0th iteration ($n = 0$) the data set y_i consists of one value, for example $y_1^{(n=0)} = 1$. In the n-th iteration, the data $y_i^{(n)}$, $i = 1, 2, \ldots, 2^n$, are obtained from

$$y_{2l-1}^{(n)} = y_l^{(n-1)} m_{2l-1}^{(n)} \quad \text{and} \quad y_{2l}^{(n)} = y_l^{(n-1)} m_{2l}^{(n)} \tag{17}$$

where the multipliers m are independent and identically distributed (i.i.d.) random numbers (see Fig. 7(b-d)). The amount of linear and nonlinear memory in the record can be triggered by varying the parameters of the distribution of the multipliers m. The mean $\bar{m}$ and the standard deviation σ_m can be tuned to best represent the multifractal record of interest. Here we have chosen normally distributed multipliers with $\bar{m} = 1$ and $\sigma_m = 0.5$

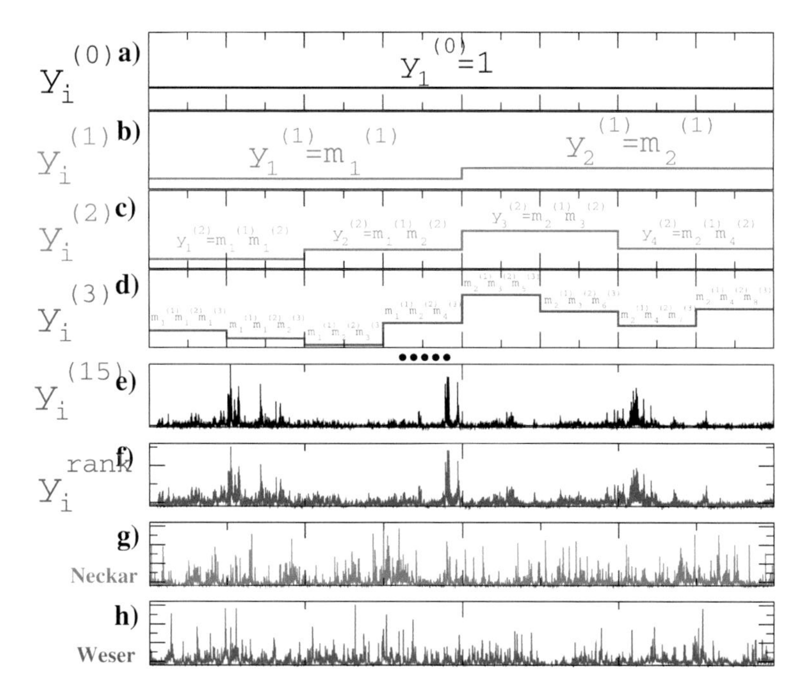

Figure 7: Multiplicative random cascade (MRC) generation algorithm. (a)-(d) The initial iterations (0-3) of the MRC algorithm. (e) A subset of the dataset obtained after 15 iterations exemplified for $\bar{m} = 1$ and $\sigma_m = 0.5$. (f) The same subset after the rankwise exchange of the synthetic data by the representative empirical river runoff data. (g), (h) Two representative runoff records, for the river Neckar at station Plochingen and for the river Weser at station Vlotho. (Figure adapted from Bogachev et al. 2009).

for a close representation of the generalized Hurst exponent distribution typical for the river runoff records.

After $n >> 1$ iterations, according to the central limit theorem, the final distribution of y_i is log-normal (Bogachev et al. 2008), see also Fig. 7(e). To emphasize the visual shape similarity between the multifractal dataset and the observational river runoff data, we next perform a rankwise exchange of the values in the multifractal datasets by the observational river runoff data (see Fig. 7(f)). Visual inspection reveals that the generated dataset contains characteristic bursty patterns closely reminiscent to those in the observational river runoff datasets, with two of their representative examples shown in Fig. 7(g, h).

Figure 8 shows the typical MF-DFA fluctuation functions for the linearly long-term correlated (monofractal) datasets obtained by the Fourier-filtering technique as well as the multifractal datasets generated by MRC. The figure shows that for the monofractal datasets the asymptotic

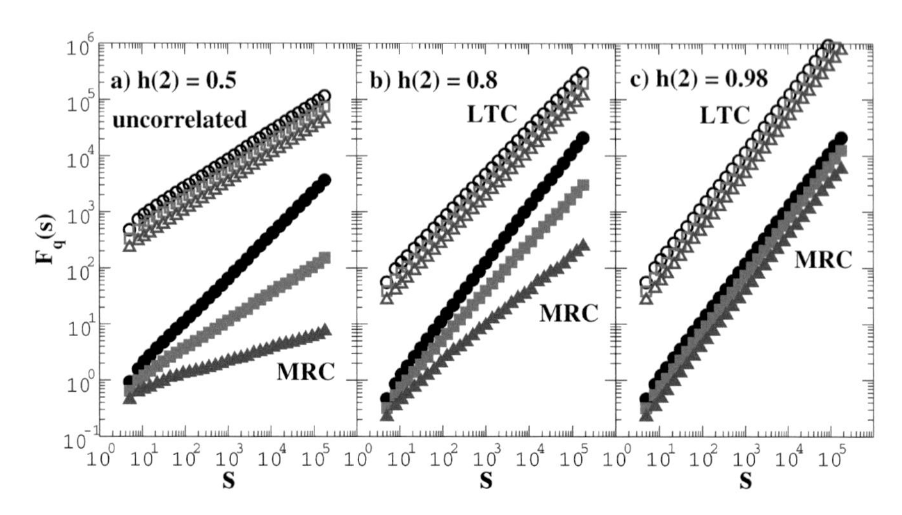

Figure 8: MF-DFA fluctuation functions $F_q(s)$ for linearly long-term correlated (LTC, open symbols) and multifractal datasets (MRC, full symbols) for (a) $h(2) = 0.5$, (b) 0.8 and (c) 0.98, for the moments $q = 1$, 2 and 5. All data are given in arbitrary units; for a better discretion, $F_q(s)$ for LTC and MRC records are separated by a vertical two decade shift. (Figure adapted from Bogachev and Bunde 2011).

$F_q(s)$ obeys the same slope for all q and thus $h(q) \equiv H$ is independent of q. In contrast, pronounced dependence on q can be observed in multifractal records. A prominent example is the MRC with zero mean and unit variance of the multipliers that leads to linearly uncorrelated but strongly nonlinearly dependent data widely used to model returns in financial markets (see Fig. 8(a)). In contrast, by choosing $\bar{m} = 1$ and adjusting σ_m one can create datasets where linear and nonlinear long-term memory are superimposed. Two particular examples for $\sigma_m = 0.5$ and 0.1 are shown in Fig. 8 (b, c). The figure also shows that at $\sigma_m << \bar{m}$ the nonlinear memory is being primarily exchanged by linear memory this way leading to roughly similar fluctuation functions for LTC and MRC in Fig. 8 (c).

5. Clustering of Extreme Events

Next we consider the consequences of linear and non-linear long-term memory on the occurrence of rare extreme events and their predictability. Understanding (and predicting) the occurrence of extreme events is one of the major challenges in science (Bunde et al. 2002). An important quantity here is the time interval between successive extreme events (see Fig. 9), and by understanding the statistics of these return intervals one aims to better understand the occurrence of extreme events.

5.1 Return Intervals

Since extreme events are, by definition, very rare and the statistics of their return intervals is poor, one usually also studies the return intervals between less extreme events, where the data exceed a certain threshold Q and where the statistics is better, and hopes to find some general "scaling" relations between the return intervals at low and high thresholds, which then allows to extrapolate the results to very large, extreme thresholds (see Fig. 9).

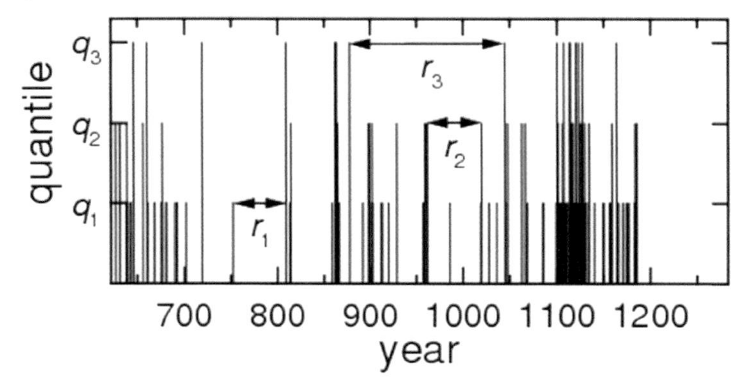

Figure 9: Illustration of the return intervals for three equidistant threshold values q_1, q_2, q_3 for the water levels of the Nile at Roda (near Cairo, Egypt). One return interval for each threshold (quantile) is indicated by arrows.

For uncorrelated data, the return intervals (with mean value R_Q) are independent of each other and their probability density function (pdf) is simply (Bunde and Havlin 1995),

$$P_Q(r) = (1/R_Q)\,(1 - 1/R_Q)^{r-1},\ r = 1, 2, \ldots \qquad (18)$$

which becomes $P_Q \cong (1/R_Q)\exp(-r/R_Q)$ for large R_Q. In this case, all relevant quantities can be derived from the knowledge of the mean return interval R_Q. Since the return intervals are uncorrelated, a sequential ordering cannot occur. There are many cases, however, where some kind of ordering has been observed where the hazardous events cluster, for example in the floods in Central Europe during the middle ages or in the historic water levels of the Nile river which are shown in Fig. 9 for 663y. Even by naked eye one can realize that the events are not distributed randomly in time but are arranged in clusters. A similar clustering was observed for extreme floods, winter storms, and avalanches in Central Europe (see Figs. 4.4, 4.7, 4.10 and 4.13 in (Pfisterer 1998), Fig. 66 in (Glaser 2001), and Fig. 2 in (Mudelsee et al. 2003)). We show below that the reason for this clustering is long-term memory.

5.2 Linear Long-term Correlations

Figure 10a shows $P_Q(r)$ for (monofractal) long-term correlated records with $H = 0.8$ (corresponding to $\gamma = 0.4$), for three values of the mean return interval R_Q (which is easily obtained from the threshold Q and independent of the correlations). Figure 10b is plotted in a scaled way, i.e., $R_Q P_Q(r)$ as a function of r/R_Q. The figure shows that all 3 curves collapse. This is a consequence of the monofractality. Accordingly, when we know the functional form of the PDF for one value of R_Q, we can easily deduce its functional form also for very large R_Q values which due to its poor statistics cannot be obtained directly from the data. This scaling is a very important property, since it allows to make predictions also for rare events which otherwise are not accessible with meaningful statistics. When the data are shuffled (black circles in Figure 10d), the long-term correlations are destroyed and the PDF becomes a simple exponential.

The functional form of the PDF is a quite natural extension of the uncorrelated case. Figure 10c suggests that $\ln P_Q(r) \sim -(r/R_Q)^{\gamma}$, i.e. simple stretched exponential behaviour (Bunde et al. 2003, Bunde et al. 2005). For

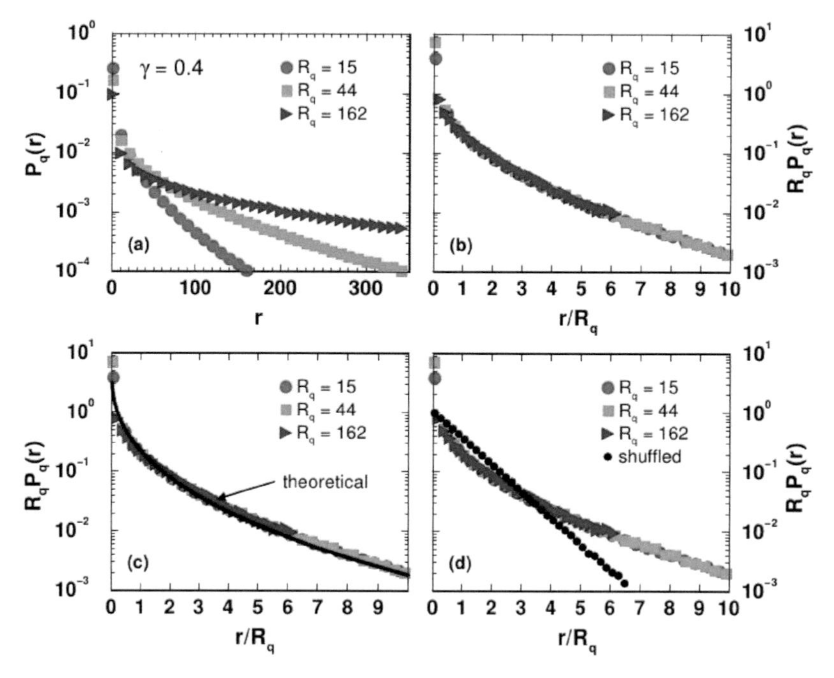

Figure 10: Probability density function $P_Q(r)$ of the return intervals in long-term correlated data with correlation exponent $\gamma = 0.4$, for three return periods R_Q, (a) plotted in the normal way, (b) plotted in a scaled way, (c) compared with a stretched exponential with exponent $\gamma = 0.4$, (full line), and (d) compared with shuffled data (black circles) (after (Bunde et al. 2005)).

γ approaching 1, the long-term correlations tend to vanish and we obtain simple exponential behaviour characteristic for uncorrelated processes. For r well below R_Q, however, there are deviations from the pure stretched exponential behaviour. Closer inspection of the data shows that for $r/R_Q<<1$ the decay of the PDF is characterized by a power law, with the exponent $\gamma - 1$. Accordingly, we have

$$P_Q(r) \sim \begin{cases} e^{-b(r/R_Q)^\gamma} & \text{for} \quad r > R_Q \\ (r/R_Q)^{\gamma-1} & \text{for} \quad r < R_Q. \end{cases} \tag{19}$$

The form of Eq. (19) suggests that $P_Q(r)$ might be written approximately as the negative derivative of a stretched exponential

$$P_Q(r) \cong -\left(\frac{d}{dr}\right)\exp\left[-b_\gamma (r/R_Q)^\gamma\right] = b_\gamma \gamma \left(\frac{r}{R_Q}\right)^{\gamma-1} \exp\left[-b_\gamma \left(\frac{r}{R_Q}\right)^\gamma\right]. \tag{20}$$

The parameter b_γ must be determined from the condition that $R_Q = \int_0^\infty dr\, r\, P_Q(r)$ which yields $b_\gamma = \left[\int_0^\infty dx \exp(-x^\gamma)\right]^{1/\gamma}$ It has been pointed out that this form, while not rigorous, can be used as a reasonable approximation for $P_Q(r)$ (Eichner et al. 2006, Fraedrich and Blender 2008). We will use it later in the next subsection to estimate the hazard function W_Q.

The stretched exponential behaviour is quite universal and does not seem to depend on the way the original data are distributed. In the cases shown here, the data had a Gaussian distribution, but similar results have also been obtained for exponential, power-law and log-normal distributions (Eichner et al. 2006). Indeed, the characteristic stretched exponential behaviour of the PDF can also be seen in long historic and reconstructed climate records (Bunde et al. 2005).

The form of the PDF indicates that return intervals both well below and well above their average value are considerably more frequent for long-term correlated data than for uncorrelated data. The distribution does not quantify, however, the way the intervals between rare events are correlated.

To study this question, (Bunde et al. 2005) and (Eichner et al. 2006) have evaluated the autocorrelation function of the return intervals in synthetic long-term correlated records. They found that the return intervals are arranged in a long-term correlated fashion, with the same exponent as in the original data. Accordingly, a long return interval is more likely to be followed by a long one than by a short one, and a short return interval is more likely to be followed by a short one than by a long one, and this leads to the clustering of extreme events above some threshold Q.

As a consequence of the long-term memory, the probability of finding a certain return interval depends on the preceding interval. This effect can easily be seen in synthetic data sets generated numerically, but not so well in climate records where the statistics is comparatively poor. To improve the statistics, we follow (Bunde et al. 2005) and only distinguish between two kinds of return intervals, "small" ones (below the median) and "large" ones (above the median), and determine the mean R_Q^+ and R_Q^- of those return intervals following a large (+) or a small (–) return interval. Due to scaling, R_Q^+/R_Q and R_Q^-/R_Q are independent of Q. Figure 11 shows both quantities (calculated numerically for long-term correlated Gaussian data) as a function of the correlation exponent γ. The lower dashed line is R_Q^-/R_Q, the upper dashed line is R_Q^+/R_Q. In the limit of vanishing long-term memory, for $\gamma = 1$, both quantities coincide, as expected. Figure 11 also shows R_Q^+/R_Q and R_Q^-/R_Q for 5 climate records with different values of γ. One can see that the data agree very well, within the error bars, with the theoretical curves.

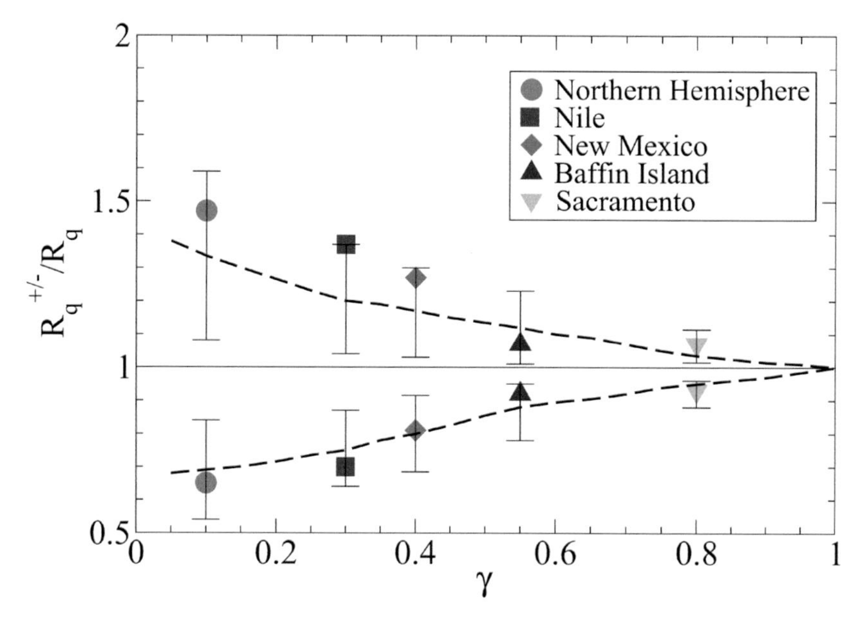

Figure 11: Mean of the (conditional) return intervals that either follow a return interval below the median (lower dashed line) or above the median (upper dashed line), as a function of the correlation exponent γ, for 5 long reconstructed and natural climate records. The theoretical curves are compared with the corresponding values of the climate records (from right to left): The reconstructed run-offs of the Sacramento river, the reconstructed temperatures of Baffin Island, the reconstructed precipitation record of New Mexico, the historic water levels of the Nile and one of the reconstructed temperature records of the Northern hemisphere (after (Bunde et al. 2005)).

5.3 Multifractal Records: Effect of the Nonlinear Long-term Correlations

Figure 12 shows the MF-DFA fluctuation functions $F_q(s)$, the PDFs of return intervals $P_Q(r)$ as well as the conditional averages of return intervals $R_Q(r_0)$ for multifractal datasets generated by MRC. The figure shows that the PDFs of the return intervals for all considered return periods, $R_Q = 10$, 70 and 500 exhibit a pronounced power-law behaviour

$$P_Q(r) \sim (r / R_Q)^{-\delta(Q)}, \ \delta\,(Q) > 1, \tag{21}$$

in marked contrast to the uncorrelated or linearly long-term correlated (monofractal) datasets. The exponent δ (shown in Fig. 12(b)) depends explicitly on the value of R_Q. Accordingly, there is no scaling, and as a consequence, the occurence of extremes cannot be estimated straightforwardly from the occurence of smaller events. When shuffling the data, i.e. destroying any temporal correlations, the scaled $R_Q P_Q(r)$ collapses to a single exponential curve, as expected (also shown in Fig. 12(b)) (Bogachev et al. 2007).

To further quantify the memory among the return intervals, we consider the conditional return intervals, i.e., we regard only those

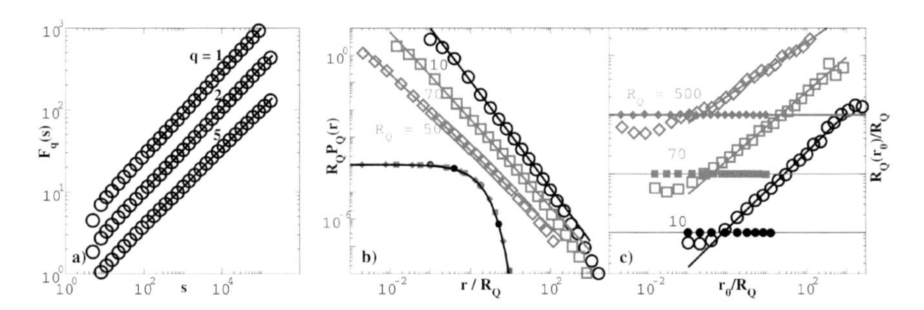

Figure 12: (a) MF-DFA fluctuation functions $F_q(s)$ of the simulated multifractal datasets obtained by MRC model with normally distributed multipliers ($\bar{m} = 0$, $\sigma_m = 1$) after 21 iterations (open symbols). Full lines show power law approximations with exponents $h(q) = 0.52$, 0.5 and 0.47 for $q = 1, 2$ and 5, respectively. (b) PDFs of the return intervals between extreme events for different return periods $R_Q = 10, 70$ and 500. For a distinct view, the data for $R_Q = 70$ and 500 have been shifted downwards by one and two decades, respectively. Full lines show the approximations by power laws $P_Q(r) \propto r^{-\delta}$ with exponents $\delta = 1.95$, 1.6 and 1.4 for $R_Q = 10$, 70 and 500, respectively. Full symbols show the same PDFs for the shuffled datasets (shifted downwards by four decades) that decay by a simple exponential. (c) Conditional averages of return intervals $R_Q(r_0)$ following intervals of a certain size r_0 for the same multifractal records (open symbols) and the same return periods R_Q. Full lines show the approximation of the asymptotic behaviour by power laws $R_Q(r_0) \propto (r_0 / R_Q)^{v}$ with exponents $v = 0.49$, 0.58 and 0.65 for $R_Q = 10$, 70 and 500, respectively. Figure adapted from (Bogachev et al. 2007).

intervals whose preceding interval is of a fixed size r_0. In Fig. 12(c) the conditional return period $R_Q(r_0)$, which is the average of all conditional return intervals for a fixed threshold Q, is plotted versus r_0/R_Q (in units of R_Q). The figure demonstrates that, as a consequence of the memory, large return intervals are rather followed by large ones, and small intervals by small ones. In particular, for r_0 values exceeding the return period R_Q, $R_Q(r_0)$ increases by a power law,

$$R_Q(r_0) \sim r_0^{\,v(Q)} \quad \text{for } r_0 > R_Q, \tag{22}$$

where the exponent v decreases with increasing value of R_Q (Bogachev et al. 2007).

We should note that theoretically the observed laws can be valid for arbitrary arguments only in infinite datasets. In real (finite) records, there exists a maximum return interval which limits the values of r as an argument of $P_Q(r)$ as well as r_0 as an argument of $P_Q(r_0)$. While in long simulated datasets nice scaling laws can still be observed, in short datasets they are largely hindered by pronounced finite size effects.

For an explicit comparison of the return interval statistics in both observational and simulated multifractal records, we next focus on the comparison of the river runoff data with multifractal datasets generated by the MRC with the same parameters $\bar{m} = 1$ and $\sigma_m = 0.5$ but with different data length. The upper panels of Fig. 13 shows (a) the MF-DFA fluctuation functions, (b) PDFs of return intervals and (c) conditional return periods averaged over 20 multifractal datasets of size 2^{20} (all shown by open symbols), with dashed lines characterizing their approximate behaviour. The figure shows that, in contrast to the datasets with vanishing linear correlations, $P_Q(r)$ exhibits pronounced deviations from power law behaviour (Bogachev et al. 2008). Additionally, the lower panels of Fig. 13 show the same characteristics obtained for the observational river runoff record (by open symbols) as well as the corresponding quantities for the simulated multifractal data created by MRC with data length 2^{15} that is comparable with typical observational river runoff record length. Figure 13(d) shows that there are relatively weak deviations of the fluctuation functions at least for $q = 1, 2$ and 5 that mainly occurs at small s indicating the contribution of additional short-term memory.

In contrast, Fig. 13(e) shows that the PDFs of the return intervals in the observational data exhibit prolonged crossovers, this way deviating from the power laws considerably, while for the artificial multifractal datasets of comparable size the expected power law behaviour can still be observed, despite of the inevitable finite size effects. This indicates that the moderate agreement between the fluctuation functions does not guarantee the reproducibility of the return interval statistics in multifractal datasets. The return interval statistics for the observational river runoff data are likely governed by a superposition of several factors such as the complex combination of linear and nonlinear long-term as well as short-term correlations, residual seasonal effects caused by the limited reproducibility of the typical annual patterns that have been subtracted from the data during its preliminary preparation, finite size effects and possibly other secondary issues. Under these circumstances, the particular contribution of each of these effects can hardly be accounted.

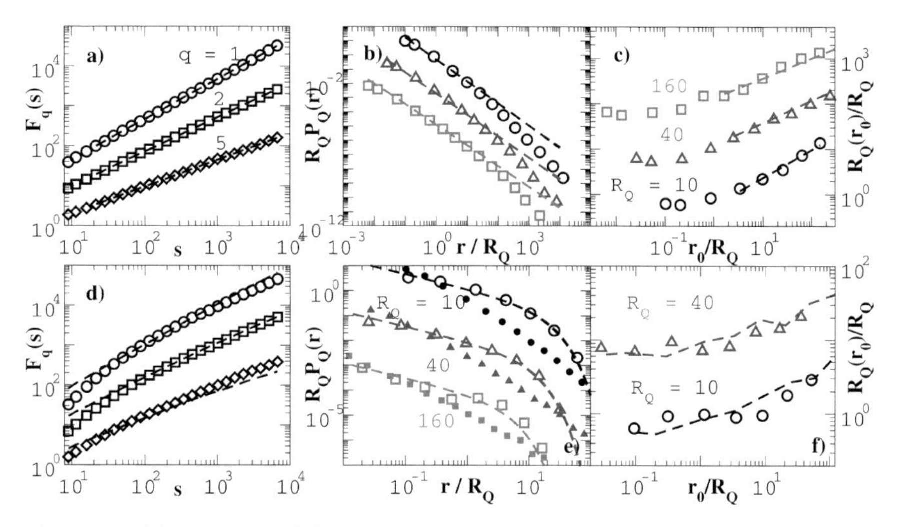

Figure 13: (a) MF-DFA of the simulated multifractal datasets obtained by MRC model with normally distributed multipliers ($\bar{m} = 1$, $\sigma_m = 0.5$) after 20 iterations (open symbols). Dashed lines show power law approximations with exponents $h(q)$ = 1.0, 0.85 and 0.67 for q = 1, 2 and 5, respectively. (b) PDFs of the return intervals between extreme events for different return periods R_Q = 10, 40 and 160. For a distinct view, the data for R_Q = 40 and 160 have been shifted down by two and four decades, respectively. The dashed lines show the approximations by power laws $P_Q(r) \propto r^{-\delta}$ with exponents δ = 2.05, 1.95 and 1.8 for R_Q = 10, 40 and 160, respectively. (c) Conditional averages of return intervals $R_Q(r_0)$ following intervals of a certain size r_0 for the same multifractal records (open symbols) for the same return periods R_Q. The dashed lines show the approximation of the asymptotic behaviour by power laws $R_Q(r_0) \propto (r_0/R_Q)^{v}$ with exponents v = 0.44, 0.49 and 0.61 for R_Q = 10, 40 and 160, respectively. (d) MF-DFA of the daily runoff of the Weser river at Vlotho, Germany (open symbols) and of the (shorter) simulated data obtained by MRC after 15 iterations to obtain comparable short-term effects as in the observational datasets (dashed lines) for q = 1, 2 and 5. (e) PDFs of the return intervals between extreme events in the same river runoff record (open symbols) and for the MRC simulated data (full symbols) for different return periods R_Q = 10, 40 and 160. Dashed lines provide reasonable approximations by a Γ-distribution (Eq. 23). For a distinct view, the data for R_Q = 40 and 160 have been shifted down by two and four decades, respectively. (f) Conditional return periods for the same river runoff record (open symbols) and for the MRC simulated data (dashed lines) for R_Q = 10 and 40. For a distinct view, the data for R_Q = 40 has been shifted upwards by one decade.

In the following, we focus on two possible practical solutions, each providing a reasonable approximation of the observational data. One possible description that takes into account deviations from the power law is its multiplication by an additional exponential term leading to the overall description by a Γ-distribution

$$P_Q(r) \propto \frac{1}{R_Q}\left(\frac{r}{R_Q}\right)^{\alpha-1} \exp\left(-\lambda \frac{r}{R_Q}\right). \tag{23}$$

While the power law dominates and thus determines the behaviour of $P_Q(r)$ at small scales, the exponential term dominates at large scales representing the deviations from the power law with asymptotically exponential behaviour (shown by dashed lines in Fig. 13(e)). The Γ-distribution has been used to also quantify the PDF of interoccurrence times between earthquakes above a certain magnitude Q in quasi-stationary seismic data (Corral 2004). In this case, the α and λ do not depend on R_Q, indicating universal behaviour of the dynamic pattern for all thresholds and seismic areas (Corral 2004).

An alternative solution is provided by the "linearization" approach that is common in applied science. The idea of this approach is to find a reasonable approximation of the (nonlinear) laws governing the behaviour of the complex system by an "effective" linear model with parameters adjusted to best fit the phenomenological observations. For short observational datasets, taking into account the finite size effect, the derivative of the Weibull distribution given by Eq. (20) with an effective exponent γ_{eff} cannot be distinguished from the Gamma-distribution given by Eq. (23). Remarkably, it has been demonstrated earlier in (Lennartz et al. 2008, Lennartz and Bunde 2011) that the seismic data can also be described properly by assuming long-term correlations in the data with an effective correlation exponent.

5.4 Estimating the Risk and Predicting Extreme Events from Long-term Correlations

Next we ask: Given that the last "extreme" event above Q occurred t time steps ago, what is the probability $W_Q(t; \Delta t)$ that in the next Δt time steps an extreme event occurs (see Fig. 14(a))? This probability is of great importance since it specifies, at any time t, the risk associated with an extreme event. It can be easily verified that this probability (which is also called "hazard function"), is related to the probability density function $P_Q(r)$ by

$$W_Q(t; \Delta t) = \int_t^{t+\Delta t} P_Q(r)dr \,/ \int_t^{\infty} P_Q(r)dr \tag{24}$$

The nominator is the probability that an extreme event occurs between t and Δt. The denominator is a normalization factor that ensures $W_Q(t,\infty)$

= 1. In the absence of memory ("white noise"), this function can be obtained analytically: $W_Q(t;\Delta t) = 1 - \exp(-\Delta t/R_Q)$ is independent of how much time has elapsed since the last event. Accordingly, for white noise records, an extreme event can never be overdue (when t exceeds the period R_Q of the event). For the most relevant case $\Delta t << R_Q$, we obtain $W_Q(t;\Delta t) = \Delta t/R_Q$.

For linear long-term correlated records, the approximation of $P_Q(r)$ by Eq. (20) yields

$$W_Q(t;\Delta t) \cong 1-\exp\left[-b_\gamma (t/R_Q)^\gamma\left(\left(1+\frac{\Delta t}{t}\right)^\gamma-1\right)\right] \cong \gamma b_\gamma \frac{\Delta t}{R_Q}\left(\frac{t}{R_Q}\right)^{\gamma-1} \quad (25)$$

for $\Delta t \ll t, R_Q$. For multifractal data, with $P_Q(r)$ following (21), the hazard function can be expressed as

$$W_Q(t;\Delta t) = 1-(1+\Delta t/t)^{1-\delta(Q)} \quad (26)$$

yielding

$$W_Q(t;\Delta t) = (\delta(Q)-1)\frac{\Delta t/R_Q}{t/R_Q} \quad (27)$$

for $\Delta t << t$ (Bogachev et al. 2009). Here, the Q- (and R_Q-) dependence is only in the prefactor.

The typical situation for precipitation and river runoff datasets is that for short and intermediate times $t \le R_Q/\lambda$ the distribution given by Eq. (23) cannot be distinguished from Eq. (20), and thus for $\Delta t \ll t \le R_Q/\lambda$ the hazard function follows Eq. (25) (with $\gamma = \alpha$ and $b = \lambda$).

For long elapsed times $t >> R_Q/\lambda$ the exponential decay dominates the power law. Thus we expect

$$W_Q(t;\Delta t) = \frac{\int_t^{t+\Delta t} P_Q(r)dr}{\int_t^\infty P_Q(r)dr} \cong \lambda\frac{\Delta t}{R_Q} \quad (28)$$

for $\Delta t << R_Q/\lambda << t$.

Figure 14 shows $W_Q(t;1)$ for the Edinburgh precipitation data (left hand side) and for the Neckar river flow data (right hand side). To obtain $W_Q(t;1)$, we fitted the PDFs of the return intervals with a Γ-distribution with $\alpha = \lambda \cong 2/3$ for the precipitation data and $\alpha = \lambda \cong 0.1$ for the runoff data. In both cases, the hazard function decays by a power law for small times t. For the precipitation data ($\alpha \cong 2/3$) the decay is slower than for the river runoff data ($\alpha \cong 0.1$). For long times t the hazard function flattens in both cases. With increasing threshold Q, R_Q also increases and the hazard function decreases as expected. In the lower panels of the figure, the hazard function is plotted for fixed $R_Q = 70$ and $\Delta t = 5$, 10 and 20. As expected, $W_Q(t;\Delta t)$ increases with increasing Δt, and the decrease of the function becomes weaker. For further increasing Δt, $W_Q(t;\Delta t)$ becomes independent of t and approaches 1 by definition.

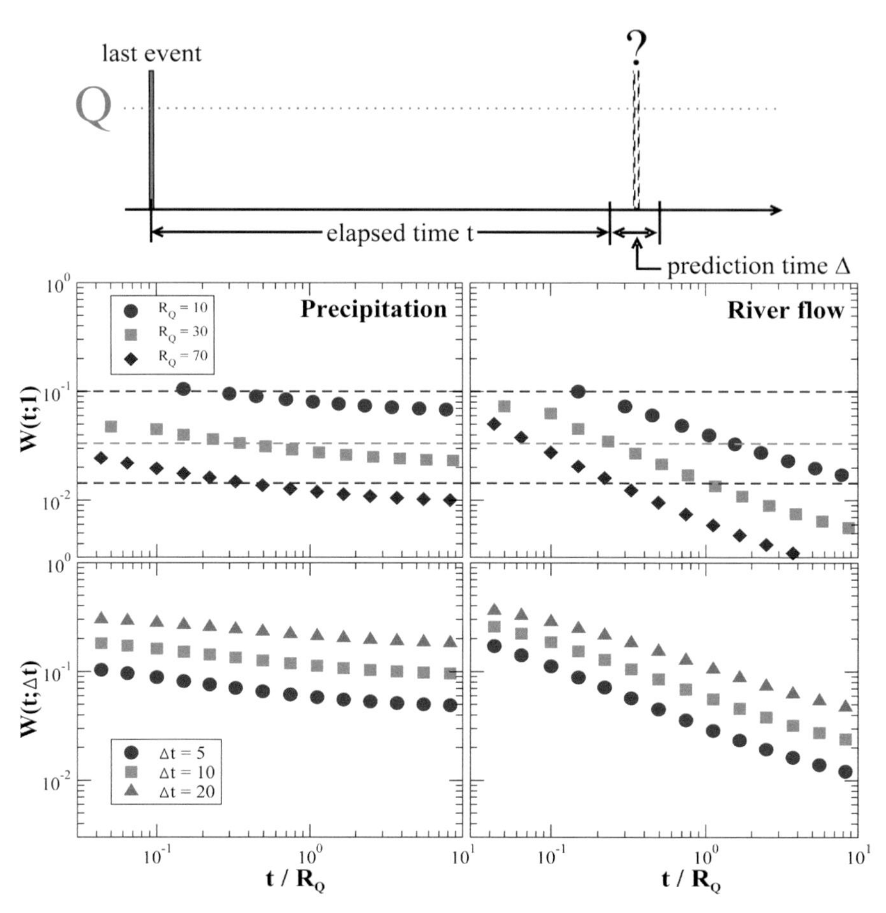

Figure 14: (a) Definition of t and Δt. (b) The hazard function $W_Q(t; 1)$ determined numerically for the gamma distribution with (i) $\alpha = \lambda \cong 2/3$ and $C \cong 0.56$, that characterizes the precipitation data (left panel) and (ii) $\alpha = \lambda \cong 0.1$ and $C \cong 0.12$ that characterizes most of the river runoff data, for $R_Q = 10$, 30 and 70 (right panel). (c) The same function $W_Q(t; \Delta t)$ but for varying $\Delta t = 5$, 10 and 20, using the best fits for precipitation (left panel) and river flow data (right panel), all at $R_Q = 70$. Figure after (Bunde et al. 2012).

Next, we discuss how the hazard function $W_Q(t; 1)$ can be used to predict events above the threshold Q.

Figure 15(a) illustrates, how this can be done in a decision making algorithm. In the algorithm, we set up a decision threshold p_d and activate an alarm whenever $W_Q(t; 1)$ exceeds p_d. In the figure, the (extreme) events above Q are shown in black, while the non-extreme events are shown in grey. The green line shows $W_Q(t; 1)$ and the red horizontal line is the decision threshold p_d. The upper line in Fig. 15(b) marks when an extreme event is correctly predicted, the second line when it is missed, the third line when a false alarm is given, and the last line when a non-extreme event is correctly predicted.

For a certain decision threshold p_d, the efficiency of the algorithm is generally quantified by the hit rate D (number of correct predictions (1st line in Fig. 15(b)) divided by the number of all extremes (sum of 1st and 2nd lines in Fig. 15(b)) and the false alarm rate μ (number of false alarms (3rd line in Fig. 15(b)) divided by the number of all non-extremes (sum of 3rd and 4th lines in Fig. 15(b))). The larger the hit rate D is for fixed false alarm rate μ, the better is the prediction provided by the algorithm. In the example of Fig. 15, the hit rate is $D = 6/9 \cong 0.67$ while the false alarm rate is $\mu = 8/16 = 0.5$.

In general D and μ depend on p_d. The overall quantification of the prediction efficiency is usually obtained from the "receiver operator characteristic" (ROC) analysis (Fawcett 2006, Bogachev and Bunde 2011), where D is plotted versus μ for all p_d between 0 and 1. By definition, for $p_d = 0$, $D = \mu = 1$, while for $p_d = 1$, $D = \mu = 0$. For $0 < p_d < 1$, the ROC curve connects the lower left and upper right corners of the ROC plot (see Fig. 16). If there is no memory in the data, $D = \mu$ and the ROC curve follow the diagonal.

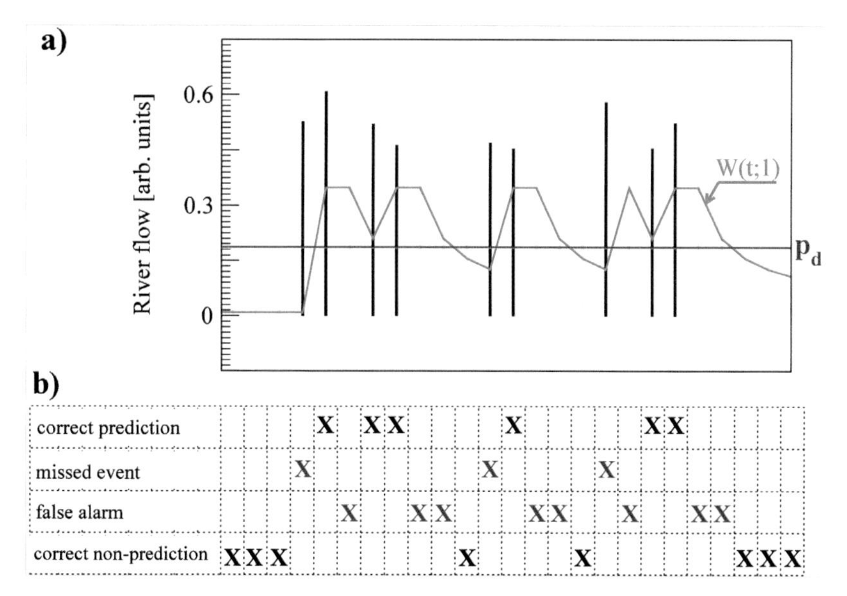

Figure 15: (a) Extreme events above Q are shown in black, while the non-extreme events are shown in grey. The full green line shows $W_Q(t; 1)$. (b) Shows when the actual event is above Q or below, and if this event has been predicted or not. The upper line gives the correct predictions of extreme events, the second line the non-predicted extremes, the third line the false alarms, and the last line the correct predicted non-extremes. Figure after (Bunde et al. 2012).

Figure 16 shows the results of the ROC-analysis for a large number of precipitation and river flow data (see (Bunde et al. 2012)) for $R_Q = 10$ (upper curves) and $R_Q = 70$ (lower curves) for false alarm rates $\mu < 0.5$. The results for the Edinburgh precipitation data and the runoffs of the Neckar river are highlighted. For comparison, the thick dashed line represents the random case. Thin vertical lines mark the 10% false alarm rate, $\mu = 0.1$. As expected, for precipitation data, the hit rate D is only slightly above the random case, due to the absence of pronounced linear and nonlinear memory. For river flow data, in contrast, due to the interplay between linear short and long-term correlations and nonlinear memory, the hit rate is quite high. On an average, for $\mu = 0.1$, the hit rate D is 0.8 for $R_Q = 10$, while for $R_Q = 70$, the hit rate decreases and the average D-value is around 0.7.

In general, it is not possible to separate the contribution of the nonlinear memory from the contribution of the linear memory. To a certain extent, one can estimate the contribution of the nonlinear memory when the linear short- and long-term memory is weak and the nonlinear memory is pronounced. This is the case for the Gaula, Susquehanna and Mary rivers.

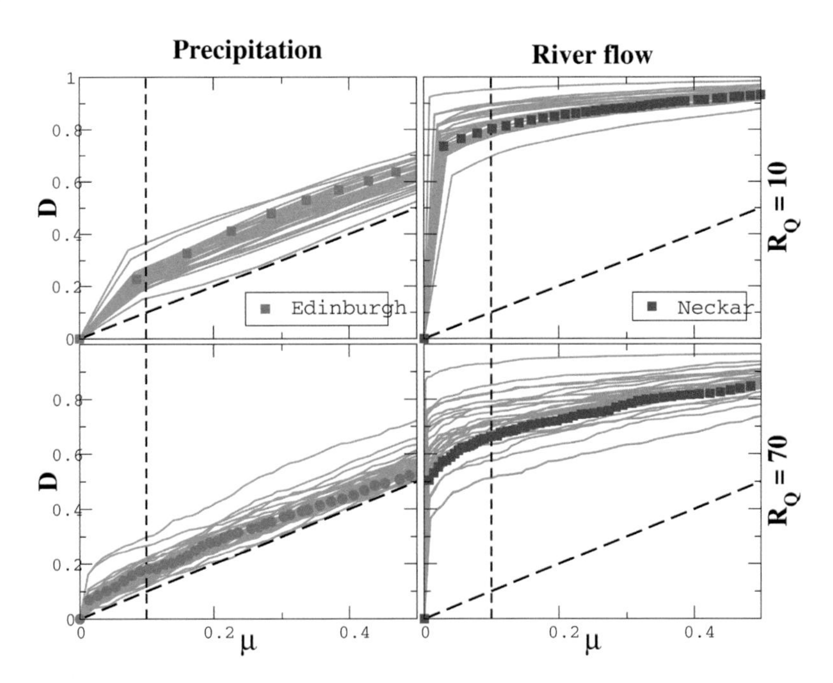

Figure 16: ROC analysis for a large number of precipitation (left hand side) and river runoff (right hand side) data for $R_Q = 10$ (upper curves) and $R_Q = 70$ (lower curves). For details, see (Bunde et al. 2012). For comparison, the thick dashed line represents the random case. Thin vertical lines mark the 10% false alarm rate $\mu = 0.1$. Figure after (Bunde et al. 2012).

In order to estimate the contribution of linear short- and long-term memory, we have determined fluctuation functions $F_2(s)$ and the autocorrelation functions $C(s)$ for these three records. We also determined $F_2(s)$ for synthetic data modeled by an autoregressive process of first order (AR1), $y_{i+1} = e^{-1/\tau} y_i + \eta_i$, where η_i is long-term correlated noise with Hurst exponent $H = 0.65$. We found that a correlation time $\tau = 3$ days models the fluctuation function best. Figure 17 (a, b) shows the $F_2(s)/s^{1/2}$ and the autocorrelation function $C(s)$ both for the river flows and for the synthetic data.

Quite obviously, the model reproduces the linear correlations nicely, on both short and long time scales. Figure 17 (c, d) show the ROC curves for the three rivers and the synthetic data for (c) $R_Q = 30$ and (d) $R_Q = 70$. For $R_Q = 30$ the results for all three rivers are well above the linear model. The highest hit rate is for the Mary river, followed by Susquehanna and Gaula. We know from previous multifractal analysis of these rivers (Koscielny-Bunde et al. 2006) that the strength of the multifractality follows the same order, with Mary river the strongest and Gaula the weakest multifractality.

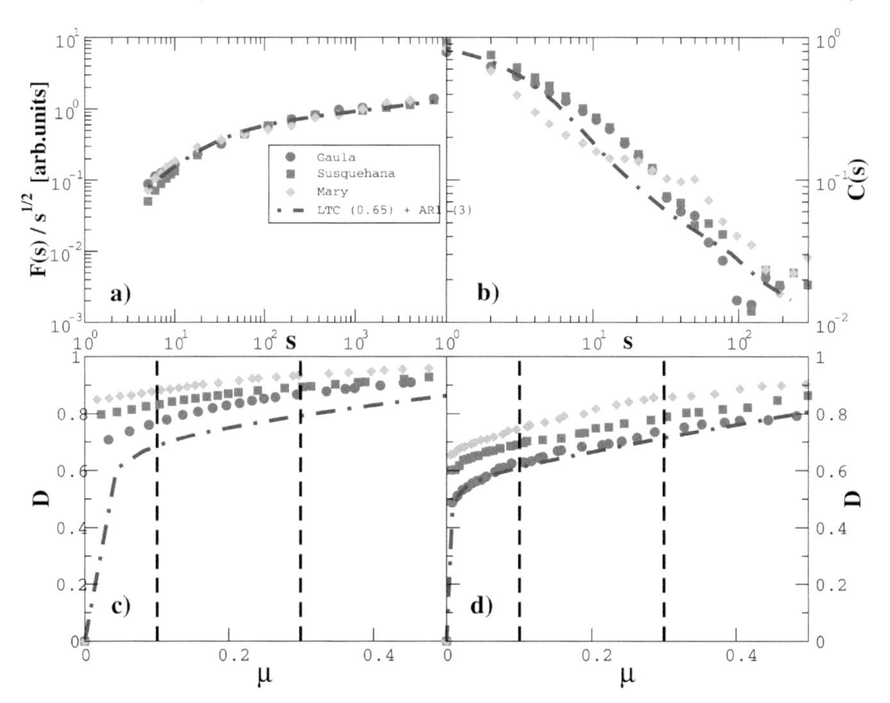

Figure 17: (a) $F_2(s)/s^{1/2}$ for three selected rivers (Gaula, Susquehana and Mary) characterized by low H values and for the fitting model (dashdot line). (b) ACF for the same data records. (c, d) ROC-curves for predictions made in the three selected rivers data (symbols) and for in fitting model data (dashdot lines) for $R_Q = 30$ and 70, respectively. Figure after (Bunde et al. 2012).

This indicates that the deviations of $D(\mu)$ from the linear model are due to nonlinear memory associated with the multifractal behaviour. For $R_Q = 70$ (Fig. 17(d)) all hit rates are lower and the results for the Gaula river are close to the linear model, while the results of the other two rivers are still well above.

5.5 Worldwide Universality of the Return Interval Properties for Hydrometeorological Data Sets

Next, to reveal whether these laws are universal for various hydrometeorological data sets, we have studied 32 daily precipitation records and 32 river runoff records from hydrometeorological stations all around the globe.

Figure 18 shows that in precipitation data for each fixed R_Q there is a reasonable data collapse for all records, indicated by small scattering of the data, except for the smallest scale $r = 1$ day and very large scales $r >> R_Q$ where the statistics is poor. For river runoff data, in contrast, the data

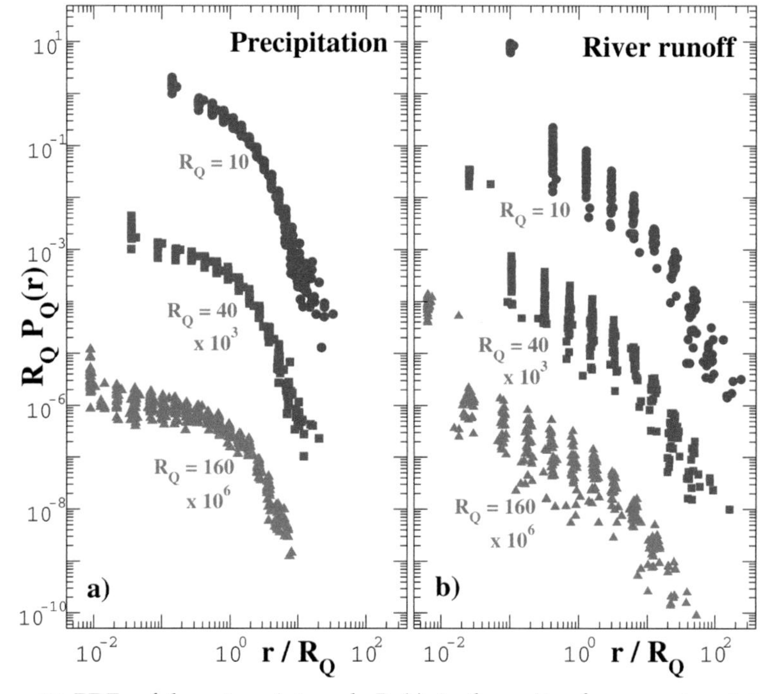

Figure 18: PDFs of the return intervals $P_Q(r)$, in the units of mean return interval R_Q for (a) precipitation and (b) river runoff data at $R_Q = 10$ (circles), 40 (squares) and 160 (triangles), top to bottom. For a better discretion, results for $R_Q = 40$ and 160 have been divided by 10^2 and 10^4, respectively. A significantly higher value at $r = 1$ shows pronounced short-term one-day memory. Figure after (Bogachev and Bunde 2012).

scatter at all scales. At $r = 1$ day, the PDF is about one order of magnitude larger than at $r = 2$ days. This feature is due to short-term memory, which is known to be pronounced in river runoffs and which shows up in the short-term patchiness in Fig. 7 (g, h). We found that this patchiness, that can be quantified by the fraction of unit return intervals, varies strongly from river to river, from 0.3 to 0.5 at $R_Q = 10$ and from 0.1 to 0.8 at $R_Q = 160$ days. Accordingly, since the PDF is normalized, the non-universal fraction of unit return intervals gives rise to the non-universal behaviour of the entire PDF (Bogachev and Bunde 2012). To reduce this unwished short-term memory effect, we now focus (for fixed R_Q) on the return intervals $\tau \geq 2$ and consider as characteristic time $\langle \tau_Q \rangle$ the average over these intervals.

Figure 19 shows the scaled PDF $\langle \tau_Q \rangle P_Q(\tau)$ as a function of $\tau / \langle \tau_Q \rangle$ for the same thresholds as in Fig. 18. One can see clearly that for fixed R_Q, the PDFs for all precipitation data collapse. It is remarkable that very different

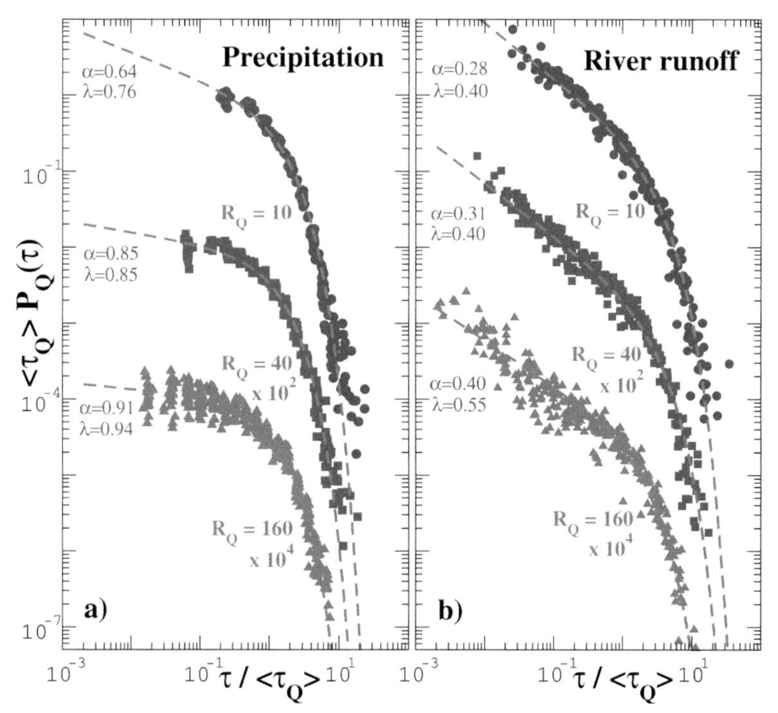

Figure 19: Probability density function $P_Q(\tau)$ of the time spans $\tau \equiv r \geq 2$, in units of the mean time span $\langle \tau_Q \rangle$ for (a) precipitation and (b) river runoff data at $R_Q = 10$ (circles), 40 (squares) and 160 (triangles), top to bottom. For visualisation, results for $R_Q = 40$ and 160 have been divided by 10^2 and 10^4, respectively. Dashed lines show optimal fits by a Γ-distribution (Eq. 23). Parameters α and λ for both kinds of data and for each R_Q are annotated in the figure. Figure after (Bogachev and Bunde 2012).

climate zones (like the Monsoon-governed area around Pusan, Korea and the continental climate zones in Europe and North America) show the same patterns. For the river runoff data, there is also a data collapse for fixed R_Q, indicating that different rivers having different basins, soil and evaporation conditions and being located in different climate zones show the same dynamic pattern. Larger scattering of the data may be due to anthropogenic influences (e.g. power stations and changes of the river bed) which for river runoffs are considerably more pronounced than for precipitation data.

As it is also shown in Fig. 19, the PDFs can be well approximated by the Γ-distribution from Eq. (23), which is plotted as dashed lines in the figure, with parameters α and λ that depend explicitely on the return period R_Q. It seems that for the precipitation data, the PDF converges to a simple exponential for large R_Q. For the river runoff data, we cannot draw conclusions on the limiting value.

Figure 20 also shows that for the precipitation and runoff data the PDFs can be described approximately by the corresponding PDFs of long-term

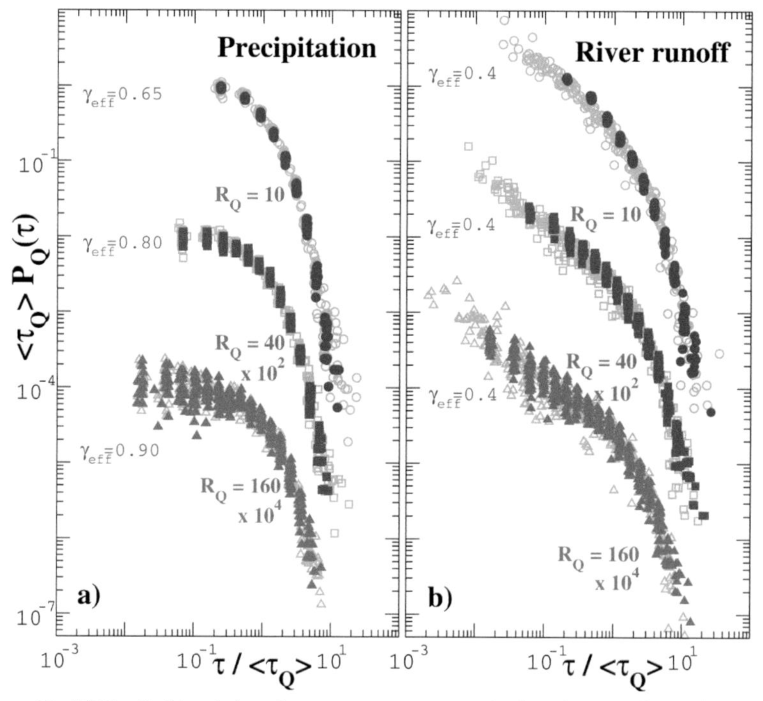

Figure 20: PDFs $P_Q(\tau)$ of the time spans $\tau \equiv r \geq 2$, for the synthetic long-term correlated records with an (effective) correlation exponent γ_{eff} (full symbols) compared to the same PDFs for the observational data (open symbols), (a) for precipitation ($\gamma_{eff} = 0.65$ for $R_Q = 10$, $\gamma_{eff} = 0.8$ for $R_Q = 40$ and $\gamma_{eff} = 0.9$ for $R_Q = 160$); (b) for river runoffs ($\gamma_{eff} = 0.4$ for all R_Q values.)

correlated data of the same length, with a (effective) correlation exponent γ_{eff}. For the precipitation data (Fig. 20 (a)), $\gamma_{eff} = 0.65(R_Q = 10)$, $\gamma_{eff} = 0.8$ ($R_Q = 40$) and $\gamma_{eff} = 0.9(R_Q = 160)$ provide fits that are nearly perfect. Scattering of the synthetic data is comparable with that one for the observational data, and thus can likely be attributed to the finite size effect. For the river runoffs (Fig. 20(b)), the agreement is reasonable, and an effective exponent $\gamma_{eff} \simeq 0.4$ can describe the observational data quite well for all considered return periods R_Q.

Scattering for synthetic and observational data is of the same order of magnitude. In the observational data, the power law regime for small arguments is more prolonged since $\langle \tau \rangle$ is larger for the same R_Q value compared to simulated data (see also Fig. 19) that is the consequence of additional short-term memory in the observational data. In the spirit of the description of the PDFs by an effective correlation exponent γ_{eff}, we can conclude that while the correlation exponent γ varies between 0.8 and 1 for precipitation and between 0.4 and 0.8 for river runoffs, the effective exponents vary between 0.65 and 0.9 for precipitaion and remain nearly unchanged ($\gamma_{eff} = 0.4$) for river runoffs. The discrepancy between γ and γ_{eff} results from the interplay between linear and nonlinear memory in the data, as well as from more pronounced finite size effects in the observational data.

6. Detection of External Trends in Long-term Correlated Records

Next we consider how the strength of an external trend in long-term correlated data sets can be estimated. This question is relevant for the estimation of the effect of anthropogenic global warming in climate data that are known to be long-term correlated. Here we follow (Lennartz and Bunde 2011, Tamazian et al. 2015). For a related method see also (Lennartz and Bunde 2009b). We start with a description of the quantities of interest.

6.1 Quantities of Interest

We consider a record $\{y_i\}$, $i = 1\ldots, L$. To estimate the increase or decrease of the data values in the considered time window of length L, one usually performs a regression analysis. From the regression line $r_i = bi + d$, one obtains the magnitude of the trend $\Delta = b(L - 1)$ as well as the fluctuations around the trend, characterized by the standard deviation $\sigma = [(1/L)\Sigma_{i=1}^{L} (y_i - r_i)^2]^{1/2}$. The relevant quantity we are interested in is the *relative trend*

$$x = \Delta/\sigma. \tag{29}$$

When a certain relative trend has been measured in a data set, the central question is, if this trend may be due to the natural variability of the data set or not ("detection problem", see, e.g. (Rybski et al. 2006)). To solve this problem, one needs to know the probability $P(x; L)dx$ that in the surrogate data with the same persistence properties as the considered data set, a relative trend between x and $x + dx$ occurs. The probability density function $P(x; L)$ is symmetric in x since the time-reversed record (y_L, y_{L-1}, ... y_2, y_1) has the same occurrence probability as the original record (y_1, y_2,... y_N), but opposite trend. In the following we consider $x > 0$. From P we derive the *trend significance*

$$S(x; L) = \int_{-x}^{x} P(x'; L)\, dx' \tag{30}$$

By definition, S is the probability that the relative trend in the record is between $-x$ and x.

The relation $S(x_{95}; L) = 0.95$ defines the upper and lower limits $\pm x_{95}$ of the 95% significance interval (also called confidence interval), see Fig. 21. If the significance of a relative trend x is above 0.95 (or 95%), one usually assumes that the considered trend cannot be fully explained by the natural

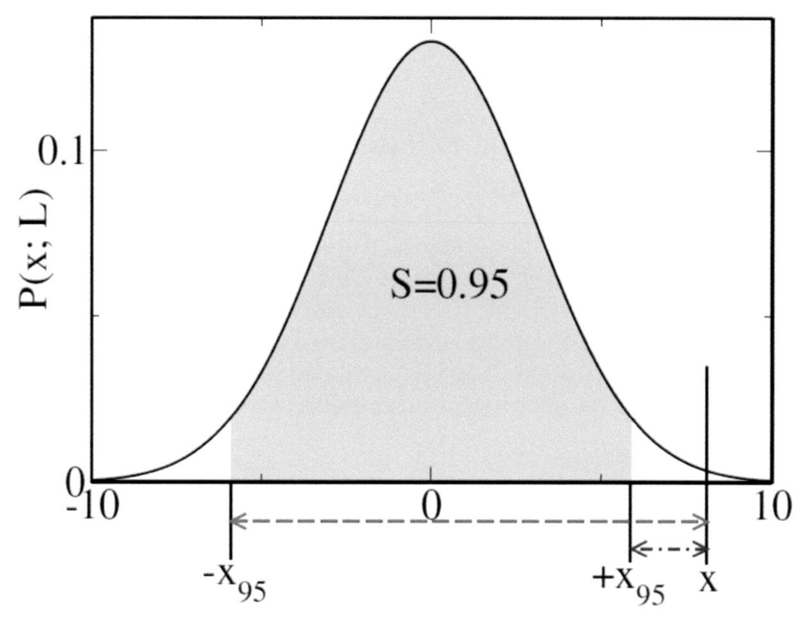

Figure 21: Sketch of the probability density function $P(x; L)$. The interval bounded by $\pm x_{95}$ is the confidence interval defined such that the filled area equals 0.95. Events inside the confidence interval are considered as natural fluctuations. Accordingly, when a certain trend $x > x_{95}$ has been measured in the record of interest, the minimum external relative trend is $x_{ext}^{min} = x - x_{95}$ (short dash-dotted line), while the maximum external relative trend is $x_{ext}^{max} = x + x_{95}$ (long dashed line).

variability of the record. In contrast, relative trends x between $-x_{95}(L)$ and $x_{95}(L)$ are regarded as natural.

If x is above x_{95} (see Fig. 21), the part $x - x_{95}$ cannot be explained by the natural variability of the record and thus can be regarded as minimum external relative trend,

$$x_{ext}^{min} = x - x_{95}. \tag{31}$$

On the other hand, the external trend cannot exceed

$$x_{ext}^{max} = x + x_{95}, \tag{32}$$

which thus represents the maximum external relative trend. By definition, x_{ext}^{min} represents the lower margin of the observed relative trend that cannot be explained by the natural variability alone, while x_{ext}^{max} is the largest possible external relative trend consistent with the natural variability of the record. According to Eqs. (31) and (32), $\pm\, x_{95}\,(L)$ can be regarded as error bars for an external relative trend in a record of length L.

For Gaussian white noise as well as for short-term correlated records generated by an AR(1) process, it has been shown (Bronstein et al. 2004, Santer et al. 2000, Tamazian et al. 2015) that $P(x; L)$ for large record length L follows a Student's t-distribution

$$P(x; L) = \frac{\Gamma\left(\frac{l(L)+1}{2}\right)}{\Gamma\left(\frac{l(L)}{2}\right)\sqrt{\pi l(L)}a}\left(1+\frac{(x/a)^2}{l(L)}\right)^{-\frac{l(L)+1}{2}} \tag{33}$$

with the degrees of freedom $l(L)$ and the scaling parameter $a = \sqrt{12/l(L)}$. Γ denotes the Γ-function. For Gaussian white noise, $l(L) = L - 2$, while for short-term correlated records, $l(L) = L[1 - C(1)]/[1 + C(1)] - 2$.

From Eqs (30) and (33) one can obtain straightforwardly the trend significance S as a function of x/a and $l(L)$,

$$S(x; L) = 2\frac{x}{a}\frac{\Gamma\left(\frac{1}{2}(l(L)+1\right)}{\sqrt{\pi l}\,\Gamma\left(\frac{l(L)}{2}\right)}\;{}_2F_1\left(\frac{1}{2},\frac{1}{2}(l(L)+1);\frac{3}{2};\frac{(x/a^2)}{l(L)}\right) \tag{34}$$

where ${}_2F_1$ is the hypergeometric function.

6.2 Significance of Trends in Long-term Correlated Records

Recently, it was shown (Lennartz and Bunde 2011) by Monte Carlo simulations that in long-term persistent data of length L, where the Hurst exponent H is determined by DFA2, the probability density $P(x; H, L)$ of the relative trend x can be reasonably approximated by a Gaussian for small x and by a simple exponential for large x. Using scaling theory, an analytic expression for $S(x; L)$ has been obtained, as a function of H, in the two x-regimes. The result was applied to a large number of climate records on the globe (Lennartz and Bunde 2011), see also (Ludescher et al. 2015, Yuan et al. 2015, Becker et al. 2014). In a more recent article it was shown (Tamazian et al. 2015) that the best approximation for P in the entire range of x is the Student's t-distribution Eq. (33), where the scaling parameter a depends on both H and L, while the effective length l depends solely on L as $l = a \ln(L)+b$, with $a = 4.05$, $b = -16.65$.

Accordingly, the significance of a relative trend in long-term persistent records is described by the same hypergeometric function as for white noise and AR1 noise, only the parameters l and a are different. Table 1 shows both parameters for the three record lengths $L = 600$, 1200 and 1800, for H values between 0.5 and 1.1. The Table shows that in the considered H range, $l(L)$ does not depend on H. For a more complete Table, with H ranging up to 1.5 and L between 400 and 2200, we refer to (Tamazian et al. 2015).

Table 1: Effective length $l(H, L)$ (left number) and scaling factor $a(H, L)$ (right number) in long term correlated data with (DFA2) Hurst exponent H and record length L

H	L=600	L=1200	L=1800
0.50	9.24/0.133	12.05/0.092	13.69/0.076
0.55	9.24/0.177	12.05/0.126	13.69/0.105
0.60	9.24/0.232	12.05/0.171	13.69/0.145
0.65	9.24/0.300	12.05/0.227	13.69/0.196
0.70	9.24/0.385	12.05/0.301	13.69/0.265
0.75	9.24/0.486	12.05/0.392	13.69/0.351
0.80	9.24/0.606	12.05/0.504	13.69/0.460
0.85	9.24/0.745	12.05/0.639	13.69/0.593
0.90	9.24/0.906	12.05/0.800	13.69/0.752
0.95	9.24/1.088	12.05/0.981	13.69/0.937
1.00	9.24/1.291	12.05/1.187	13.69/1.151
1.05	9.24/1.517	12.05/1.420	13.69/1.390
1.10	9.24/1.776	12.05/1.681	13.69/1.657

6.3 Example: The Antarctic Byrd Record

It is straightforward to apply this methodology to observational data. Important applications are climate data (e.g., river flows, precipitation, and temperature data) where one likes to know the significance of trends due to anthropogenic climate change. When considering climate data, it is important to use monthly data where additional short-term dependencies have been averaged out and seasonal trends can be better eliminated than in daily data (Lennartz and Bunde 2011). To be specific, let us consider monthly temperatures T_i.

First, we subtract the monthly seasonal mean $\langle T_i \rangle$ from the data, i.e. $y_i = T_i - \langle T_i \rangle$. Then a regression analysis is performed to obtain the trend Δ, the standard deviation σ, and the relative trend $x = \Delta/\sigma$. From the DFA2 analysis of the seasonally detrended data we obtain the Hurst exponent H. From x and H we can estimate the significance S of the temperature trend from Table 1 as well as the boundary $\pm \Delta_{95} = x_{95}\sigma$ of the 95% significance interval which represent the error bars of the trend.

For a specific application, we follow (Tamazian et al. 2015) and consider the monthly (corrected) Byrd record between 1957 and 2013 (684 months) that was recently reconstructed by Bromwich et al. (Bromwich et al. 2014). An earlier version (Bromwich et al. 2013) of the Byrd record has been discussed in (Bunde et al. 2014). The Byrd station is located in the center of West Antarctica which is one of the fastest warming places on Earth. It is obvious that the question of the significance of Antarctic warming is highly relevant, since the warming trend influences the melting of the West Antarctic Ice Shelf and thus contributes to future sea level rise.

Figure 22(a) shows the temperature anomalies of the monthly Byrd record. The regression analysis yields $\Delta = 2.015$ °C and $\sigma = 3.09$, leading to $x = 0.654$. Figure 22(b) shows the result of the DFA2 analysis. In the double logarithmic plot, the DFA2 fluctuation function $F(s)$ follows a straight line with exponent $H = 0.65$ between $s = 10$ and $L/4$. Accordingly, the data are long-term persistent and our methodology applies. The exponent $H = 0.65$ is in agreement with earlier estimates (Bunde et al. 2014).

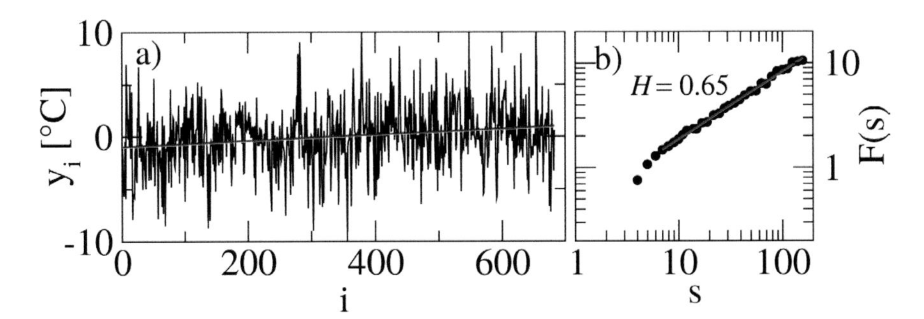

Figure 22: (a, b) Monthly temperature anomalies of the Byrd record between 1957 and 2013 (black lines) and corresponding DFA2 fluctuation function $F(s)$. The red line in (a) is the regression line (after (Tamazian et al. 2015)).

For obtaining the degrees of freedom l and the scaling factor a for $H = 0.65$ and $L = 684$, we consider the 4th line in Table 1 and 2 of (Tamazian et al. 2015) and use the respective values for $L = 500$, 600, 700 and 800 for a cubic interpolation. This gives $l = 9.78$ and $a = 0.286$. Inserting these values into Eq. (34) gives $S = 0.954$ and $x_{95} = 0.639$, leading to $\Delta_{95} = 1.975$ °C. Accordingly, the significance of the warming trend is 95.4%. The minimum external trend is 0.04 °C, while the maximum external trend is 3.99 °C.

When (Bromwich et al. 2013) determined these quantities, they used the (incorrect) conventional hypothesis (Stocker et al. 2013) that the *annual* linearly detrended temperature data follow an AR(1) process. For the annual detrended data, they obtained $C(1) = 0.075$ and thus $l = 49.05$ and $a = 0.546$. Inserting these values into Eq. (34) yields $S = 0.999$ and $x_{95} = 0.997$. As a consequence, the minimum external trend is 0.87 °C, while the maximum external trend is 3.09 °C. Accordingly, by the conventional hypothesis the trend significance and the minimum external trend are strongly overestimated, while the maximum external trend is considerably underestimated. For a more extensive discussion of Antarctic temperature trends, we refer to (Ludescher et al. 2015, Yuan et al. 2015).

7. Conclusions

In this review, we studied fractals and multifractals in climate and hydrological time series and quantified their relation to linear and nonlinear long-term memory. We discussed several methods for detecting linear as well as nonlinear long-term memory, also in the presence of polynomial external trends which is an important task in the context of global warming. We showed explicitely how the occurrence of rare events can be quantified in these records and how this can be effectively used in risk estimation. We also showed how the significance of trends can be estimated in (monofractal) long-term correlated temperature records and applied the method to the temperature data of Central West Antarctica.

To a first order approximation, one may apply this method also to the (multifractal) precipitation data and river flow data, being characterized by nonlinear correlations that, among others, show up in the bursty behaviour of the data sets (Kantelhardt et al. 2002, Kantelhardt et al. 2006, Koscielny-Bunde et al. 2006). Accordingly, the formalism described here to estimate the significance of a trend cannot be rigorously applied to these data sets, and we consider it as an interesting and important challenge to develop a method that allows to estimate the significance of trends also for systems with nonlinear long-term memory.

Acknowledgements

We would like to thank all our coworkers in this field for valuable discussions, in particular Eva Koscielny-Bunde, Shlomo Havlin, Jan Eichner, Jan Kantelhardt, Sabine Lennartz, Josef Ludescher, Diego Rybski, John Schellnhuber and Hans von Storch. M.I.B. would also like to thank the Russian Science Foundation (RSF) grant 16-19-00172 for financial support during work on this review.

REFERENCES

Arneodo, A., B. Audit, N. Decoster, J.-F. Muzy and C. Vaillant. 2002. Wavelet Based Multifractal Formalism: Applications to DNA Sequences, Satellite Images of the Cloud Structure, and Stock Market Data. pp. 27-102. In: The Science of Disasters. A. Bunde, J. Krupp and H.J. Schellnhuber (eds). Springer, Berlin, Heidelberg.

Barabasi, A.-L. and T. Viscek. 1991. Multifractality of self-affine fractals. *Phys. Rev.* A 44: 2730–2733.

Becker, M., M. Karpytchev and S. Lennartz-Sassinek. 2014. Long-term sea level trends: Natural or anthropogenic? *Geophys. Res. Lett.* 41: 5571–5580.

Beretta, A., H.E. Roman, F. Raicich and F. Crisciani 2005. Long-time correlations of sea-level and local atmospheric pressure fluctuations at trieste. *Physica A* 347: 695–703.

Blender, R., C. Raible and F. Lunkeit. 2015. Non-exponential return time distributions for vorticity extremes explained by fractional Poisson processes. *Quart. J. Roy. Meteorol. Soc.* 141: 249–257.

Blender, R. and K. Fraedrich. 2003. Long time memory in global warming simulations. *Geophys. Res. Lett.* 30: 1769.

Blender, R., K. Fraedrich and F. Sienz 2008. Extreme event return times in long-term memory processes near 1/f. *Nonlin. Processes in Geophys.* 15: 557–565.

Bloomfield, P. and D. Nychka. 1992. Climate spectra and detecting climate change. *Climatic Change* 21: 275–287.

Bogachev, M.I., J.F. Eichner and A. Bunde. 2007. Effect of Nonlinear Correlations on the Statistics of Return Intervals in Multifractal Data Sets. *Phys. Rev. Lett.* 99: 240601.

Bogachev, M.I., J.F. Eichner and A. Bunde 2008. On the occurence of extreme events in long-term correlated and multifractal data sets. *Pure Appl. Geophys.* 165: 1195–1207.

Bogachev, M.I., I.S. Kireenkov, E.M. Nifontov and A. Bunde 2009. Statistics of return intervals between long heartbeat intervals and their usability for online prediction of disorders. *New J. Phys.* 11: 063036.

Bogachev, M.I. and A. Bunde. 2009. On the occurrence and predictability of overloads in telecommunication networks. *EPL* 86: 66002.

Bogachev, M.I. and A. Bunde. 2011. On the predictability of extreme events in records with linear and nonlinear long-range memory: Efficiency and noise robustness. *Physica* A390: 2240–2250.

Bogachev, M.I. and A. Bunde. 2012. Universality in the precipitation and river runoff. *EPL* 97: 48011.

Bromwich, D.H. et al. 2013. Central West Antarctica among the most rapidly warming regions on earth. *Nature Geoscience* 6: 139.

Bromwich, D.H. et al. 2014. Reply to 'How significant is West Antarctic warming?' *Nature Geoscience* 7: 247.

Bronstein, I. et al. 2004. Handbook of Mathematics, 4th edition. Springer, Berlin.

Bunde, A., M.I. Bogachev and S.L. Lennartz. 2012. Precipitation and River Flow: Long-Term Memory and Predictability of Extreme Events. In: Complexity and Extreme Events in Geosciences. S. Sharma, A. Bunde, D. Baker and V. Dimri (eds). AGU Geophysical Monograph Series 196: 139–152.

Bunde, A. and S. Havlin (eds.). 1995. Fractals and Disordered Systems. Springer, Berlin.

Bunde, A., S. Havlin, J. Kantelhardt, T. Penzel, J.-H. Peter and K. Voigt. 2000. Correlated and uncorrelated regions in heart-rate fluctuations during sleep. *Phys. Rev. Lett.* 85: 3736.

Bunde, A., J. Kropp and H.J. Schellnhuber (eds.). 2002. The science of disasters – Climate disruptions, heart attacks, and market crashes. Springer, Berlin.

Bunde, A., J. Eichner, S. Havlin and J. Kantelhardt. 2003. The effect of long-term correlations on the statistics of rare events. *Physica* A330: 1.

Bunde, A., J. Eichner, S. Havlin and J. Kantelhardt. 2005. Long-term memory: A natural mechanism for the clustering of extreme events and anomalous residual times in climate records. *Phys. Rev. Lett.* 94: 048701.

Bunde, A., U. Buentgen, J. Ludescher, J. Luterbacher and H. von Storch 2013. Is there memory in precipitation? *Nature Climate Change* 3: 360.

Bunde, A., J. Ludescher, C. Franzke and U. Büntgen. 2014. How significant is West Antarctic warming? *Nature Geoscience* 7: 246–247.

Cohn, T. and H. Lins 2005. Nature's style: Naturally trendy. *Geophys. Res. Lett.* 32: L23402.

Corral, A. 2004. Long-Term Clustering, Scaling, and Universality in the Temporal Occurrence of Earthquakes. *Phys. Res. Lett.* 92: 108501.

Dangendorf, S. et al. 2014. Evidence for long-term memory in sea level. *Geophys. Res. Lett.* 41: 5530–5537.

Eichner, J., J. Kantelhardt, A. Bunde and S. Havlin. 2006. Extreme value statistics in records with long-term persistence. *Phys. Rev.* E93: 016130.

Eichner, J., E. Koscielny-Bunde, A. Bunde S. Havlin and H.J. Schellnhuber. 2003. Power-law persistence and trends in the atmosphere: A detailed study of long temperature records. *Phys. Rev.* E68: 046133.

Fatichi, S., S.M. Barbosa, E. Caporali and M.E. Silva 2009. Deterministic versus stochastic trends: Detection and challenges. *J. Geophys. Res.* 114: D18121.

Fawcett, T. 2006. An introduction to ROC analysis. *Patt. Recogn. Lett.* 27: 861–874.

Feder, J. 1982. The Fractal Geometry of Nature. Freeman, San Francisco.

Feder, J. 1988. Fractals. Springer, NY.

Fraedrich, K. and R. Blender 2003. Scaling of atmosphere and ocean temperature correlations in observations and climate models. *Phys. Rev. Lett.* 90: 108501.

Franzke, C. 2010. Long-range dependence and climate noise characteristics of Antarctic temperature data. *J. Climate* 23: 6074.

Franzke, C. 2012. Nonlinear trends, long-range dependence, and climate noise properties of surface temperature. *J. Climate* 25: 4172–4183.

Franzke, C. 2013. A novel method to test for significant trends in extreme values in serially dependent time series. *Geophys. Res. Lett.* 40: 1391–1395.

Giese, E., I. Mossig, D. Rybski and A. Bunde. 2007. Long-term analysis of air temperature trends in Central Asia. *Erdkunde* 61: 186–202.

Gil-Alana, L.A. 2005. Statistical modeling of the temperatures in the northern hemisphere using fractional integration techniques. *J. Climate* 18: 5357–5369.

Glaser, R. 2001. Klimageschichte Mitteleuropas. Wissenschaftliche Buchgesellschaft, Darmstadt.

Govindan, R., D. Vjushin, S. Brenner, A. Bunde, S. Havlin and H.J. Schellnhuber 2002. Long-range correlations and trends in global climate models: Comparison with real data. *Phys. Rev. Lett.* 89: 028501.

Halley, J. 2009. Using models with long-term persistence to interpret the rapid increase of earth's temperature. *Physica* A388: 2492–2502.

Hurst, H.E. 1951. Long-term storage capacity of reservoirs. *Trans. Am. Soc. Civil Eng.* 116: 770–808.

Kantelhardt, J., E. Koscielny-Bunde, H.H.A. Rego, S. Havlin and A. Bunde. 2001. Detecting long-range correlations with detrended fluctuation analysis. *Physica* A295: 441–454.

Kantelhardt, J., S. Zschiegner, E. Koscielny-Bunde, A. Bunde, S. Havlin and H. Stanley. 2002. Multifractal detrended fluctuation analysis of nonstationary time series. *Physica* A316: 87–114.

Kantelhardt, J.W., E. Koscielny-Bunde, D. Rybski, P. Braun, A. Bunde and S. Havlin. 2006. Long-term persistence and multifractality of precipitation and river runoff records. *J. Geophys. Res. Atmosphere* 111: D01106.

Király, A., I. Bartos and I. Jánosi. 2006. Correlation properties of daily temperature anomalies over land. *Tellus* A5: 593.

Koscielny-Bunde, E., A. Bunde, S. Havlin and Y. Goldreich 1996. Analysis of daily temperature fluctuations. *Physica* A231: 393–396.

Koscielny-Bunde, E., A. Bunde, S. Havlin, H.E. Roman, Y. Goldreich and H.J. Schellnhuber. 1998. Indication of a universal persistence law governing atmospheric variability. *Phys. Rev. Lett.* 81: 729.

Koscielny-Bunde, E., J.W. Kantelhardt, P. Braun, A. Bunde and S. Havlin 2006. Long-term persistence and multifractality of river runoff records: Detrended fluctuation studies. *J. Hydrol.* 322: 120–137.

Koutsoyiannis, D. 2003. Climate change, the hurst phenomenon, and hydrological statistics. *Hydrological Sciences J.* 48: 3–24.

Koutsoyiannis, D. 2006. A toy model of climatic variability with scaling behaviour. *J. Hydrology* 322: 25–48.

Lennartz, S., A. Bunde and D.L. Turcotte. 2008. Missing data in aftershock sequences: Explaining the deviations from scaling laws. *Phys. Rev.* E78: 041115.

Lennartz, S. and A. Bunde. 2009a. Eliminating finite-size effects and detecting the amount of white noise in short records with long-term memory. *Phys. Rev.* E79: 066101.

Lennartz, S. and A. Bunde 2009b. Trend evaluation in records with long-term memory: Application to global warming. *Geophys. Res. Lett.* 36: L16706.

Lennartz, S. and A. Bunde. 2011. Distribution of natural trends in long-term correlated records: A scaling approach. *Phys. Rev.* E84: 021129.

Livina, V., Y. Ashkenazy, P. Braun, A. Monetti, A. Bunde and S. Havlin. 2003. Nonlinear volatility of river flux fluctuations. *Phys. Rev.* E67: 042101.

Lovejoy, S. and D. Schertzer 2013. The Weather and Climate: Emergent Laws and Multifractal Cascades. Cambridge University Press, Cambridge.

Ludescher, J. and A. Bunde. 2014. Universal behavior of the interoccurrence times between losses in financial markets: Independence of the time resolution. *Phys. Rev.* E90: 062809.

Ludescher, J., A. Bunde, C. Franzke and H.J. Schellnhuber. 2015. Long-term persistence enhances uncertainty about anthropogenic warming of Antarctica. *Clim. Dyn.* 46: 263–271.

Lux, T. and M. Ausloos. 2002. In: Bunde, A., J. Kropp and H.J. Schellnhuber (eds.). The science of disasters – Climate disruptions, heart attacks and market crashes. Springer, Berlin.

Mudelsee, M., G.T. Börngen and U. Grünwald. 2003. No upward trends in the occurrence of extreme floods in Central Europe. *Nature* 425: 166–169.

Malamud, B.D. and D.L. Turcotte. 1999. Long-range persistence in geophysical time series. *Advances in Geophysics*, 40: 1.

Mandelbrot, B.B. 2002. Gaussian Self-Affinity and Fractals. Springer, New York, Berlin, Heidelberg.

Mandelbrot, B.B. and J.R. Wallis. 1968. Noah, Joseph, and operational hydrology. *Water Resources Research* 4: 909–918.

Monetti, R., S. Havlin and A. Bunde. 2003. Long-term persistence in the sea surface temperature fluctuations. *Physica* A320: 581–589.

Montanari, A. 2003. Long-range dependence in hydrology. In: Doukhan et al. (eds). Theory and Application of Long-range Dependence. pp. 461–472. Birkhäuser, Basel.

Montanari, A., R. Rosso and M.S. Taqqu. 2000. A seasonal fractional ARIMA model applied to the nile river monthly flows at Aswan. *Wat. Resources Res.* 36: 1249–1259.

Mudelsee, M. 2007. Long memory of rivers from spatial aggregation. *Wat. Resources Res.* 43: W01202.

Muzy, M., E. Bacry and A. Arneodo. 1991. Wavelets and multifractal formalism for singular signals: Application to turbulence data. *Phys. Rev. Lett.* 67: 3515–3518.

Peitgen, H.-O., H. Jürgens and D. Saupe 1991. Chaos and Fractals. Springer, Heidelberg.

Pelletier, J.D. and D.L. Turcotte. 1997. Long-range persistence in climatological and hydrological time series: Analysis, modeling and application to drought hazard assessment. *J. Hydrol.* 203: 198–208.

Peng, C.-K., J. Mietus, J. Hausdorff, S. Havlin, H. Stanley and A. Goldberger. 1993. Long-range anticorrelations and non-Gaussian behavior of the heartbeat. *Phys. Rev. Lett.* 70: 1343–1346.

Pfisterer, C. 1998. Wetternachhersage: 500 Jahre Klimavariationen und Naturkatastrophen 1496–1995. Verlag Paul Haupt, Bern.

Rybski, D. and A. Bunde 2009. On the detection of trends in long-term correlated records. *Physica* A388: 1687–1695.

Rybski, D., A. Bunde, S. Havlin and H. von Storch. 2006. Long-term persistence in climate and the detection problem. *Geophys. Res. Lett.* 33: L06718.

Rybski, D., A. Bunde and H. von Storch. 2008. Long-term memory in 1000-year simulated temperature records. *J. Geophys. Res. Atmosph.* 113: D02106.

Santer, B.D. et al. 2000. Statistical significance of trends and trend differences in layer-average atmospheric temperature time series. *J. Geophys. Res. Atmosph.* 105: 7337–7356.

Santhanam, M. and H. Kantz. 2005. Long-range correlations and rare events in boundary layer wind fields. *Physica* A345: 713–721.

Stocker, T.F. et al. (eds). 2013. Climate Change 2013: The Physical Science Basis. Contribution of Working Group I to the Fifth Assessment Report of the Intergovernmental Panel on Climate Change. Cambridge University Press, Cambridge.

Talkner, P. and R. Weber 2000. Power spectrum and detrended fluctuation analysis: Application to daily temperatures. *Phys. Rev.* E62: 150.

Tamazian, A., J. Ludescher and A. Bunde. 2015. Significance of trends in long-term correlated records. *Phys. Rev.* E91: 032806.

Tessier, Y., S. Lovejoy, B. Hubert, D. Schertzer and S. Pecknold. 1996. Multifractal analysis and modeling of rainfall and river flows and scaling, causal transfer functions. *J. Geophys. Res. Atmosph.* 101: 26427–26440.

Turcotte, D. 1997. Fractals and Chaos in Geology and Geophysics 2nd ed. Cambridge University Press, Cambridge.

Vjushin, D., I. Zhidkov, S. Brenner, S. Havlin and A. Bunde. 2004. Volcanic forcing improves atmosphere-ocean coupled general circulation model scaling performance. *Geophys. Res. Lett.* 31: L10206.

Weber, R. and P. Talkner 2001. Spectra and correlations of climate data from days to decades. *J. Geophys. Res.* 106: 20131–20144.

Yuan, N., Z. Fu and S. Liu. 2014. Extracting climate memory using fractional integrated statistical model: A new perspective on climate prediction. *Scientific Reports* 4: 6577.

Yuan, N. et al. 2015. On the long-term climate memory in the surface air temperature records over Antarctica: A non-negligible factor for trend evaluation. *J. Climate* 28: 5922–5934.

Zorita, E., T. Stocker and H. v. Storch. 2008. How unusual is the recent series of warm years? *Geophys. Res. Lett.* 35: L24706.

CHAPTER

10

Multi-fractal Random Walk and Its Application in Petro-physical Quantities

Gholamreza Jafari[1*], Soheil Vasheghani Farahani[2], Zahra Koohi Lai[3] and Seyed Mohammad Sadegh Movahed[1]

[1] Department of Physics, Shahid Beheshti University, G.C., Evin, Tehran 19839, Iran

[2] Department of Physics, Tafresh University, Tafresh 39518 79611, Iran

[3] Department of Physics, Firoozkooh Branch, Islamic Azad University, Firoozkooh, Iran

1. Introduction

In order to produce an increase in the oil and gas production from rock reservoirs, it is essential to study the geological structure of large-scale porous media, which are at some length scales highly heterogeneous. In addition, further studying and modeling porous media requires investigating the flow of unwanted industrial constituents (known as leakage) in groundwater and accordingly capturing the distribution of the petro physical characteristics (e.g., permeability) in the porous media. However, due to the lack of sufficient data and their non-stationary characteristics, the results obtained by most models are vague. The reason that the data is non-stationary is because of the fluctuations of the data over the long length-scale. Note that in characterizing laboratory-scale data this problem is not faced (Muller 1992, Adler 1992, Adler 1999, Sahimi 1993).

For a large-scale porous medium (e.g., a reservoir) the spatial distribution of the porosity together with the spatial distribution of the permeability must be either measured directly or determined indirectly.

*Corresponding author: g_jafari@sbu.ac.ir

The porosity at different depths and locations is routinely measured in the well-logging process (Jensen et al. 2000), while the permeability distributions could be found either by laboratory measurements (Sahimi 1993, Jensen et al. 2000) or by in situ nuclear magnetic resonance (Mair et al. 1999). The method described here aims to analyze the porosity logs, which also works for the permeability distributions, temperature, gamma-ray, and resistivity logs, in addition to stochastic time series.

Here we analyze some properties of the porosity of rocks based on gamma radiation, sound transmission, and neutron porosity. These quantities are continuously plotted versus the well depth. The resulting curves are referred to as well-logging of the petro-physical quantities in terms of the American Petroleum Institute (API). Well-logging is the process of recording physical, chemical, and electrical properties of the rocks or other fluids that constitute the structure of a well. In oil well-logging, the measured physical properties of the compounds are shown on one axis while the depth of the well is shown on the other axis of the graph.

Studying well-logs is useful in the evaluation of oil production and resources, particularly by characterizing the physical properties of the material around the wells that are obtained by physically plotting the well on a map. Well-logging methods have recently been greatly developed. In addition to their applications in petroleum engineering, they are used for evaluating groundwater and water wells, mining engineering, rock mechanics, and environmental and geophysical studies.

The nonlinear dynamics of geological strata due to external forces and internal instabilities fall into the category of complex systems. Experimental observations show that petro-physical data are of non-Gaussian form due to the existence of fat tails in the probability density function. The reason for this behavior relates to the large fluctuations in the system at extreme values. To analyze petro-physical parameters, some techniques have been invoked in the field of statistical physics, namely; physics of fractals (Muller 1992, Xie et al. 2007), percolation algorithm (Andrade et al. 1995, King et al. 1999), artificial intelligence (Corso et al. 2003, Christie et al. 2006), Tsalis entropy (Koohi Lai et al. 2012) and light scattering-intensity fluctuations (Shayeganfar et al. 2009, 2010, 2013).

An obvious consequence of a non-Gaussian behavior is the existence of multi-fractal properties in the data. In order to model non-Gaussian petro-physical data, the multi-fractal random walk (MRW) method may be implemented, which takes into account the correlation between local variances in the time series. So, at the location of the reservoir, non-Gaussian characteristics and as a result multi-fractal features are observed in the data, which can be studied using the MRW approach. On the other hand, reciprocal correlations between the petro-physical series exist which is due to the existence of porosity in the location where hydrocarbons and water accumulate, and is studied by the NP and STT series affected by inorganic and organic content, particularly in shale. Layers composed of various types of rocks and heterogeneity in the reservoir are detected by the GR series. But there is another issue in these petro-physical data, that is the inter-dependency (correlation) among these data, which appear in the data set as reciprocal correlations. This leads us to modify the MRW approach to the joint MRW approach in order to obtain clearer information. We shall see for example that shale layers in the reservoir create negative correlations between the large fluctuations of the Gamma radiation series and the neutron porosity, and positive correlations between the Gamma radiation and the sound transmitting time.

2. Well Log Indicators

The quality of each well is evaluated by indicators that are mostly defined by the porosity of the medium. Since in this section the aim is to provide a technical analysis of the data, we first define three applied indicators (i.e., measured natural gamma, neutron porosity logs and sonic logging) before showing how the multi fractal random walk approach works in this context.

2.1 Neutron Porosity Logs

The use of neutron-detected porosity in well-logging has a long history. The initial application of neutrons was to characterize the porous structures in the environment. Nowadays, in addition to porosity, neutrons are used as pulses in order to find their absorption rate in structures, where after performing spectroscopy, information about the pore space and solid matrix are obtained. The neutron sources are the consequence of precise nuclear reactions. When the structure is bombarded with neutrons several different reactions can occur between neutrons and the nuclei of atoms which make this a practical technique in well-logging.

The maximum deceleration and energy of neutrons takes place when the mass of the nucleus and the neutron are equal. Among all of the elements, hydrogen has the greatest effect in the neutron speed reduction, i.e., a factor ten greater than any other atoms. Hydrogen is highly relevant in the context of oil and its derivatives. Any void space filled by water contains hydrogen. The deceleration phase is highly dependent on the hydrogen density, which is very high in organic material like oil. Therefore, hydrogen could be a good indicator for the detection of oil reservoirs and the quality of wells.

2.2 Sonic Logging

In an audio apparatus one transmitter and one or two receivers are used. The transmitter, which is influenced by an electromagnetic field controlled from the surface, creates a sound wave with a frequency between 20 and 40 kHz. Sonic-logging, in simple words, records the time needed for sound to pass through one foot depth of the structure, named as the transmission time (Δt), inversely proportional to the speed. The transmission time for a porous rock depends on its solid matrix and void space. Sound propagation is a complex phenomenon and is influenced by the mechanical properties of several distinct sound fields that include: earth column, structure, and the device itself. The sound emitted from the transmitter hits the wall of the well and creates longitudinal and transverse waves, which penetrate into the structure. These waves are surface and body waves that propagate in the walls of the well.

Factors affecting the speed of sound are discussed as following.

2.2.1 Solid Matrix

The speed of sound depends on the type of minerals such as quartz, dolomite, and feldspar. The effect of minerals on speed depends on the density and elastic properties of the minerals. Δt is routinely measured for some minerals. In the case of complex rocks, the effect of each mineral depends on the speed of sound and the volume percentage of minerals in the solid matrix.

2.2.2 Fluids and Porosity

The speed of sound waves is dependent on the value of the medium porosity and the kind of fluid in the pores. A higher porosity and lower density of fluid typically means a lower sound speed. Generally speaking, the speed of sound is higher in water than in oil and in oil is higher than in gas. So the speed of sound is worth analysis to obtain useful information about the properties of well-log.

2.2.3 Grain Arrangement

The arrangement and placement of the grains and pores affects the speed of sonic waves. It has been shown that the type, size and distribution of pores (inter-granular, voids and fractures) all influence the speed of sound. The sound wave length fits in a specific range. This implies that there is a specific size for the pore spaces such that shorter wavelength sound waves would enter the rock and be influenced by the elastic properties of the rock. The limiting porous spaces are directly proportional to the sound wave-length. This means that by adjusting the wave-length, we could extract information on the grain arrangement of the porosity. This point helps better understand the continuous porosity phase at different scales. In low-porosity structures, the void space is more or less isolated and is distributed randomly. If the voids are mostly separate from each other, the structure of the medium is in the continuous phase. This is logical since the fastest wave would propagate in this phase and avoid the voids. In contrast, if the voids are connected to each other, as seen in an uncondensed shale or shallow sand (over 48 to 50 percent porosity), it is the fluid that makes a continuous phase, and what is measured is the propagation time of sound in the fluid.

2.3 Measured Natural Gamma

Almost all of the gamma existing on Earth is produced by the decay of three radioactive elements namely, Uranium, Thorium and Potassium. Clay and shale are known to contain some amounts of thorium, uranium and potassium. In addition, shale contains large amounts of phosphates, organic materials (enriched uranium) and radioactive minerals (feldspar, mica and heavy minerals). The clay and shale are nicely arranged in the porous earth, and therefore the measured gamma rays somehow provide an indication of the amount of shale and clay, and hence the porosity of the earth and well.

The intensity of the gamma rays is not only a function of the radioactivity and density of the rock, but also a function of well characteristics (e.g., diameter), casing and tubing, size and location of the device. The reason is that all of the listed items are located between the receiver and the porous medium, which changes the amount of gamma rays that are detected.

The existence of shale in hydrocarbon reservoirs highly influences the estimation of the supply and production capacity of the reservoir. The presence of clay in the shale makes it difficult to estimate the porosity and saturation of the reservoir. The existence of even a small amount of clay in the medium strongly affects the permeability. Without knowing enough about the type of clay minerals in the rock, the use of inappropriate mud-drilling may cause the permeability of the structure to be severely affected.

Clay has an impact on all log measurements. But how is clay different from the other parts of the rock? Clay has hydrogen, and therefore affects the measurement of neutron porosity. Whenever the neutron beam intensity damps, a change in the environment is faced. This is due to the existence of hydrogen. The reason for this deduction is that hydrogen is a strong damper of the neutron beam intensity. Now where does the hydrogen come from? The answer is in oil components or water located in the porosity of clay.

3. Univariate Multi-fractal Model

Consider a spatially (h) dependent stochastic process (x) represented by $x(h)$, where at space lag l its increment or in other words its fluctuations, $\Delta_1 x(h)$, is defined as

$$\Delta_l x(h) = x(h + l) - x(h) \tag{1}$$

Before proceeding let us discuss some facts. A one-dimensional stochastic process studied with fractal calculus ($x(h)$) is actually configured by its exponents $\xi(q)$ that directly scales the power-law behavior of the absolute moments, m, of the fluctuations (Kiyono et al. 2007) where we have

$$m(q, l) = K_q l^{\xi q}. \tag{2}$$

Now if one takes

$$m(q, l) = \sum_h | x(h+l) - x(h) |^q, \tag{3}$$

the type of fractality can be distinguished. In particular, if the exponent ξ_q linearly depends on q, $x(h)$ would be mono-fractal, while if the exponent ξ_q is nonlinearly dependent on q then $x(h)$ would be multi-fractal (Bacry et al. 2001, Brachet et al. 2000, Sornette 2009). Our focus here is on the multi-fractal aspects of stochastic behaviors. An approach very suitable for studying multi-fractal stochastic processes is based on the cascade model. As a matter of fact Muzy et al. (2000) evolved cascade model counts as a multi-fractal random walk (Muzy et al. 2000, 2001).

In this approach we focus on the probability distribution and assume that the cascading process starts from a large scale, which eventually tends to small scales. Thus, there would be a relationship between the increment of the indicator $\Delta_l x(h) = x(h + l) - x(h)$, where $x(h)$ could be the value of any indicator in well-log. Since these indicators are non-Gaussian, we model this non-Gaussianity by a white noise $\xi_l(h)$ and a noise whose logarithm is white noise $e^{\omega_l(h)}$. In a multi-fractal random walk (MRW) the increment of fluctuation is represented by

$$\Delta_l x(h) \equiv \xi_l (h)\, e^{\omega_l(h)} \tag{4}$$

where $\xi_l(h)$ and $\omega_l(h)$ are independent Gaussian distributions having means equal to zero and variances $\sigma^2(l)$ and $\lambda^2(l)$, respectively (Bacry et al. 2001, Castaing et al. 1990, Arneodo et al. 1997). This multiplication ensures that we have a non-Gaussian process. The variance of the $\omega_l(h)$ process ($\lambda^2(l)$) controls the non-Gaussianity.

3.1 The Non-Gaussian Probability Density Function

In order to study the multi-fractal random walk characteristic of a variable, we must first introduce the working function. The multi-fractal characteristic of a stochastic field is due to the existence of long-range correlations in the field. These long-range correlations may be described by the concept of non-Gaussianity. Therefore, we proceed by understanding the form of the probability density function and its dependence on its width, which is represented by the parameter $\lambda^2(l)$ (Kantelhardt et al. 2002, Grahovac and Leonenko 2014). Consider a non-Gaussian probability density function (PDF), $P_l(\Delta_l x)$, with heavy tails as (Castaing et al. 1990)

$$P_l(\Delta_l x) = \int G_l(\ln \sigma_l) \frac{1}{\sigma_l} F_l(\frac{\Delta_l x}{\sigma_l}) d \ln \sigma_l, \tag{5}$$

in which

$$G_l (\ln \sigma_l) = \frac{1}{\sqrt{2\pi}\lambda(l)} \exp\left(-\frac{\ln^2(\sigma_l)}{2\lambda^2(l)}\right), \tag{6}$$

and

$$F_l\left(\frac{\Delta_l x}{\sigma_l}\right) = \frac{1}{\sqrt{2\pi}} \exp\left(-\frac{\Delta_l x^2}{2\sigma^2(l)}\right). \tag{7}$$

As a special case, if $\lambda^2(l)$ tends to zero, the probability density function, $P_l(\Delta_l x)$, converges to a Gaussian function. The parameter $\lambda^2(l)$ thus measures the magnitude of the non-Gaussianity. As the tail of the profile starts to fatten the non-Gaussianity becomes more pronounced. This means that a large value of $\lambda^2(l)$ is a sign for a high probability of finding large fluctuations in a data set. This recalls the criticality in a system stated in (Kiyono et al. 2006).

3.2 The Relation between $\lambda^2(l)$ and the qth Order Absolute Moments

To estimate the non-Gaussian parameter $\lambda^2(l)$, we can vary the variances of two Gaussian noises so that the PDF of this process matches the PDF of real data. One suitable fitting algorithm could be the χ^2 statistic. But whenever a condition applies so that $\lambda^2(l) < 0.01$, finding the minimum χ^2 statistic is non-trivial. Therefore, Kiyono et al (2007) introduced a method that results in an explicit relation between the non-Gaussianity and the qth order absolute moments of the fluctuations. From Eq. (5) we have

$$\left\langle |\Delta_l x|^q \right\rangle = \int_{-\infty}^{+\infty} |\Delta_l x|^q P_l(\Delta_l x) dx = \frac{2^{q/2}\Gamma(\frac{q+1}{2})}{\sqrt{\pi}} \exp\left(\frac{q(q-2)\lambda_q^2(l)}{2}\right) \tag{8}$$

where the brackets represent the statistical average, and Γ is the gamma function satisfying $q > -1$. From Eq. (8) the non-Gaussianity is obtained as

$$\lambda_q^2(l) = \frac{2}{q(q-2)} \left| \ln\left(\frac{\sqrt{\pi}|\Delta_l x| q}{2^{q/2}}\right) - \ln\Gamma\left(\frac{q+1}{2}\right) \right| \tag{9}$$

Kiyono et al. (2007) evaluated their method in order to check whether $\lambda_q^2(l)$ would determine correctly the value of $\lambda^2(l)$ by numerical analysis of stochastic process data set:

$$\Delta_l x(h) \equiv \xi_l(h) e^{\lambda(l)\omega_l(h) - \lambda^2(l)} \tag{10}$$

where $\xi_l(h)$ and $\omega_l(h)$ are both Gaussian white noise processes with zero mean and unit variance, and independent of each other. Using Eq. (9) $\lambda_q^2(l)$ is estimated for Eq. (11) according to Fig. 1.

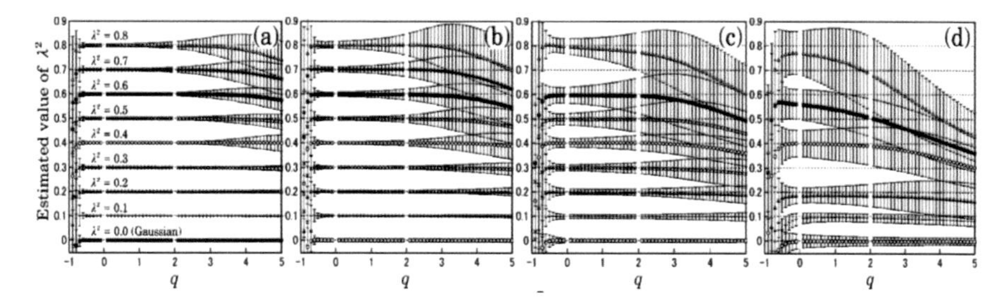

Figure 1: The sample mean of the estimator λ_q^2, where the observe time series $x(t)$ is standardized before computing λ_q^2. The sample mean was estimated from 200 samples. The theoretical values of λ^2 are shown in panel (a). The error bars indicate the sample standard deviation. (a) Data length $n = 106$, (b) $n = 105$, (c) $n = 104$ and (d) $n = 103$ (Kiyono et al. 2007).

3.3 Review of Some Applications on Multi-fractal Random Walk

The multi-fractal random walk (MRW) method is based on a log-normal deviation from a Gaussian PDF. This non-Gaussianity leads to multi-fractality in the system. Thus, determining the non-Gaussian parameter, $\lambda^2(l)$, provides useful information about the multi-fractality and the occurrence of large fluctuations with heterogeneity of local variance. Recently, many systems, as will be mentioned in the following, have been investigated by estimating $\lambda^2(l)$ at various scales l.

Fully-developed turbulent flows: Ghashghaie et al. (1996) showed that in turbulent flows λ^2 depends on the scale so that λ^2 decreases as scale increases (see Fig. 2).

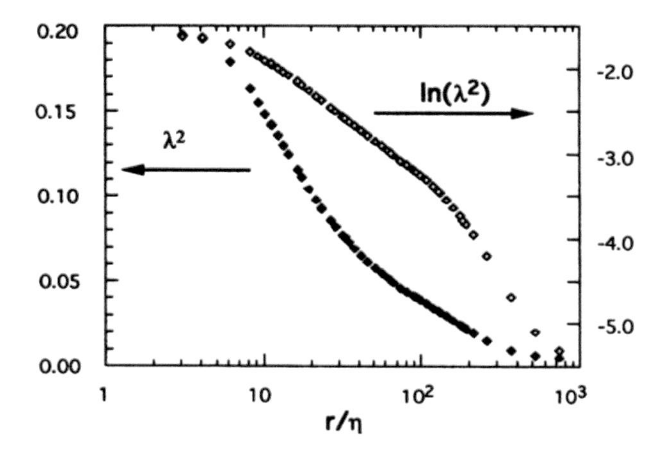

Figure 2: Dependence of λ^2 on scale for Reynolds number $R_\lambda = 328$. The solid diamonds correspond to a logarithmic linear presentation, whereas the opened diamonds correspond to a double logarithmic presentation (Chabaud et al. 1994).

Foreign exchange markets: Ghashghaie et al. (1996) observed scaling dependence of λ^2 in the foreign exchange rate. Their study indicated that non-Gaussianity of PDF was established only at small scales (see Fig. 3).

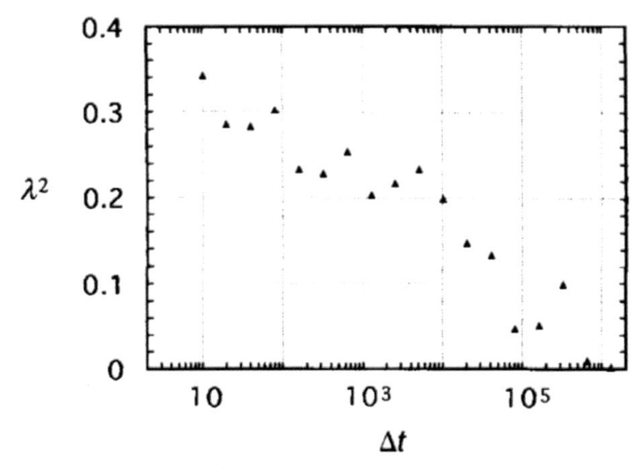

Figure 3: Dependence of λ^2 vs. scale in seconds (Ghashghaei et al. 1996).

Identification of phase transition in a healthy human heart rate: The human heart rate in healthy individuals undergoes a dramatic breakdown of criticality characteristics, reminiscent of continuous second order phase transitions. The hallmark of criticality is observed only during usual daily activity, and a breakdown of these characteristics occurs in prolonged, strenuous exercise and sleep (see Fig. 4).

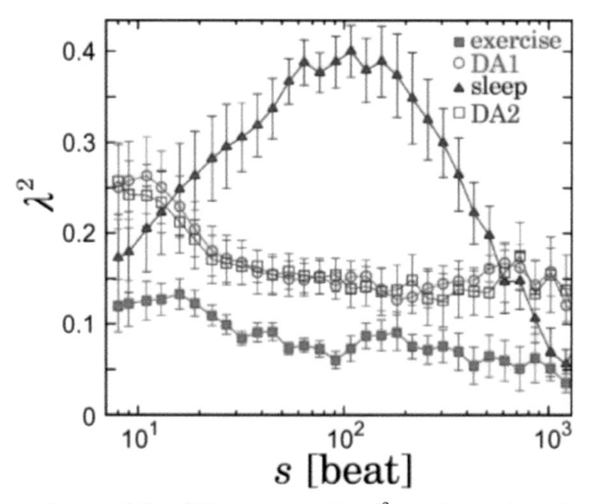

Figure 4: Dependence of the fitting parameter λ^2 on the scale s during constant exercise, usual daily activity after the exercise (DA1), sleep, and usual daily activity the next morning (DA2) (Kiyono et al. 2005).

The existence of criticality and a phase transition in stock-price fluctuations: A high value for λ^2 is a sign for the occurrence of rare or great events which proves adequate to call the situation "critical". This critical-like behavior could be readily noticed by the large fluctuations shown by the red lines in Fig. 5. Kiyono et al. 2006 showed (by red lines) the scaling behavior of the stock-price leading to the black Monday crash of 1987, where the value of λ^2 stays quite high. However, the green fluctuations are linked to its aftermath, where the descend of the value of λ^2 means that the conditions are healing.

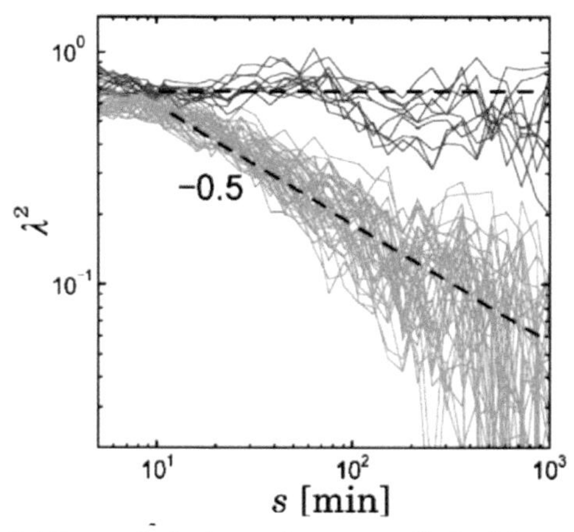

Figure 5: The scale independence of λ^2. Red lines correspond to the data around the black Monday and the green lines are related to data before black Monday (after Kiyono et al. 2006).

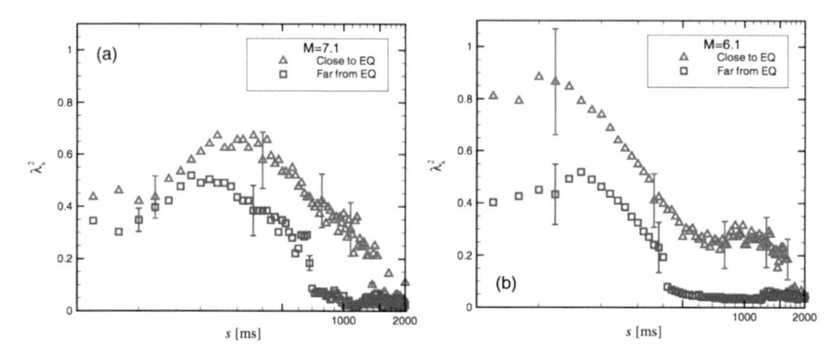

Figure 6: Scale dependence of λ_s^2 vs. scale (in milliseconds). (a) The magnitude M = 7.1 event, far from [data set (I) denoted by squares] and close to [data set (II) shown by triangles] the earthquake. For data set (I) and s > 700 ms, $\lambda_s^2 \to 0$, implying that the increments' PDF is Gaussian, but, for data set (II), λ_s^2 deviates strongly from 0 for 700 ms < s < 1500 ms. (b) The same as in (a), but for the M = 6.1 event. When $\lambda_s^2 \to 0$, the error bars are about the same size as the symbols (Manshour et al. 2009).

Turbulence-like behavior of seismic time series: The analysis of Earth's vertical velocity time series revealed a pronounced transition in their probability density function from Gaussian to non-Gaussian (Manshour et al. 2009). The transition occurs 5–10 hours prior to a moderate or large earthquake (see Fig. 6).

Stochastic qualifier of gel and glass transitions in Laponite suspensions: Based on the multiplicative log-normal cascade models, a criterion to distinguish gels from glasses was identified (Shayeganfar et al. 2010). The non-Gaussian parameter λ_s^2 behaves differently for gel and glass samples. Therefore, it can be employed as a criterion for distinguishing the gel transition from the glass transition. Shayeganfar et al. (2010) found that in both ergodic and non-ergodic regimes of aging, λ_s^2 was larger for the glassy samples compared to that of gel samples (see Fig. 7).

In the following section, the applications of the multi-fractal random walk method in petro-physical and well-logging data are presented.

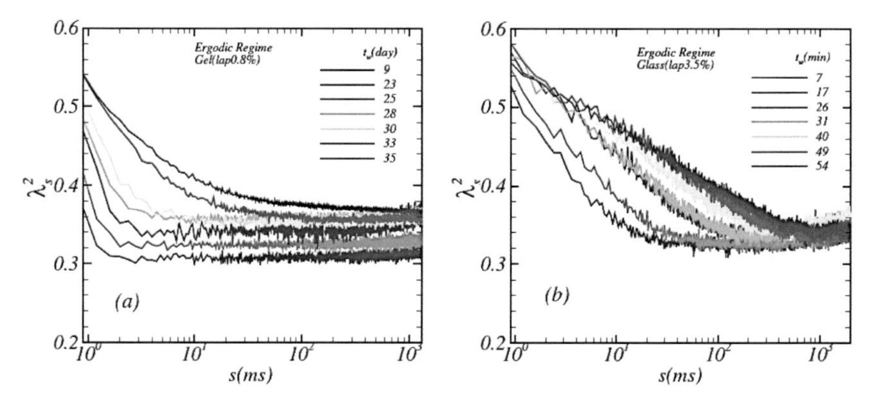

Figure 7: Scale dependence of λ_s^2 vs. s (in milli-seconds) for ergodic regime, (a) gel samples and (b) glass samples (Shayeganfar et al. 2010).

4. Multi-fractal Random Walk Applications in Well-log Quantities

The petro-physical data used in this study were collected from four wells in Maroon reservoir located in Khuzestan Iran. Each log contains between 1200 and 20,000 data points. The attributes of each well are as follows:

- Well number one: Depth from 2760 m to 3334 m
- Well number two: Depth from 3504 m to 3946 m
- Well number three: Depth from 3669 m to 4214 m
- Well number four: Depth from 2958 m to 3557 m

Each data set includes measurements recorded every 15.4 cm for GR, STT, and NP logs. The logged interval contains the Asmari region

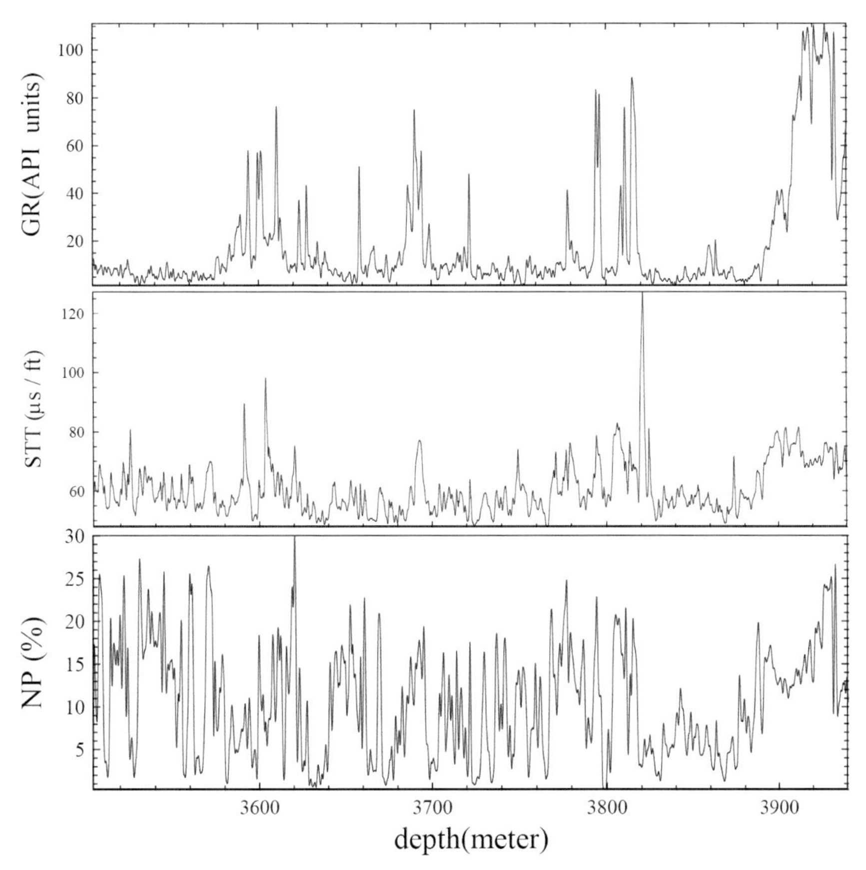

Figure 8: Three well logs, upper panel: the gamma ray (GR), middle panel: sonic transient time (STT) and the lower panel: neutron porosity (NP), versus the depth recorded every 15.4 cm over the depth interval 3504 m-3946 m for well number 2.

formation, including mostly fractured carbonate, some sandstone, shaly sand, and some signs of anhydrate. The data corresponding to the samples extracted from well number 2 are plotted in Fig. 8 showing the spatial heterogeneity of the reservoir at distinct depths and length scales (Jafari et al. 2001, Ferreira et al. 2009, Koohi Lai et al. 2013).

4.1 Non-Gaussianity of Well-log Data

Based on the data collected from the four wells studied here, the amount of non-Gaussianity of the well-log data could be quantified. Figure 9 shows the non-Gaussian parameter $\lambda^2(l)$ as a function of the scale l. By implementing the multi-fractal formalism and carrying out the χ^2 test, the best value for $\lambda^2(l)$ is achieved.

It can be readily observed from Fig. 10 that the scale dependence of the non-Gaussian parameter, $\lambda^2(l)$, does not follow the same pattern for the three indicators: GR (square), STT (triangle) and NP (right triangle). By browsing Fig. 9 one notices that the non-Gaussian parameter $\lambda^2(l)$ obtained from the GR indicator is large and scale invariant for all wells but not remarkable in well # 3. The large value of $\lambda^2(l)$ indicates a non-Gaussian PDF, which is due to the occurrence of large fluctuations of the gamma rays received probably from the shaly layers in the reservoir. The scale invariant characteristic of $\lambda^2(l)$ is an indication of the critical-like behavior of GR. This property is a consequence of porosity and permeability variations in rock reservoirs and large scales.

Regarding the STT data, Fig. 9 shows that the small value of $\lambda^2(l)$ is a consequence of the shale effects in the media. But, contrary to that observed for a fully developed turbulence, for the four oil wells studied here the non-Gaussian parameter does not decay logarithmically. The presence of

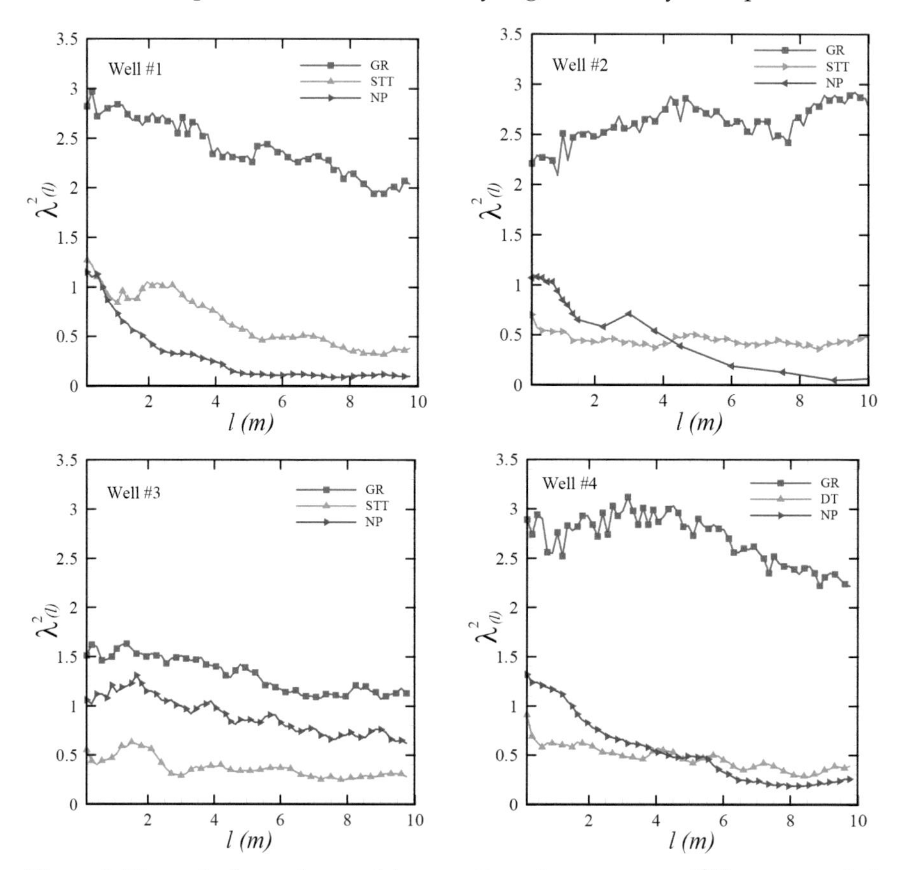

Figure 9: The scale dependence of the non-Gaussian parameter $\lambda^2(l)$ versus scale l (depth) for three well-log data, GR (square), STT (triangle) and NP (right triangle) plotted for the four wells studied here. DT and NP in well # 4 (bottom right plot) are the same as STT and NP in other plots. Note that the data points corresponding to the non-Gaussian regime in the scale between 15 cm and 10 m have been shown.

local correlations in the indicators increases the scale invariance features of the non-Gaussian parameter $\lambda^2(l)$. About the NP data, Fig. 9 shows that the non-Gaussian parameter $\lambda^2(l)$ is large at smaller scales, and tends to zero at larger scales for all four wells. A large value of $\lambda^2(l)$ indicates that the NP data possesses a non-Gaussian behavior accompanied by long-ranged correlations at small scales.

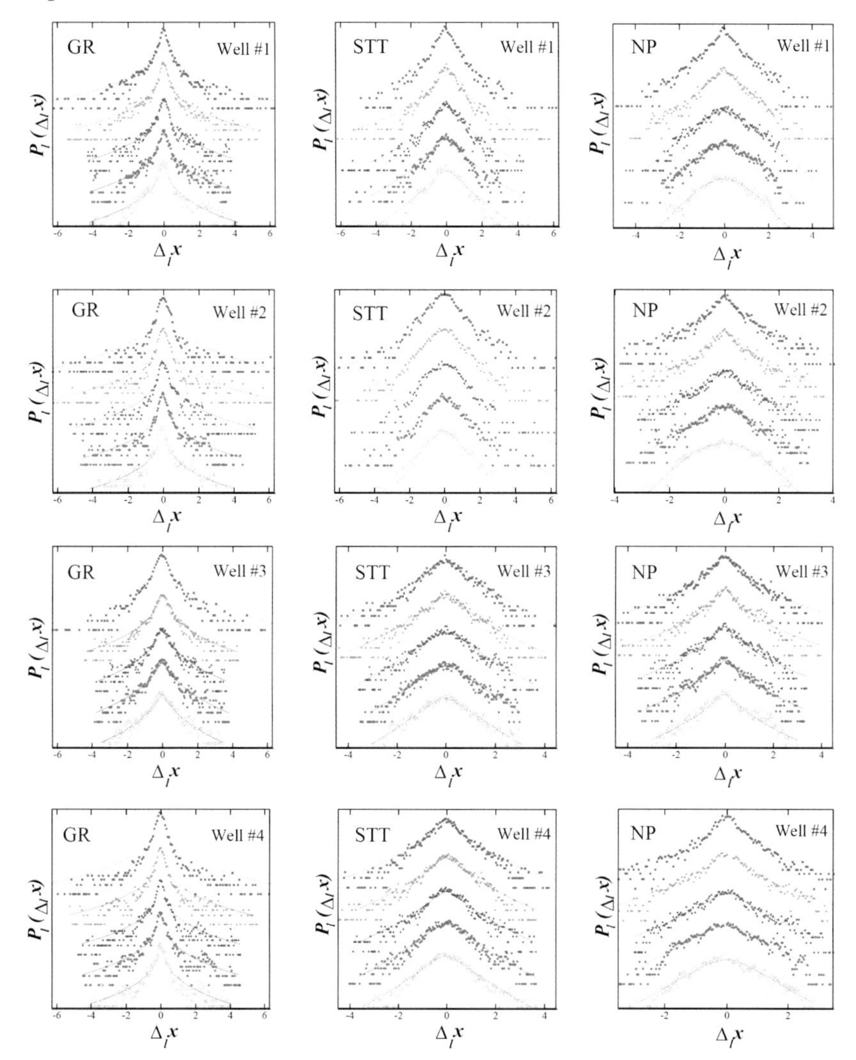

Figure 10: Deformation of the PDFs in scales for the three indicators GR (left panels), STT (middle panels) and NP (right panels) from four oil wells. From top to bottom we have PDFs at scales l = 15, 75, 225, 450, 900 cm measured at a space interval of 15 cm from the four oil wells under consideration located at the southwest of Iran. Solid lines are theoretical estimations obtained by fitting the well-log data to Eq. (5).

Figure 10 shows the increment PDFs of the well-log data sets for the four wells at different scales. The well-log data sets have been fitted to Eq. (5). It could readily be seen that the probability density function of GR is strongly non-Gaussian for all wells at all scales. The non-Gaussianity for the STT data shows that the non-Gaussianity is smaller than that for GR. The PDF of NP has a non-Gaussian behavior at small scales, while at large scales it becomes Gaussian.

It is interesting to observe that the scale invariance or data collapse exists for the probability density function of GR and STT and not for NP that exhibits Gaussian behavior at large scales. This scale-invariant feature of the probability density function resembles the phase transition of the indicators in the reservoir region.

4.2 Correlation Function of Logarithm of Stochastic Variances

In the presence of correlations in the logarithm of the stochastic variances, $\omega_l(h)$, the non-Gaussian parameter of the probability density function experiences an increase in $\lambda^2(l)$. This correlation function is defined as

$$C_l(\tau) \equiv \left\langle \left[\bar{\omega}_l(i) - \langle \bar{\omega}_l \rangle\right]\left[\bar{\omega}_l(i+\tau) - \langle \bar{\omega}_l \rangle\right]\right\rangle, \tag{11}$$

in which

$$\bar{\omega}_l(i) = \frac{1}{2}\ln \sigma^2(l,i), \tag{12}$$

and

$$\sigma^2(l, i) = \frac{1}{l}\sum_{j=1+(i-1)l}^{il} \Delta_l x^2(j). \tag{13}$$

Figure 11 shows the correlation function of the well-log data according to Eq. (11) at scales $l = 3$ m, $l = 6$ m and $l = 9$ m. It can easily be seen that GR shows long-range correlations at all scales, particularly at larger ones. This results in a high non-Gaussian parameter for GR and confirms that large fluctuations of Gamma ray of shaly layers exist probably everywhere in the reservoir. The correlation function of NP is long-ranged due to the non-Gaussianity at small scales. The correlation of large fluctuations for STT is weaker than GR and NP for all the well data except for well number 1. This can be interpreted as a result of shale effects in the reservoir.

5. Multivariate Multi-fractal Random Walk

In order to extract the most information out of an oil/gas reservoir, one must analyze the appropriate dependent indicators. Since these indicators which are named as Gamma radiation (GR), sonic transient time (STT) and Neutron porosity (NP) are interdependent, by measuring the influence that these indicators have on each other, the best information from the well can be extracted. Depending on the depth of the well, the

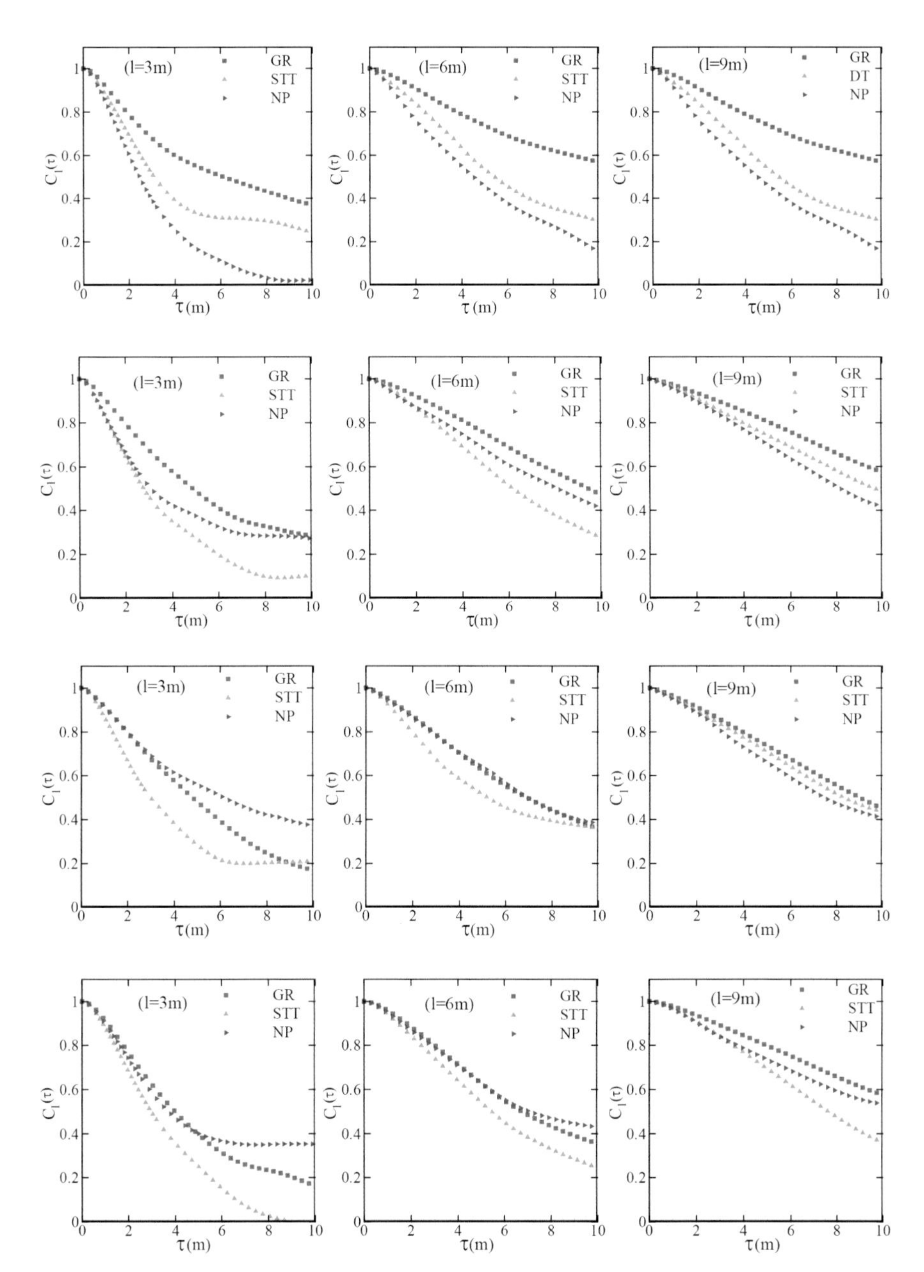

Figure 11: Correlation function of log stochastic variances, $C_l(\tau)$ versus the space lag τ at scales l = 3 m, l = 6 m and l = 9 m for wells 1 to 4 from top to bottom for the three well-log data GR (square), STT (triangle) and NP (right triangle). DT and NP in the top right plot are the same as STT and NP in other plots.

probability density function of these indicators possesses a non-Gaussian behavior, which is due to the gradient caused by different materials and local heterogeneities in the reservoir. Here we shed light on the way these petro-physical quantities may influence each other and highlight the scale at which each of these quantities are affected from the gradients within the rock reservoir. The reason to focus on these quantities is because the behavior of each provides nontrivial insights in the oil/gas production for each well. The assessment carried out by Koohi Lai et al. (2015) gives a good illustration of the way the petro-physical quantities react to the changes in the reservoir. They stated for their data under consideration that GR had a strong multi-fractal behavior with non-Gaussianity, while STT showed almost mono-fractal behavior. Non-Gaussianity (λ_l^2) in the context of well logging is a result of changes in the medium. This change while deepening the well is an indication of nearing a reservoir. The results obtained by the four studied wells shows that GR possesses the strongest multi-fractality features. This means that GR is most sensitive to medium changes (nearing a reservoir) compared with the other three indicators. But interestingly in addition to the fruitful information about the oil/gas well given by the magnitude of the non-Gaussianity of the probability density function, the type of correlation between the indicators also plays an important role on the information extraction from the well. This clearly requires studying the cross correlation between the petro-physical quantities. Results of Koohi Lai et al. (2015) showed that between GR and NP there existed a negative correlation, while between GR and STT a positive correlation. We should point out that the information based on the non-Gaussianity and the cross-correlations is unique for each well. Another important aspect of the non-Gaussianity in the probability density function of the indicators is that it has a direct proportionality with the quality of the well, while the strength of correlation between the indicators is inversely proportional to the information extracted from the well.

In order to provide a clear illustration of the model we present two non-stationary coupled processes whose PDFs are non-Gaussian. The model could then be extended for multi-coupled series.

The aim here is to study the coupling between two non-stationary processes, which usually exists between neighbors. For this purpose, the multi-fractal random walk approach is generalized to model the correlation between processes based on the lognormal cascade method (Muzy et al. 2001). This generalization is a multivariate multifractal model that describes the scale invariance of joint statistical properties of processes across various scales. Similar to what was done for a one-dimensional multi-fractal random walk, a non-Gaussian data set is generated by multiplying a Gaussian noise by a noise whose logarithm is also Gaussian. By controlling their variances one could measure the strength of their non-Gaussianity.

Suppose $\mathbf{x}(h) \equiv \{x_1(h), x_2(h)\}$ represents a bivariate process representing two non-Gaussian indicators of the well logs, for the increment $\Delta_l x(h) \equiv x(h + l) - x(h)$. The non-Gaussian behavior of these processes, as of the case of a single process, are controlled by two Gaussian noises represented by ξ_1 and ξ_2. Note that generally two processes are not independent of each other. Thus, the cross correlation between two non-Gaussian processes should be represented by two matrix; one consisting of the cross correlation between ξ_1 and ξ_2 and the other consisting of the cross correlation between ω_1 and ω_2. The bivariate version of the multi-fractal random walk (BiMRW) is (Muzy et al. 2001)

$$\Delta_l \mathbf{x}(h) \equiv (\xi_l^{(1)}(h) e^{\omega_l^{(1)}(h)}, \xi_l^{(2)}(h) e^{\omega_l^{(2)}(h)}), \tag{14}$$

where each increment at the space lag l has a unique ξ and ω, and the bivariate processes $\left|\xi_l^{(1)}, \xi_l^{(2)}\right|$ and $\left|\omega_l^{(1)}, \omega_l^{(2)}\right|$ are independent of each other. Note that both processes have a joint Gaussian distribution with a zero mean. In this case the covariance matrices Σ_l and Λ_l are respectively

$$\Sigma_l \equiv \begin{pmatrix} \Sigma_l^{11} & \Sigma_l^{12} \\ \Sigma_l^{21} & \Sigma_l^{22} \end{pmatrix} = \begin{pmatrix} \left\langle \xi_l^{(1)}(h)\xi_l^{(1)}(h) \right\rangle & \left\langle \xi_l^{(1)}(h)\xi_l^{(2)}(h) \right\rangle \\ \left\langle \xi_l^{(2)}(h)\xi_l^{(1)}(h) \right\rangle & \left\langle \xi_l^{(2)}(h)\xi_l^{(2)}(h) \right\rangle \end{pmatrix}, \tag{15}$$

$$\mathbf{\Lambda}_l \equiv \begin{pmatrix} \Lambda_l^{11} & \Lambda_l^{12} \\ \Lambda_l^{21} & \Lambda_l^{22} \end{pmatrix} = \begin{pmatrix} \left\langle \omega_l^{(1)}(h)\omega_l^{(1)}(h) \right\rangle & \left\langle \omega_l^{(1)}(h)\omega_l^{(2)}(h) \right\rangle \\ \left\langle \omega_l^{(2)}(h)\omega_l^{(1)}(h) \right\rangle & \left\langle \omega_l^{(2)}(h)\omega_l^{(2)}(h) \right\rangle \end{pmatrix}. \tag{16}$$

This non-diagonal matrix shows that the two processes (indicators) are not independent of each other.

The four diagonal elements of these matrices, namely $\Sigma_l^{11} \equiv \sigma_1^2(l)$, $\Sigma_l^{22} \equiv \sigma_2^2(l)$, $\Lambda_l^{11} \equiv \lambda_1^2(l)$, and $\Lambda_l^{11} \equiv \lambda_1^2(l)$ are defined for the two individual processes 1 and 2. Since these matrices are symmetric, we have $\Sigma_l^{12} = \Sigma_l^{21} = \Sigma_l \sigma_1(l)\, \sigma_2(l)$ and $\Lambda_l^{12} = \Lambda_l^{21} = \Lambda_l \lambda_1(l) \lambda_2(l)$, in which Σ_l is known as the *"Markowitz matrix"* (Markowitz 1959). The Markowitz matrix measures the variance and correlation of the ξ 's, and $\mathbf{\Lambda}_l$ represents the *multi-fractal* matrix, which determines the non-linearity of the ω's. The shape of the joint-PDF would consequently look like

$$P_l(\Delta_l x_1, \Delta_l x_2) = \left| d(\ln \sigma_1(l)) \right| d(\ln \sigma_2(l)) \\ G_l(\ln \sigma_1(l), \ln \sigma_2(l)) \frac{1}{\sigma_1(l)\, \sigma_2(l)} F_l\left(\frac{\Delta_l x_1}{\sigma_1(l)}, \frac{\Delta_l x_2}{\sigma_2(l)} \right), \tag{17}$$

where G_l ($\ln \sigma_1(l)$, $\ln \sigma_2(l)$) and $F_l\left(\frac{\Delta_l x_1}{\sigma_1(l)}, \frac{\Delta_l x_2}{\sigma_2(l)} \right)$, are the probability density functions of the bivariate processes $(\omega_l^{(1)}, \omega_l^{(2)})$ and $(\xi_l^{(1)}, \xi_l^{(1)})$, respectively. By considering the cross correlation between the two systems, their joint-probability density function is given by

$$G_l(\ln\sigma_1(l), \ln\sigma_2(l)) = \frac{1}{2\pi\lambda_1(l)\lambda_2(l)\sqrt{1-\Lambda_l^2}} \times$$

$$\exp\left(-\frac{1}{2(1-\Lambda_l^2)}\right)\left(\frac{\ln^2(\sigma_1(l))}{\lambda_1^2(l)}\right) - 2\Lambda_l\left(\frac{\ln\sigma_1(l)}{\lambda_1(l)}\right)\left(\frac{\ln\sigma_2(l)}{\lambda_2(l)}\right),$$

and

$$F_l\left(\frac{\Delta_l x_1}{\sigma_1(l)}, \frac{\Delta_l x_2}{\sigma_2(l)}\right) = \frac{1}{2\pi\sqrt{1-\Sigma_l^2}} \times$$

$$\exp\left(-\frac{1}{2(1-\Sigma_l^2)}\left|\left(\frac{\Delta_l x_1^2}{\sigma_1^2(l)}\right) + \left(\frac{\Delta_l x_2^2}{\sigma_2^2(l)}\right) - 2\Sigma_l\left(\frac{\Delta_l x_l}{\sigma_1(l)}\right)\left(\frac{\Delta_l x_2}{\sigma_2(l)}\right)\right|\right), \tag{18}$$

If the cross correlation coefficients Λ_l and Σ_l were zero, G_l and F_l would simply be the product of the two independent processes. This confirms that any deviation from the product of these two independent processes is a result of coupling between the two processes, though experimental inferences are less straightforward. According to the covariance matrix defined by Eq. (18), the parameter Λ_l controls the strength of the joint multi-fractality of the two processes (Muzy et al. 2001, Koohi Lai et al. 2015).

The parameter Λ_l at scale l could be evaluated by implementing Bayesian statistics (Bickel and Doksum 2001). Therefore, the multivariate Gaussian function would be

$$L(P_{\text{data}}(y) \mid P_{\text{theory}}(y; \mathbf{\Sigma}_l, \mathbf{\Lambda}_l)) = \frac{\sqrt{\text{Det}\{F\}}}{(2\pi)^{N/2}} \exp\left(-\frac{\Delta^T.F.\Delta}{2}\right), \tag{19}$$

where $\Delta \equiv P_{\text{data}}(y) - P_{\text{theory}}(y; \mathbf{\Sigma}_{l,} \mathbf{\Lambda}_{l,})$ is a column vector, F is the Fisher information matrix determined keeping in mind $F^{-1} = \langle\Delta(y)\Delta(y')\rangle$, the parameters $(\mathbf{\Sigma}_l, \mathbf{\Lambda}_{l,})$ are model free, and y is an independent parameter. $P_{\text{data}}(y)$ is computed directly from the data sets while $P_{\text{theory}}(y; \mathbf{\Sigma}_l, \mathbf{\Lambda}_l)$ is estimated from Eq. (17). Since a cross correlation between $\Delta(y)$'s for different y's is absent, the Fisher matrix is diagonal. Note that if there exist cross correlations in the covariance matrix, the proper similarity transformation could be implemented to diagonalize the matrix. By maximizing the likelihood function, one optimizes the best-fit values for the model-free parameters of scale l. In this case the chi-square, χ^2, tends to its global minimum

$$\chi^2(\mathbf{\Lambda}_l; \mathbf{\Sigma}_l) = \mathbf{\Sigma}_y \frac{\left[P_{\text{data}}(y) - P_{\text{theory}}(y; \mathbf{\Lambda}_l, \mathbf{\Sigma}_l)\right]^2}{\sigma_{\text{data}}^2(y) + \sigma_{\text{theory}}^2(y; \mathbf{\Lambda}_l, \mathbf{\Sigma}_l)} \tag{20}$$

where $\sigma^2_{\text{data}}(y)$ and $\sigma^2_{\text{theory}}(y;\, \mathbf{\Lambda}_l\, \mathbf{\Sigma}_l)$ are the mean standard deviation of $P_{\text{data}}(y)$ and $P_{\text{theory}}(y;\, \mathbf{\Lambda}_l,\, \mathbf{\Sigma}_l)$, respectively. By integrating over $\mathbf{\Sigma}_l$ we obtain

$$\chi^2(\mathbf{\Lambda}_l) = \mathbf{\Sigma}_y \int d\, \mathbf{\Sigma}_l \left(\frac{\left[P_{\text{data}}(y) - P_{\text{theory}}(y; \Lambda_l, \Sigma_l \right]^2}{\sigma^2_{\text{data}}(y) + \sigma^2_{\text{theory}}(y; \Lambda_l, \Sigma_l)} \right). \tag{21}$$

The best-fit values for the non-Gaussian parameters $\mathbf{\Lambda}_l$ are determined systematically by searching in the landscape of marginalized chi-square. The cross correlation function of the processes is obtained by

$$C_l^{\text{joint}}(\tau) \equiv \left\langle \left| \bar{\omega}_l^{(1)}(i) - \left\langle \bar{\omega}_l^{(1)} \right\rangle \right| \left| \bar{\omega}_l^{(2)}(i+\tau) - \left\langle \bar{\omega}_l^{(2)} \right\rangle \right| \right\rangle, \tag{22}$$

where τ is the space lag and is greater than scale $l(l < \tau)$. Since the vector process is stationary, the cross correlation only depends on τ. The local variance parameter $\bar{\omega}_l^{(\Diamond)}(i)$ is defined as,

$$\bar{\omega}_l^{(\Diamond)}(i) = \frac{1}{2} \ln \sigma^2_{(\Diamond)}(l, i) \tag{23}$$

in which the magnitude is given by

$$\sigma^2_{(\Diamond)}(l, i) = \frac{1}{l} \sum_{j=1+(i-1)l}^{il} \Delta_l x^2_{\Diamond}(j). \tag{24}$$

The symbol ($\Diamond$) could stand for either (1) or (2) according to the appropriate data set (Kiyono et al. 2006, Arneodo et al. 1998).

6. Implementation of BiMRW on Oil and Gold Markets

It could be useful to investigate the reciprocal effects of the oil and gold markets via BiMRW using daily recordings from 1995 to 2012 (www.goldratefortoday.org). For a BiMRW, the joint multi-fractal parameter, Λ_l, describes the coupling of large fluctuations of the two processes (Muzy et al. 2001, Koohi Lai et al. 2015). A large value of Λ_l, refers to a robust joint multi-fractality, which results in a coupled criticality or an uncertainty state in the system. The scaling parameter, Λ_l, that plays an important role in $G_l(\ln \sigma_1(l), \ln \sigma_2(l)$, is written as (Koohi Lai et al. 2015)

$$\Lambda_l = \frac{\Lambda_l^{(12)}}{\lambda_1(l)\, \lambda_2(l)} = \frac{\langle \ln \sigma_1(l) \ln \sigma_2(l) \rangle}{\lambda_1(l)\, \lambda_2(l)} \tag{25}$$

where $\langle \ldots \rangle$ denotes the ensemble average of all windows of size l. It can be observed from Eq. (25) that the non-Gaussian parameters λ's, and the cross correlation of the stochastic variances $\langle \ln \sigma_1(l), \ln \sigma_2(l) \rangle$, influence the scaling parameter, Λ_l. Now since Λ_l is scale-dependent, the scale at which Λ_l ascends must be determined.

In fact, a large value of Λ_l implies the emergence of coupled criticality or uncertainty. In the following we explain cases in which Λ_l is probable.

1. If the two systems under consideration are uncorrelated, the independency of the associated individual uncertainty (measured by λ) implies no coupling in the system. Hence no uncertainty between them arises.
2. If the two systems are correlated, coupled criticality emerges when at least one of the systems possesses a Gaussian distribution. This makes the λ's zero, which resembles a state of resonance. However, in this normal state, the weak cross correlation of large fluctuations decreases $\Lambda_{l.}$
3. If the two systems are correlated and both are in their high criticality states, as the λ's increase, the coupled uncertainty decreases.

Given that we presented the coupling between two systems, we should now discuss the uncertainty of their coupling before proceeding with the direction of their coupling. First let us ascertain facts; a system which is in a higher uncertainty state compared to others, with large λ's, would possess higher amplitude fluctuations. This causes a state of uncertainty in the neighboring system. This implies that the two systems are correlated. The result would be that the coupled uncertainty is amplified. The amplification of the coupled uncertainty is accompanied

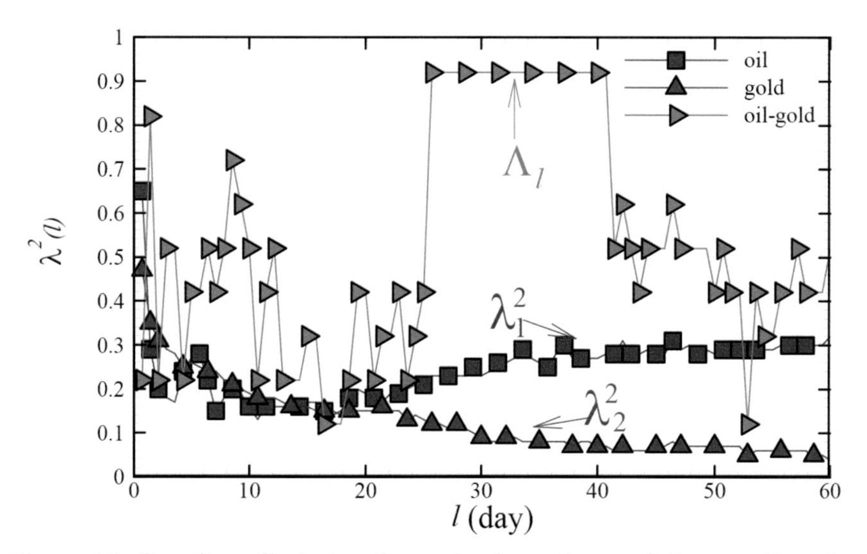

Figure 12: Sampling illustrates the scale dependence of the non-Gaussian parameter, $\lambda^2(l)$, for oil and gold markets recorded daily at the time interval between 1995 and 2012. The coupling multi-fractal parameter $\Lambda_l$for the joint markets has also been plotted versus scale, l (Koohi Lai et al. 2015).

by an increase in Λ_l. It could now be understood that the uncertainty induced by the neighboring system is responsible for the behavior of Λ_l. In contrast, if a system in the normal state causes large fluctuations in the neighboring system in a higher uncertainty state, the cross correlation of large fluctuations decreases resulting in a decrease in Λ_l.

A question that arises here is whether the occurrence of an uncertainty state in one system produced by the other system is a directed phenomenon or not. In order to answer this question, we must first show how the occurrence of large fluctuations and a high criticality in one system is influenced by large fluctuations in the other system. This requires evaluating the variance of the conditional distribution for the stochastic local magnitudes. The variance of conditional distribution of the local magnitudes is defined as

$$Var\left(\bar{\omega}_l^{(1)}\middle|\bar{\omega}_l^{(2)}\right) \equiv \sum_i \left|\bar{\omega}_l^{(1)}(i) - \left\langle\bar{\omega}_l^{(1)}\right\rangle\right|^2 P\left(\bar{\omega}_l^{(1)}(i)\middle|\bar{\omega}_l^{(2)}(i)\right),$$

where the conditional distribution function $P\left(\bar{\omega}_l^{(1)}(i)\middle|\bar{\omega}_l^{(2)}(i)\right)$ is

$$P\left(\bar{\omega}_l^{(1)}(i)\middle|\bar{\omega}_l^{(2)}(i)\right) = \frac{P\left(\bar{\omega}_l^{(1)}(i), \bar{\omega}_l^{(2)}(i)\right)}{P\left(\bar{\omega}_l^{(2)}(i)\right)}, \tag{26}$$

and $P\left(\bar{\omega}_l^{(1)}(i), \bar{\omega}_l^{(2)}(i)\right)$ is the joint distribution function of the stochastic local magnitude, $\bar{\omega}_l^{(\lozenge)}$, for the two processes under consideration. The large conditional variance is due to the occurrence of large fluctuations in a system due to the fluctuations in the other system. However, a system that is in a higher criticality state (large $\lambda^2_{(\lozenge)}(l)$) would cause large fluctuations accompanied by a high conditional criticality in the other system. This phenomenon is due to a strong cross correlation of large fluctuations between the two systems.

It is worth noting that when dealing with coupled indicators, they do not always have a symmetric effect on each other. In such a case, instead of a cross correlation matrix, we recommend a conditional cross correlation matrix formalized as (Koohi Lai et al. 2015)

$$\begin{aligned} C_l^{(\text{conditional})}(\tau) &\equiv \left\langle\left|\bar{\omega}_l^{(1)}(i) - \left\langle\bar{\omega}_l^{(1)}\right\rangle\right|\left|\bar{\omega}_l^{(2)}(i+\tau) - \left\langle\bar{\omega}_l^{(2)}\right\rangle\right|\right\rangle \\ &\quad \times P\left(\bar{\omega}_l^{(1)}(i)\middle|\bar{\omega}_l^{(2)}(i)\right) \end{aligned} \tag{27}$$

This discussion helps understand the anti-symmetrical action between two interacting systems. This phenomenon comes from the fact that when two systems or bodies work or interact with each other, depending on their weights, one's effect on the other is not similar the other way round. Consider the coupling between two systems non-balanced, or e.g. in social life a one-sided relation, such as one-sided love. This causes one person to be in an uncertainty or criticality state against the other, while this is not the case the other way round. In order to investigate this directed coupled uncertainty, the conditional cross correlation function of processes (1) and (2) at time/space scale l against time/space lag $\tau(l < \tau)$ must be defined. Using the formulation of Eq. (27), it is possible to detect such unbalanced correlations in well-logging variables.

7. Conclusion

For accurate analysis of oil/gas wells and reservoirs, due to the high level of complexity, a large number of petro-physical parameters and indicators are required. Spatio-temporal dependency of such indicators at reservoir scales causes changes e.g., non-stationary and non-Gaussianity in the behavior of those indicators. More specifically, local heterogeneity forces the distribution function of those indicators to become non-Gaussian with heavy tails, resulting in a multi-fractal behavior. Non-Gaussianity means that the frequency of such indicators at small and large scales (tails of the distribution function) are more probable when compared with the normal (Gaussian) function. The multi-fractal random walk (MRW) approach is a promising method to analyze these types of data. However, one should keep in mind that those indicators might not be independent from each other and thus analyzing them independently might result in data misinterpretation. Since the large events (in the tail of the distribution function) play an important role in these solidarities, the Joint MRW method seems a more appropriate technique of analysis.

Acknowledgments

The authors are grateful to the Editors, Behzad Ghanbarian (University of Texas at Austin) and Allen Hunt (Wright State University), for their constructive comments and proof reading.

REFERENCES

Adler, P.M. 1992. Porous Media: Geometry and Transport. Butterworth-Heinemann, Stoneham, MA.

Adler, P.M. and J.-F. Thovert. 1999. Fractures and Fracture Networks. Kluwer, Dordrecht.

Andrade Jr., J.S., D.A. Street, T. Shinohara, Y. Shibusa and Y. Arai. 1995. Percolation disorder in viscous and nonviscous flow through porous media. *Phys. Rev.* E51: 5725–5731.

Arneodo, A., S. Roux and J.F. Muzy. 1997. Experimental Analysis of Self-Similarity and Random Cascade Processes: Application to Fully Developed Turbulence Data. *J. Phys. II France* 7: 363–370.

Arneodo, A., E. Bacry, S. Manneville and J.F. Muzy. 1998. Analysis of Random Cascades Using Space-Scale Correlation Functions. *Phys. Rev. Lett.* 80: 708–711.

Bacry , E., J. Delour and J.F. Muzy. 2001. Multifractal random walk. *Phys. Rev.* E64: 026103-1:4.

Bickel , P.J. and K.A. Doksum. 2001. Mathematical Statistics. Vol. 1: Basic and Selected Topics (Second edition 2007).

Brachet, M.-E., E. Taflin and J.-M. Tcheou. 2000. Scaling transformation and probability distributions for financial time series. *Chaos, Solitons and Fractals* 11: 2343–2348.

Castaing, B., Y. Gagne and E. Hopfinger. 1990. Velocity probability density functions of high Reynolds number turbulence. *Physica* D46: 177–200.

Chabaud, B., A. Naert, J. Peinke, F. Chilla, B. Castaing and B. Hebel. 1994. Transition toward developed turbulence. *Phys. Rev. Lett.* 73: 3227–3230.

Christie, M., V. Demyanov and D. Erbas. 2006. Uncertainty quantification for porous media flows. *J. Comput. Phys.* 217: 143–158.

Corso, G., P.S. Kuhn, L.S. Lucena and Z.D. Thom'e. 2003. Seismic ground roll time–frequency filtering using the gaussian wavelet transform. *Physica* A318: 551–561.

Ferreira, R.B., V.M. Vieira, I. Gleria and M.L. Lyra. 2009. Correlation and complexity analysis of well logs via Lyapunov, Hurst, Lempel–Ziv and neural network algorithms. *Physica* A388: 747–754.

Ghashghaie, S., W. Breymann, J. Peinke, P. Talkner and Y. Dodge. 1996. Turbulent cascades in foreign exchange markets. *Nature* (London) 381: 767–770.

Grahovac, D. and N.N. Leonenko. 2014. Detecting multifractal stochastic processes under heavy-tailed effects. *Chaos, Solitons and Fractals* 65: 78–89.

Jafari, G.R., M. Sahimi, M.R. Rahimi Tabar and M.R. Rasaei. 2011. Analysis of porosity distribution of large-scale porous media and their reconstruction by Langevin equation. *Phys. Rev.* E83: 026309-1:7.

Jensen, J.L., L.W. Lake, P.W.M. Corbett and D.J. Goggin. 2000. Statistics for Petroleum Engineers and Geoscientists (2nd Edn). Prentice Hall, Englewood Cliffs, NJ.

Kantelhardt, J.W., S.A. Zschiegner, E. Koscielny-Bunde, S. Havlin, A. Bunde and H.E. Stanley. 2002. Multifractal detrended fluctuation analysis of nonstationary time series. *Physica* A316: 87–114.

King, P.R., J.S. Andrade Jr., S.V. Buldyrev, N. Dokholyan, Y. Lee, S. Havlin and H.E. Stanley. 1999. Predicting oil recovery using percolation. *Physica* A266: 107–114.

Kiyono, K., Z.R. Struzik, N. Aoyagi, F. Togo and Y. Yamamoto. 2005. Phase Transition in a Healthy Human Heart Rate. *Phys. Rev. Lett.* 95: 058101-1:4.

Kiyono, K., Z.R. Struzik and Y. Yamamoto. 2006. Criticality and Phase Transition in Stock-Price Fluctuations. *Phys. Rev. Lett.* 96: 068701-1:4.

Kiyono, K., Z.R. Struzik and Y. Yamamoto. 2007. Estimator of a non-Gaussian parameter in multiplicative log-normal models. *Phys. Rev.* E76: 041113-1:8.

Koohi Lai, Z., S. Vasheghani Farahani and G.R. Jafari. 2012. Non-Gaussianity of petro-physical parameters using q-entropy and a multifractal random walk. *Physica* A391: 5076 –5081.

Koohi Lai, Z. and G.R. Jafari. 2013. Non-Gaussianity effects in petro-physical quantities. *Physica* A392: 5132–5137.

Koohi Lai, Z., S.V. Farahani, S.M.S. Movahed and G.R. Jafari. 2015. Coupled uncertainty provided by a multifractal random walker. *Phys. Lett.* A379: 2284–2290.

Manshour, P., S. Saberi, M. Sahimi, J. Peinke, A.F. Pacheco and M.R. Rahimi Tabar. 2009. Turbulence like Behavior of Seismic Time Series. *Phys. Rev. Lett.* 102: 014101-1:4.

Markowitz, H. 1959. Portfolio selection: Efficient diversification of investments. John Wiley and Sons, New York.

Mair, R.W., G.P. Wong, D. Hoffmann, M.D. Hurlimann, S. Patz, L.M. Schwartz and R.L. Walsworth. 1999. Probing Porous Media with Gas Diffusion NMR. *Phys. Rev. Lett.* 83: 3324–3327.

Muller, J. 1992. Multifractal characterization of petro-physical data. *Physica* A191: 284–288.

Muzy, J.F., J. Delour and E. Bacry. 2000. Modelling fluctuations of financial time series: From cascade process to stochastic volatility model. *Eur. Phys. J.* B17: 537–548.

Muzy, J.F., D. Sornette, J. Delour and A. Arneodo. 2001. Multifractal returns and hierarchical portfolio theory. *Quantitative Finance* 1: 131–148.

Sahimi, M. 1993. Flow phenomena in rocks: From continuum models to fractals, percolation, cellular Automata, and simulated annealing. *Rev. Mod. Phys.* 65: 1393.

Shayeganfar, F., S. Jabbari-Farouji, M.S. Movahed, G.R. Jafari and M.R.R. Tabar. 2009. Multifractal analysis of light scattering-intensity fluctuations. *Physical Review* E80(6): 061126-1:8.

Shayeganfar, F., S. Jabbari-Farouji, M.S. Movahed, G.R. Jafari and M.R. Rahimi Tabar. 2010. Stochastic qualifier of gel and glass transitions in Laponite suspensions. *Phys. Rev.* E81: 061404-1:7.

Shayeganfar, F., M.S. Movahed and G.R. Jafari. 2013. Discrimination of Sol and Gel states in an aging clay suspension. *Chemical Physics* 423: 167–172.

Sornette, D. 2009. Critical phenomena in natural sciences: Chaos, fractals, self-organization and disorder: Concepts and tools (2nd Edn). Springer.

Xie, S.Y., Q.M. Cheng, G. Chen, Z.J. Chen and Z.Y. Bao. 2007. Application of local singularity in prospecting potential oil/gas targets: Nonlinear process. *Geophy* 14: 285–292. http://www.goldratefortoday.org.

CHAPTER

11

Combining Fractals and Multifractals to Model Geoscience Records

Carlos E. Puente[1]*, Mahesh L. Maskey[1] and Bellie Sivakumar[1,2]

[1] Department of Land, Air & Water Resources, University of California, Davis, 1 Shied Avenue, Davis, CA, 95616

[2] UNSW Water Research Centre, School of Civil and Environmental Engineering, The University of New South Wales, Vallentine Annexe (H22), Level 1, Kensington Campus, Sydney, Australia

1. Introduction

Understanding hydrological systems is crucial not only for the proper management of water resources but also for elucidating possible climate change impacts. There is not just a single variable playing a vital role in this regard but rather several geoscience signatures, such as precipitation, temperature, streamflow and others.

With the advancement of computational capabilities, various geophysical-hydrological models have been proposed that try to account for the physical processes behind available data and for the seemingly stochastic character of the records. Irrespective of how this is attempted, the overall objective of these efforts is to try to understand the nonlinear and, hence, complex dynamics involved. Beyond the simplified nature of the assumed notions, these models end up requiring meaningful geophysical parameters, which, as they are often hard to acquire, may lead to misinterpretations. Despite the fact that we are in a computational era, fully understanding and, hence, fully recognizing the intrinsic variability of signals is still challenging. This is the case as records contain intricate, "chaotic," and, altogether, convoluted details.

Given that natural time series are typically erratic, noisy and intermittent, it has become natural to model them using stochastic (fractal)

*Corresponding author: cepuente@ucdavis.edu

theories, that also account for statistical self-similarity (e.g., Mandelbrot 1982, Feder 1988). Specifically, in order to model complex natural signals several efforts have been made towards implementing stochastic theories (e.g., Rodríguez-Iturbe 1986), stochastic-fractal theories (e.g., Gupta and Waymire 1990, Lovejoy and Schertzer 2013), and chaotic analysis (e.g., Sivakumar 2000, 2004). These kinds of efforts, however, result in representations that are quite difficult to condition. As such, they may miss the specific convoluted details present in a given natural set, such as the exact locations of major peaks.

Given the intrinsic limitation in assuming that what is seen is a realization of a stochastic process, a geometric approach, combining both fractal and multifractal notions, and known as the Fractal-Multifractal (FM) method was introduced (Puente 1996). Such a deterministic method models random-looking natural sets as fractal transformation of multifractal measures and results in a host of patterns that indeed share the same geometric and statistical features of natural records (Puente 2004). In fact, the FM approach may be conditioned to not only preserve key statistical indicators (e.g. moments, autocorrelation function, power spectrum and multifractal spectrum) but also the inherent textures of data sets.

The present chapter focuses on the application of such a notion in the context of geosciences research, especially in hydrology. From its inception, the FM procedure has advanced from encoding to understanding to predicting the complexity of natural records. In this spirit, the following sub-sections summarize our efforts: (a) encoding of mildly and highly intermittent records, (b) simulating mildly and highly intermittent sets, (c) disaggregating or downscaling such geophysical records, and (d) classifying sets geometrically aiming at the prediction of future scenarios. In what follows, the geophysical sets are limited to rainfall, streamflow, and water temperature, but, as will become apparent, the FM approach may be used to model several other sets.

2. The Fractal-multifractal Method

Puente (1996) combined fractal functions and multifractal measures in order to model geophysical records in a holistic manner. This section briefly reviews the so-called Fractal-Multifractal (FM) method and also some of its variants, which have been found suitable to encode natural sets.

2.1 Original Approach

A *fractal interpolating function* $f: x \rightarrow y$, passing through $N + 1$ ordered points on the plane $\{(x_n, y_n) \mid x_0 < x_1 < ... < x_N\}$, is defined as the unique fixed point of N affine maps with the structure (Barnsley 1988):

$$w_n\begin{pmatrix} x \\ y \end{pmatrix} = \begin{pmatrix} a_n & 0 \\ c_n & d_n \end{pmatrix}\begin{pmatrix} x \\ y \end{pmatrix} + \begin{pmatrix} e_n \\ f_n \end{pmatrix} \quad n = 1, ..., N, \tag{1}$$

with the x component decoupled from the y component ($b_n = 0$), so that the graph $G = \{(x, f(x) \mid x \in [0, 1]\}$ satisfies $G = w_1(G) \cup w_2(G) \cup ... w_N(G) \cup$. While the parameters d_n are vertical scalings, $|d_n| < 1$, the other parameters, a_n, c_n, e_n and f_n, are evaluated from contractive initial conditions that guarantee the existence of function f, namely:

$$w_n\begin{pmatrix} x_0 \\ y_0 \end{pmatrix} = \begin{pmatrix} x_{n-1} \\ y_{n-1} \end{pmatrix}, \qquad w_n\begin{pmatrix} x_N \\ y_N \end{pmatrix} = \begin{pmatrix} x_n \\ y_n \end{pmatrix}. \tag{2}$$

These equations give rise to simple linear systems of equations for the "other" parameters in terms of the vertical scalings and the coordinates by which the function passes. At the end, a convoluted "wire" function f is generated whose graph has a fractal dimension D between 1 and 2.

A fractal function may be produced via a point-wise sampling of the "attractor" G iterating successively the affine maps. This process, known as the "chaos game" (Barnsley 1988), starts at a point within G (say, an interpolating point) and progresses guided by arbitrary "coin tosses" that assign distinct usage proportions to the N maps. Allowing enough time for the iterations, the process not only paints a graph G point-by-point, but also induces a unique invariant measure over G, which typically contains intermittencies and possesses multifractal properties.

Seen over the coordinate x, such invariant measures define *deterministic multinomial multifractals* – with length scales given by the placements of the interpolating points in x and with the proportions for the "coin tosses" defining the intermittencies – (Mandelbrot 1989, Puente 1996). When seen over the coordinate y, the invariant measures over G define deterministic projections, which turn out to encompass some of the irregular shapes encountered in nature, such as rainfall, streamflow, temperature and other sets.

Figure 1 illustrates how a binomial multifractal measure dx may be naturally combined with a fractal interpolating function to define an interesting output dy. When the two maps

$$w_1\begin{pmatrix} x \\ y \end{pmatrix} = \begin{pmatrix} 0.80 & 0 \\ 2.91 & -0.72 \end{pmatrix}\begin{pmatrix} x \\ y \end{pmatrix} + \begin{pmatrix} 0 \\ 0 \end{pmatrix} \tag{3}$$

and

$$w_2\begin{pmatrix} x \\ y \end{pmatrix} = \begin{pmatrix} 0.20 & 0 \\ -0.65 & -0.54 \end{pmatrix}\begin{pmatrix} x \\ y \end{pmatrix} + \begin{pmatrix} 0.80 \\ 2.19 \end{pmatrix}, \tag{4}$$

are iterated according to proportions 68–32%, they: (a) generate a fractal wire f passing through $\{(0, 0), (0.80, 2.19), (1, 1)\}$ (at the center of circles) and with dimension $D = 1.31$; and (b) induce a simple binomial multifractal dx over x – with length scales 0.8 and 0.2 and redistributions 68–32% – and also a unique and, hence, deterministic derived measure dy over y – exhibiting non-trivial variability.

In a practical setting and requiring little effort in the implementation as explained in the Online Appendix http://puente.lawr.ucdavis.edu/omake/fractal_geosciences_2016/fractal_geosciences_2016.html, dy is found as a histogram of all chaos game points over y, adding all "events" over x corresponding to the crossings of function f for a given value of y i.e., $dy = f^{-1}(dx)$. By varying the parameters, and as shown on the far right graph in Fig. 1 via local integration, objects dy or dy_s turn out to be "random-looking" sets that do resemble geoscience records (Obregón et al. 2002b, Puente 2004). The aforementioned Online Appendix includes a demo that allows the reader to reproduce Fig. 1.

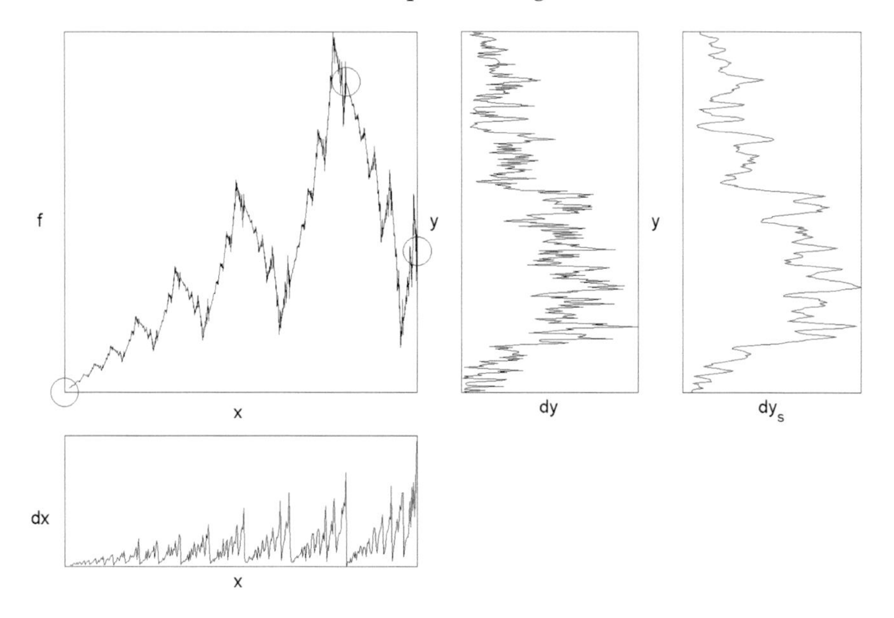

Figure 1: The FM approach: from a multifractal dx to a projection dy, via a fractal interpolating function f, a "wire" from x to y. dy_s is a smoothed version of dy.

The FM procedure, besides transforming multifractals relevant in the study of turbulence, i.e., when dx has equal length scales and redistributions 70–30% (Meneveau and Sreenivasan 1987), may be assigned a more direct physical interpretation (Cortis et al. 2013). This is the case as the measure dy, although entirely deterministic, may be interpreted as a "realization" of a non-trivial but conservative multiplicative cascade, as customarily done via stochastic cascades of tracers (e.g., Lovejoy and Schertzer 2013).

In summary, the modeling of geoscience records via the original FM method requires the solution of an inverse problem for suitable geometric parameters, namely: (a) the interpolating points by which a fractal function passes, (b) the vertical scaling d_n, (c) the frequencies used in chaos game calculations, and (d) a smoothing parameter depending upon the nature of a target set. When using two or three maps, the FM approach requires 6 or 10 parameters, hence leading to sizable compressions.

2.2 Two Extensions

Instead of defining a fractal "wire," the procedure may be modified to generate more general attractors. Such may be done using N maps as in Eq. (1), but using new contractive initial conditions,

$$w_n\begin{pmatrix}x_0\\y_0\end{pmatrix}=\begin{pmatrix}x_{2n}\\y_{2n}\end{pmatrix},\ w_n\begin{pmatrix}x_{2N-1}\\y_{2N-1}\end{pmatrix}=\begin{pmatrix}x_{2n+1}\\y_{2n+1}\end{pmatrix},\qquad n=1,\ldots,N, \tag{5}$$

such that the range of map w_n in x becomes the more general interval $[x_{2n}, x_{2n+1}]$, for $x_0 \leq x_{2n} < x_{2n+1} \leq x_{2N-1}$. When such sub-intervals contain gaps, the resulting attractors are Cantorian in nature (Maskey et al. 2015) and when such ranges overlap or when the corresponding end-points in y do not match as in Eq. (2), the resulting attractors are not functions but instead are shaped as interesting "leaves" (Huang et al. 2013).

Figure 2 presents an example of an FM-based derived measure dy based on a Cantorian case, and Fig. 3 does so for a leafy attractor. While both representations use two maps, the former uses as end-points {(0, 0), (0.39, – 1.54)} and {(0.77, – 5.0), (1, 1)} and the latter {(0, 0), (0.23, 5.0)} and {(0.19, 0.01), (1, 1)}, as marked by circles. Whereas the scaling parameters and iteration frequencies for the first case are 0.28 and –0.47 and 66–34%, for the second case they are 0.55 and –0.89 and 15–85%, respectively.

As seen in Fig. 2, the iterations of such maps generate a Cantorian attractor, which, as before, induces projections dx and dy. While dx contains multiple spikes and gaps reflecting the original hole of size (from 0.39 to 0.77, as marked by the second and third open circles), the derived distribution dy also exhibits holes, but not in an obvious repetitive fashion as dx, hence becoming itself useful to model highly intermittent sets. In this spirit, the other entity in Fig. 2 (in the far right), named dy_v, represents yet an additional adaptation found by using a vertical threshold (φ_v) in the construction, so that (after proper normalization) new sets with an increased number of zeros may be found.

As seen in Fig. 3, the iteration of the corresponding maps for the second case yields other interesting patterns dx and dy. Now the overlap, also seen by the second and third open circles, results in a leafy attractor

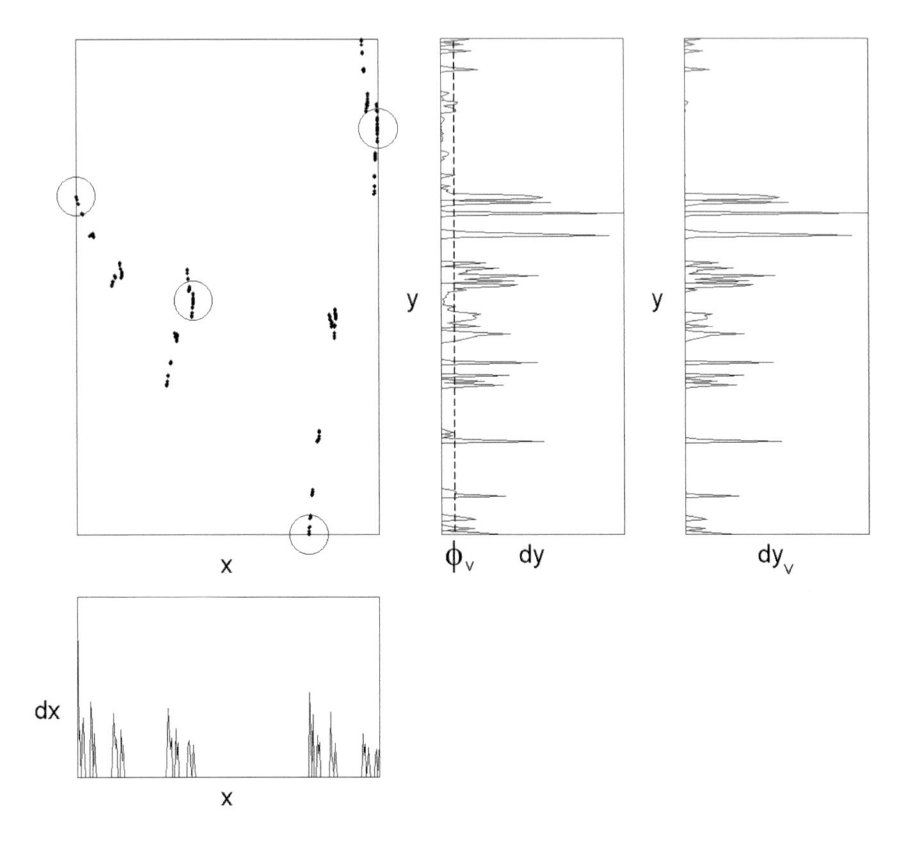

Figure 2: A generalization of the *FM* approach: from a Cantorian texture *dx*, to a derived *dy*, via a disperse attractor. dy_v is found pruning *dy* below a threshold φ_v and renormalizing.

that induces yet another representation *dy* whose smoothened version (not shown) may resemble, say, the diurnal or yearly variations of air or water temperature.

For these extensions of the FM approach, the following are the key parameters needed for a suitable representation: (a) the end-points that define the more general attractor, (b) the vertical scalings d_n, (c) the iteration frequencies, and (d) a smoothing or a threshold parameter, if needed. When using two or three maps, these FM approaches require 8 or 15 geometric parameters, leading still to sizable compressions. For more information on these extensions, the reader may access the aforementioned Online Appendix, which besides including pertinent Matlab codes also allows the reader to interact with the notions leading to Figs 2 and 3.

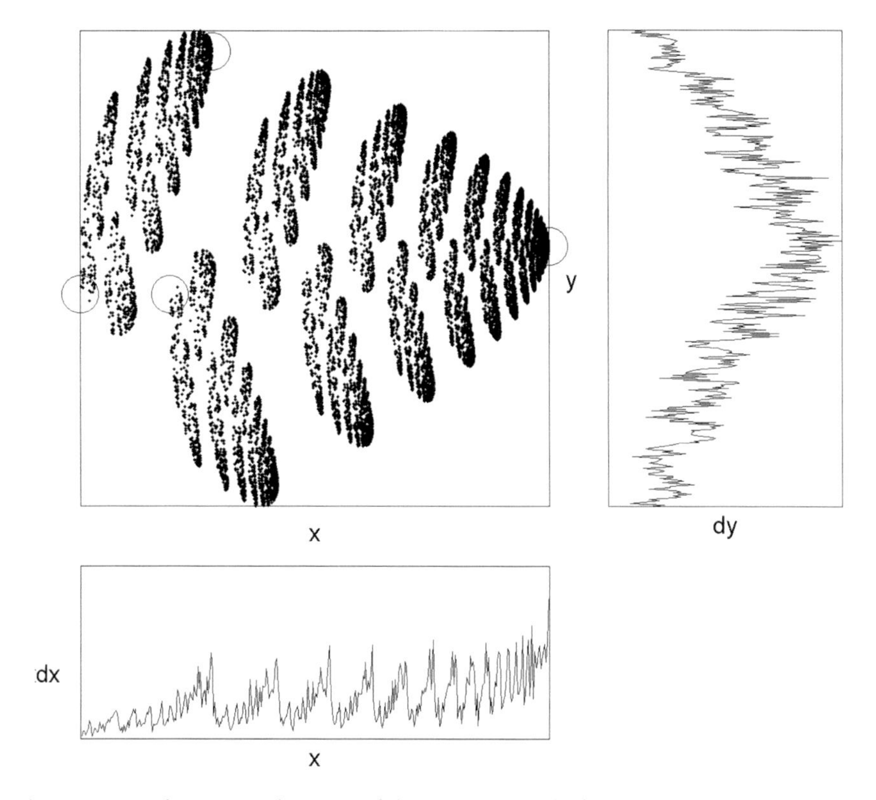

Figure 3: Another generalization of the FM approach: from an input texture dx, to a projection dy, via a "leafy" attractor.

3. General Strategy

Even though the FM methodology is ultimately simple and computationally efficient – once a set of parameters is known – the finding of a suitable representation for a given set is not a trivial task. As there are neither analytical formulas for attractors nor for the derived measures dy, only a numerical solution is possible. Unfortunately, such is hampered by the dimensionality of the problem, the choice of the objective function, and the optimization algorithm used (Huang et al. 2013), which reflect a highly complex parameter space where local minima may happen based on distinct initial conditions.

The results to be presented herein reflect our own experience throughout the years and are based on a generalized particle swarm optimization (GPSO) algorithm and an objective function that minimizes squared errors of suitable attributes, either the accumulated data sets from beginning to end when encoding sets or statistical qualifiers for simulation purposes.

Inspired by the collective social behavior of animals and the notion that all members of the swarm ought to be leaders, a "cloud" version of the particle swarm optimization (PSO) algorithm (Fernández-Martínez et al. 2010) is adopted – with 300 initial swarm populations used for generating FM parameters after 100 successive iterations. As just mentioned, the L^2 norm (i.e., the root mean square error) of statistical attributes of the records is used to define an objective function. As an example, for encoding purposes such a function is

$$\in_{ac} = \sqrt{\frac{1}{N}\sum_{i=0}^{N}(r_i - \hat{r}_l)^2} \tag{6}$$

where N is the number of data points; and r_i and $\hat{r}_i$ are the ith accumulated values of the original record and FM fit, respectively. In order to ensure that solutions share similar geometrical features with the target set, various penalties are also imposed on the objective function so that: (i) the maximum deviations on accumulated sets, $\in_{mx}$, at each point would not exceed 10%, and (ii) the length of the FM fit and the target set would not differ by more than 10%. At the end, results shown herein do satisfy such constraints.

The evaluation of the resulting FM models is made using various statistics not included in the objective function. For this purpose, various statistical attributes, such as autocorrelation, histogram, and Renyi entropy functions, are evaluated in terms of Nash-Sutcliffe efficiencies (Nash and Sutcliffe 1970).

4. Encoding Geophysical Records

The present section illustrates the applicability of the FM approach for encoding complex geoscience records, which may be categorized into two types: (1) mildly intermittent sets, such as rainfall events and typical streamflow and water temperature records gathered daily; and (2) highly intermittent sets, exhibiting considerable swings of activity and inactivity and a large numbers of zeros, such as daily rainfall records gathered over a year. These types of data sets are now considered step-by-step.

4.1 Mildly Intermittent Sets

4.1.1 Rainfall Events

Faithful FM encodings of high-resolution rainfall events in Boston and Iowa City, lasting for a few hours, (not shown here) have been found. As reported, the FM method, coupled with a suitable search procedure, is capable of preserving not only the overall shapes of the records but also their main statistical and multifractal properties. In regards to the Boston event, gathered every 15 seconds, various FM representations based on three to five affine maps yield, for the accumulated objective function, ϵ_{ar} in Eq. (6), values smaller than 0.4% and maximum deviations in accumulated sets, ϵ_{max}, that are less than a mere 1.4% (Puente and Obregón 1996, Obregón et al. 2002a, Cortis et al. 2009, Huang et al. 2013). For four distinct events gathered in Iowa City, every five seconds, results are comparably good in terms of geometrical and statistical features, with ϵ_{max} values that are consistently below 2.5%, for alternative variants of the FM method (Obregón et al. 2002b, Huang et al. 2012a, b).

4.1.2 Daily Streamflow

Having obtained excellent encodings of rainfall events, the FM approach and its extensions are tested here as a modeling method for streamflow records gathered daily over a year. As these mildly-intermittent sets visually share similar "complexity" as the rainfall events, the FM approaches turn out to also yield faithful descriptions. This is illustrated in the following for 64 years of daily records gathered at the Sacramento River (USGS station 11447650 near Freeport) and for even smoother, decadal records for a total of 55 years, where a constant base flow is subtracted from the raw data and a water year (labeled by the end year) starts on October 1st and ends on the following September 30th. Prior to encoding, such records are normalized so that the accumulated volume becomes unity, as required for the FM methodology, which is used in Fig. 1, with a constant smoothing of 5 days.

Figure 4 shows an example of two FM encodings (in gray) of the streamflow set corresponding to the water year ending in 1965 (black). While representation A corresponds to a wire generated via two maps, B emanates from a wire generated via three maps. As seen, the overall geometry of the streamflow set is captured well in both cases, as the corresponding accumulated sets are hard to distinguish from the actual set. Such is corroborated by the small values of the root mean square and maximum errors in accumulated sets (ϵ_{ar} and ϵ_{max}) – below 2.1 and 4.0%, respectively – and the high Nash-Sutcliffe efficiency for the records, η_d, always above 66%, as reported in the figure. As may be expected, given its additional degrees of freedom (10 parameters), the encodings for representations B results in reasonable fittings of the autocorrelation, histogram and entropy functions, as implied by high Nash-Sutcliffe efficiencies (η_c η_h and η_e) – equal to 95, 95 and 96%, respectively, also as included in the figure.

After encoding all 64 streamflow years, from 1951 to 2014, Fig. 5 shows the implied performance for case B (a fractal wire based on the iteration of three maps), when FM encodings are upgraded by the yearly

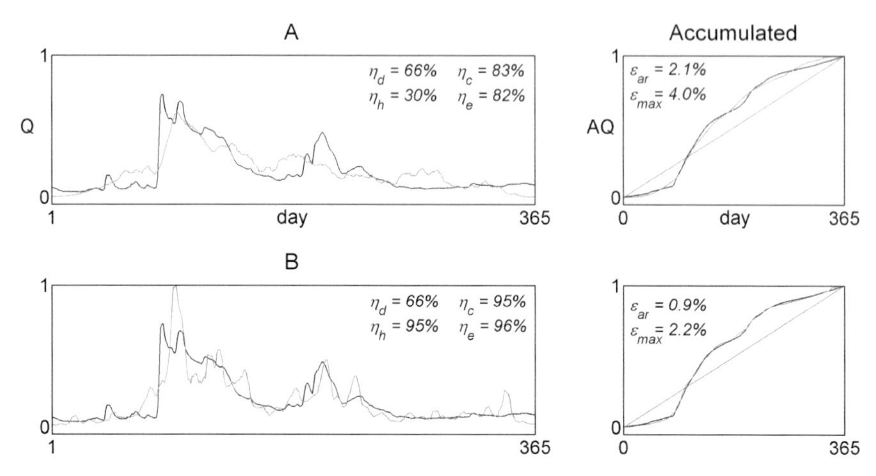

Figure 4: Measured streamflow at the Sacramento River for water year 1965 (black) and two FM representations (gray), corresponding to wires based on two and three maps.

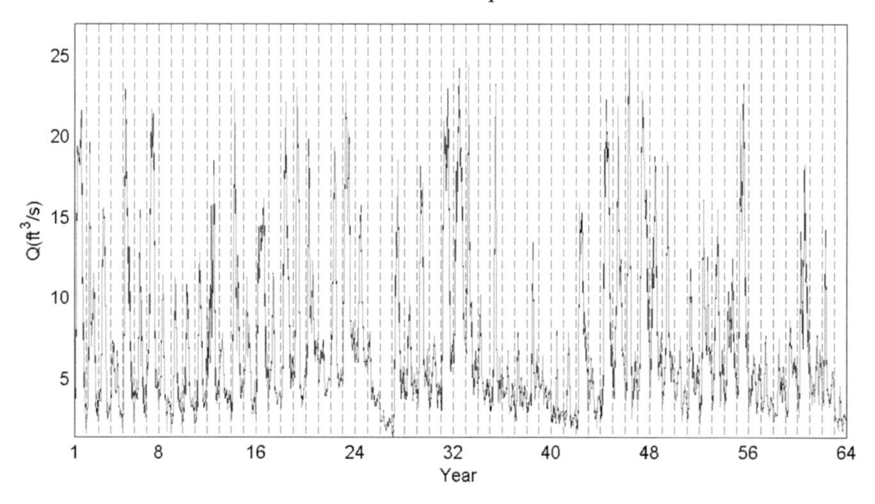

Figure 5: Observed (dark) and FM encoded (gray) streamflow for 64 years in the Sacramento River, from water year 1951 till 2014. The FM method uses wires based on three maps. The vertical scale of the graph is in 100,000.

volumes and then the constant base flow is added. As seen, the overall FM representation of streamflow is excellent visually (in gray) and also statistically, as the qualifiers $\in_{ar}$, $\in_{max}$, and η_d have means plus or minus standard deviations of $0.8 \pm 0.3\%$, $1.8 \pm 0.5\%$ and $64 \pm 19\%$, respectively.

As climate change studies often rely on information averaged over a decade, Fig. 6 includes the FM analysis of a typical decadal streamflow set at the Sacramento River (black) for two variants: a wire model based on

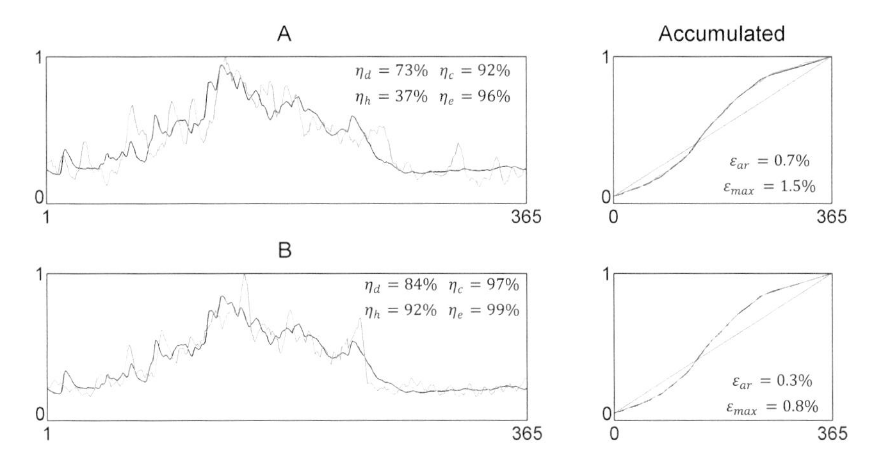

Figure 6: Measured streamflow at the Sacramento River for water decade 1966 (black) and two FM representations (gray), corresponding to wires based on two and three maps.

two maps (A) and a wire representation that uses three maps (B) (both in gray). Notice how the smooth decadal records are also well represented by the FM notions, as expected. The optimization exercise itself turns out to be easier this time than that for yearly sets (in part because parameters for the previous decade may be used as initial conditions for a future search) and such a fact gets reflected in a higher range of statistical qualifiers. As included in Fig. 6, for the decade presented, the ϵ_{ar} and ϵ_{max} values are found to be 0.5 and 1.1% for FM representation A (relying on 6 parameters) and 0.3 and 0.8% for FM variant B (employing 10 parameters), which are rather small numbers, smaller than the typical ones reported in Fig. 4.

As done before, the overall performance over the entire period of 55 streamflow decades is shown in Fig. 7, for the wire model based on three maps. Clearly, the agreement between data (black) and FM sets (gray) is again excellent, and such is reflected by small numbers for the attributes ϵ_{ar}, ϵ_{max} (0.5 ± 0.1, 1.0 ± 0.3, respectively), and noteworthy values of η_d (76 ± 12). As seen contrasting Figs. 5 and 7, the fits are better for the smoother sets.

As seen in Fig. 8, the FM representations in Figs. 5 and 7 (gray) result in very close preservations of the Spring flows at the Sacramento River (black), for the months of March, April and May (bottom to top). This fact is particularly useful as the FM encodings may be used to properly represent the main volumes in the river.

4.1.3 Daily River Water Temperature

Having faithful FM representations of streamflow sets motivates encoding water temperature at the Sacramento River. Since the lowest temperatures

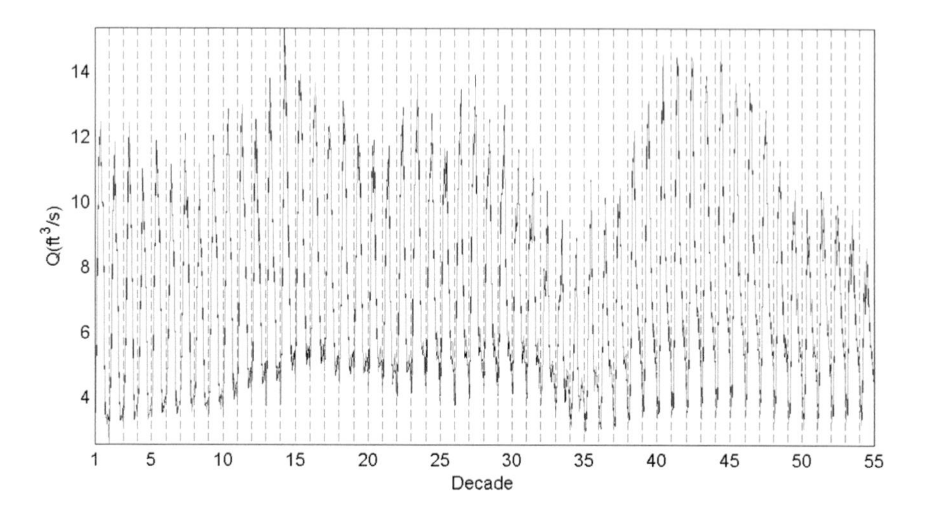

Figure 7: Observed (dark) and FM encoded (gray) streamflow for 55 decades in the Sacramento River, from water decade 1959 till 2014. The FM method uses wires based on three maps. The vertical scale of the graph is in 100,000.

over the year happen around December and January, instead of the water year cycle, records from January to December are considered, taking out a minimum temperature and normalizing, as done in case of streamflow. Once the records are defined, the FM approach with a 7-day internal smoothing is used for 51 years spanning 1962 to 2012.

A sample outcome of such an effort is shown in Fig. 9 for the year 1968, which includes close fittings of the accumulated temperature sets based on a wire, defined using two maps (gray). As seen, this set has excellent $\in_{ar}$ and $\in_{max}$ values of 0.2 and 0.4% and close to one Nash-Sutcliffe values η_d, η_c and η_e. Fig. 10 presents the corresponding overall temperature records as encoded via the wire representation based on a total of 6 parameters. As seen, the FM representations (gray) are truly excellent and such is reflected by rather high Nash-Sutcliffe efficiency for the records of 91 ± 4%. Compared to streamflow sets, temperature records are smoother and, at the end, easier to represent using the FM method. As illustrated, this is accomplished with less number of parameters.

4.2 Highly Intermittent Records

The deterministic FM method, in its Cantorian version coupled with a vertical threshold (Fig. 2), may also be used to encode daily rainfall records exhibiting noticeable intermittency. In what follows this is illustrated using 20 years of records gathered at Laikakota, Bolivia.

As an example, Fig. 11 in black shows rainfall records for water year 1966 (lasting nine months from September 1st of the previous year) and

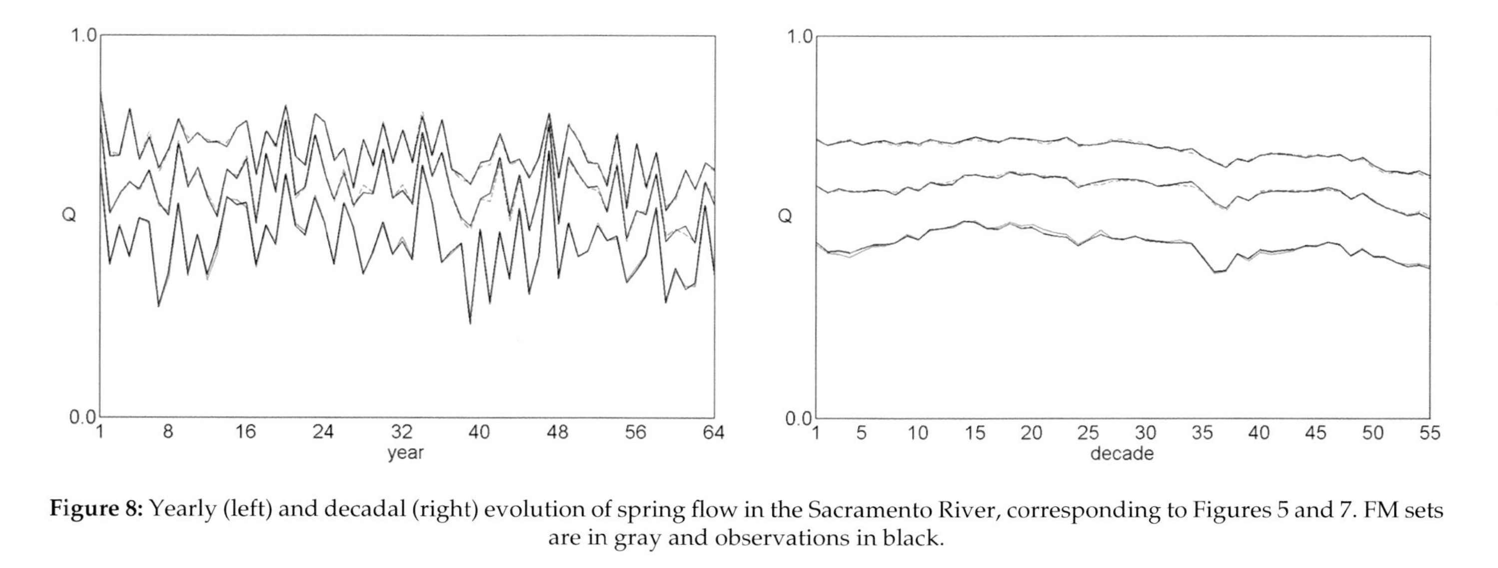

Figure 8: Yearly (left) and decadal (right) evolution of spring flow in the Sacramento River, corresponding to Figures 5 and 7. FM sets are in gray and observations in black.

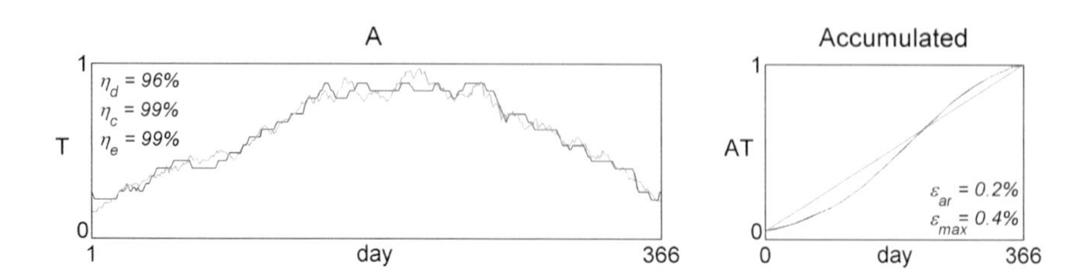

Figure 9: Measured water temperature in the Sacramento River for the year 1968 (black) and an FM representation (gray), corresponding to a wire based on two maps.

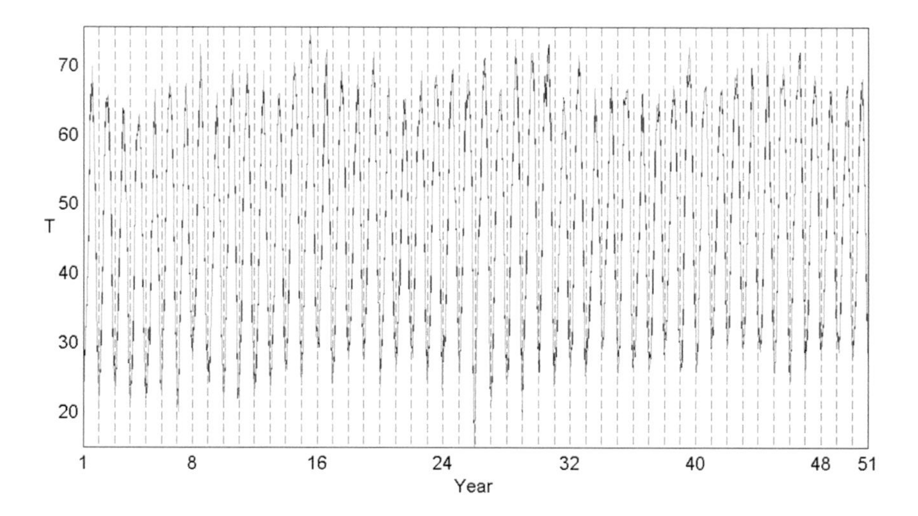

Figure 10: Observed (dark) and FM encoded (gray) water temperature records for 51 calendar years in the Sacramento River, from 1962 till 2012. The FM method uses wires based on two maps.

the corresponding accumulated mass set (top), and two alternative FM encodings (and accumulated sets) in gray, labeled A and B, found via Cantorian inputs and subsequent pruning via a threshold; see Maskey et al. (2015) for further details. These representations rely on the iteration of three maps and approximate the accumulated set closely as they have mean square and maximum errors in ϵ_{ar} and ϵ_{max} that are (as seen in the graph) less than 1.7 and 5%, respectively. Although the precise locations of major peaks on this complex data set are not perfectly captured, the FM encodings do preserve the volume of the most massive second peak, which may be seen in the corresponding steepest section of the accumulated profile. As seen, the FM sets not only share similar textures and overall

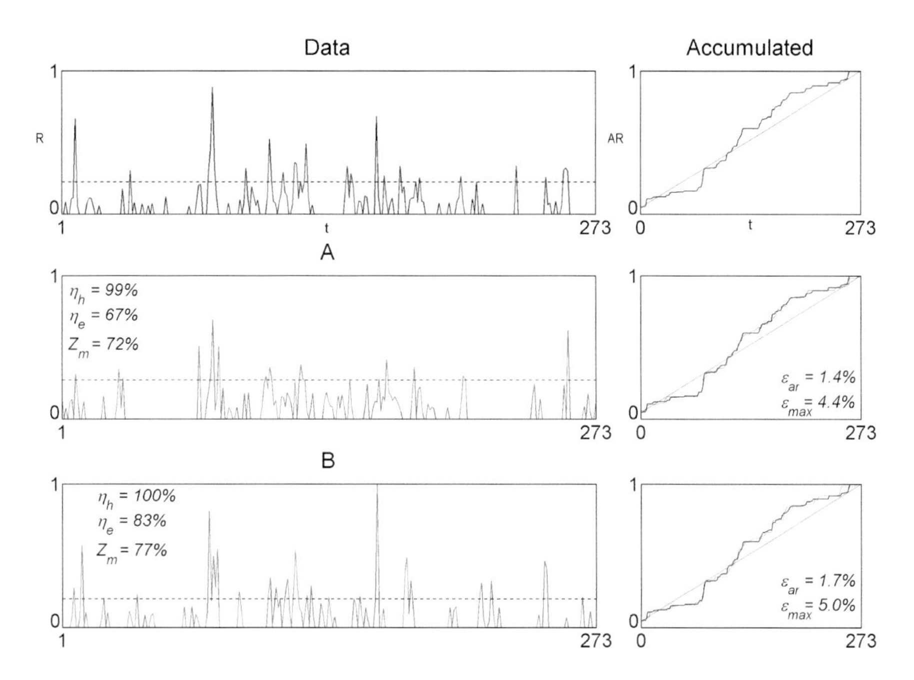

Figure 11: Measured rainfall at Laikakota, Bolivia from September 1965 to May 1966 (top-black) and two FM representations (bottom-gray), corresponding to Cantorian constructions based on three maps and a threshold.

distribution of rain throughout the season, but also closely capture the extremes in the set, as indicated by the shown horizontal dashed lines at 90% of the mass, especially the one named B. The goodness of the encodings may also be corroborated by the rather high Nash-Sutcliffe efficiencies on the histogram and entropy of the data set: η_h and η_e always above 99 and 67%, and by the percent of zeros matched by the encodings Z_m above 72%, as reported in the graph.

Figure 12 shows the results of applying the FM approach (gray) on each of 20 years of data at the Laikakota site (black, from 1963 to 1984), but using a representation that is best in maximum accumulated error $\in_{max}$ while using only two maps and a threshold. Although the obtained approximation is not as faithful as the one based on three maps, the overall behavior over the years remains good, as optimized accumulated values, $\in_{ar}$, range from 1.3 to 2.2% and maximum accumulated errors, $\in_{max}$, span 3.5 to 8.8%. Notice how the real set and the FM encoding share similar (although non-identical) characteristics which, by having the mass properly accommodated within each individual year, result in almost indistinguishable overall accumulated rain over the 20 years, as seen in Fig. 12(b). It is quite remarkable that the FM approach, relying on only 9 parameters, may accomplish such a feat.

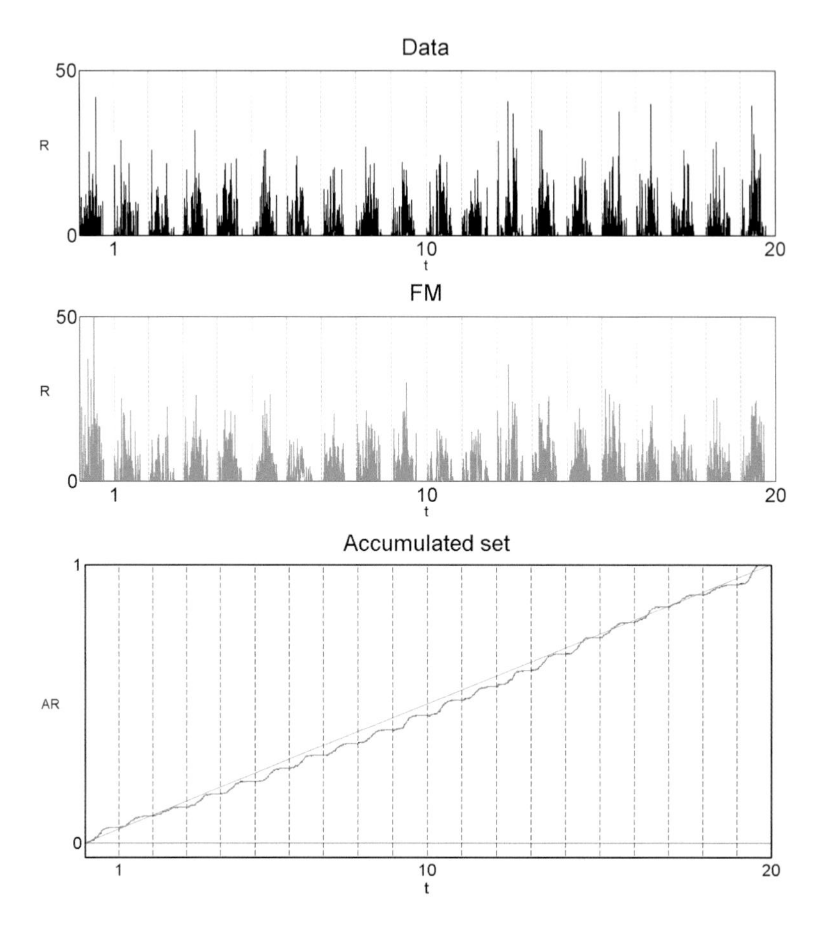

Figure 12: (a) Measured rainfall at Laikakota, Bolivia from 1964-1965 to 1983-1984 (top-black) and a FM representation (bottom-gray), corresponding to Cantorian constructions based on two maps and a threshold; and (b) corresponding accumulated rainfall sets over the period (After Maskey et al., 2015).

4.3 Remarks

The above results have clearly illustrated how the FM approach and its variants (depending on 6 to 15 parameters) may be coupled with a heuristic optimization scheme, especially the GPSO, in order to approximate the specific geometry of a host of geophysical patterns. As seen, the more the intermittency on a set the harder it is to encode it, with daily water temperature patterns being the easiest followed by daily streamflow records and daily rainfall sets, which are the most difficult. Certainly, the more complex-looking a given target is, the higher the number of maps required in the iterations and, hence, the higher the number of parameters, leading to more involved searches, which are compounded if a threshold and penalties, involving an inherent discontinuity, are required. Although some calculations may be accomplished on a standard personal computer, calculations for alternative sets require running the optimization procedure from scratch, as distinct sets have variable geometries.

As both rainfall and streamflow records are not measured perfectly (e.g., Lanza and Vuerich 2009, Hammel et al. 2006), the encodings shown, within 3% in accumulated records, represent sensible approximations of the natural phenomena. These results, providing geometries that match the overall shapes and textures present in the sets, clearly substantiate the notion that complex geophysical patterns may be wholly characterized in a deterministic way (Puente and Sivakumar 2007). As the information in the sets is now compressed into the FM parameters of subsequent sets, an assessment of such evolutions shall be presented later in Section 7.

5. Simulation of Geophysical Records

Having shown that the FM method may produce faithful encodings of geophysical records, this section shows that the deterministic parsimonious approach may also be used to obtain simulations having increased levels of complexity, namely, rainfall events, daily streamflow sets, and daily rainfall sets. This is accomplished by replacing the objective function (Eq. 6), so that instead of accumulated sets it now uses suitable statistical information, such as autocorrelations, histograms, entropies, and numbers of zero values.

5.1 Mildly Intermittent Sets

To illustrate the notions with mildly intermittent sets, one rainfall event in Iowa City and one year of daily streamflow records gathered at the Sacramento River, California, are analyzed. In both cases, alternative plausible simulations are found by minimizing the root mean square errors for correlations ϵ_c and histograms ϵ_h, employing two maps that generate either a wire or a leaf (as in Figs. 1 and 3). While the simulated streamflow records rely on a smoothing of the obtained FM set (Figs 1 and 3), the rainfall set does not require of a local average of the output derived projection (Fig. 2). As a consequence, the numbers of FM parameters for the wire and leaf representations of rainfall are 5 and 7, while for streamflow they are 6 and 8.

5.1.1 Rainfall Events

Figure 13 displays a high-resolution storm event gathered in Iowa City in 1990 and lasting 11.4 hours (Georgakakos et al. 1994) together with the records' autocorrelation function and histogram (top-black),

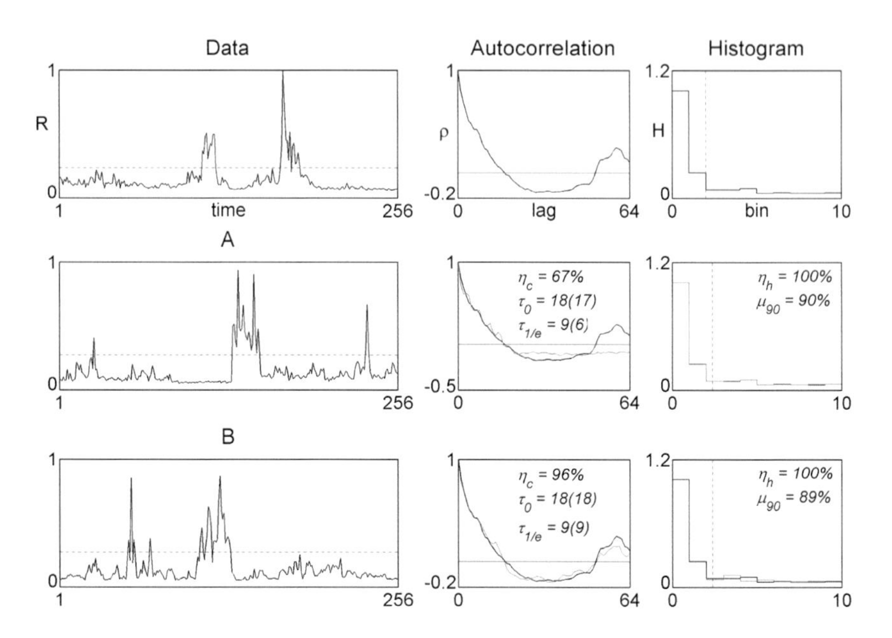

Figure 13: A rainfall event in Iowa City (top) and two suitable FM simulations (bottom), based on a wire and a leaf representation using two maps each. While A is based on fitting the histogram, B optimizes the autocorrelation.

followed by two FM simulations that preserve either the histogram or the autocorrelation function of the records (bottom-gray). As seen, while the original storm contains two prominent regions of activity, the simulations yield sensible sets having similar-looking geometries that spread the mass and peaks at different locations.

As can be appreciated via the statistics in the graph, both representations (A from a wire fitting the histogram only and B via a leaf seeking the autocorrelation only) preserve very well both the autocorrelation and histogram. Nash-Sutcliffe values for the histogram η_h are rather close to 100% and the location of the extreme 90% values μ_{90} match very well, as seen by the close horizontal and vertical dashed lines in the graphs. Although Nash-Sutcliffe statistics for the autocorrelation for case A is not as high as in case B, 67% vs. 96% in η_c notice the close agreement on the decay of the function for case A not optimizing such an attribute, as indicated by the number of lags when zero and $1/e$ correlations are found, τ_0 and $\tau_{e^{-1}}$.

5.1.2 Streamflow

Figure 14 shows a year of daily streamflow record gathered at the aforementioned station of the Sacramento River, in water year 2003,

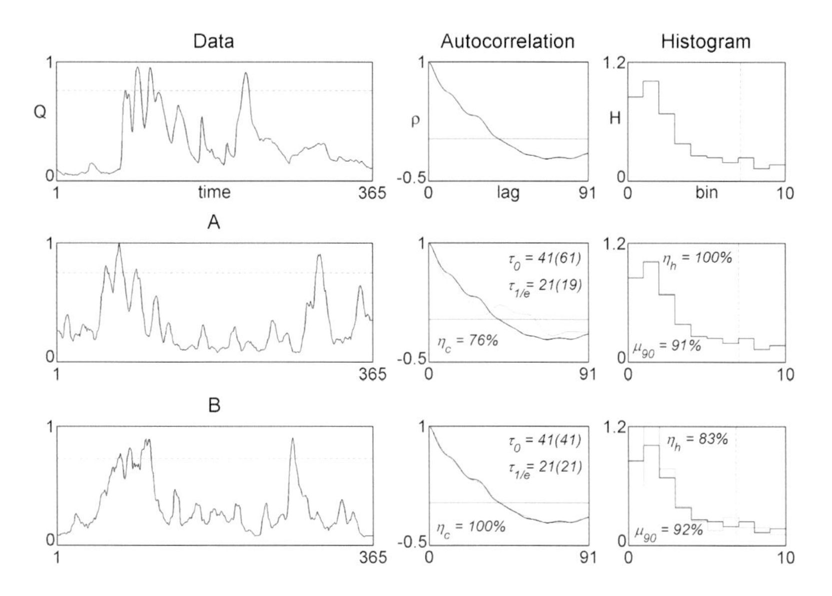

Figure 14: A streamflow set at the Sacramento River from October 2002 to September 2003 (top) and two suitable FM simulations (bottom), based on a wire and a leaf representation using two maps each. While A is based on fitting the histogram, B optimizes the autocorrelation.

together with the record's autocorrelation function and histogram (top-black), followed by two FM simulations that, once again, preserve either the histogram or the autocorrelation function of the record (bottom-gray). As can be observed, while the original record and the simulations do look alike geometrically, the latter spread the mass in different ways, hence yielding peaks at various locations.

As done for the rainfall events just described, while simulation A uses a wire based on two maps in order to preserve the record's histogram, the one labeled B emanates from a leaf based on two maps that seeks to fit the data's autocorrelation function. Unlike what was done for rainfall, though, a smoothing parameter of 7 days is used to generate the simulated sets. As may be seen, the shown simulations preserve almost perfectly the statistic included in the objective function: while simulation A matches the histogram, simulation B very closely preserves the autocorrelation function. Such may be corroborated by the corresponding Nash-Sutcliffe statistics close to 100%, rather close values of μ_{90} close to 90%, and perfect fittings on the decay of the autocorrelation function via the lags τ_0 and $\tau_{e^{-1}}$. Although preserving one attribute does not imply close fittings of the other statistic, notice how the shown simulations do reasonably well on a statistic not included in the objective function, with η_c and η_h values being substantially large at values of 76 and 83%.

5.2 Highly Intermittent Records

To illustrate that the FM approach may also be used to simulate highly intermittent sets, a year of daily rainfall gathered at Tinkham Creek, Washington State, USA is analyzed. As autocorrelations of rain decay towards zero rather quickly, alternative plausible simulations are found by minimizing the root mean square errors for entropy ϵ_e and histograms ϵ_h, employing two maps that generate either a wire or a Cantorian attractor, both coupled with a threshold (as in Figs. 1 and 2). As such, the numbers of FM parameters for the wire and Cantorian simulations are 6 and 8, respectively.

Figure 15 displays rainfall record at Tinkham Creek for water year 2001 together with the record's histogram, Renyi entropy, and autocorrelation function (top-black), followed by two FM simulations that preserve either the entropy function or the histogram of the record (bottom-gray). The original set and the two simulations share similar looking intermittent geometries that, as found for mildly intermittent sets, spread the mass throughout the year giving rise to distinct peaks.

While simulation A uses a wire based on two maps and uses a threshold in order to preserve the record's entropy, the one labeled B

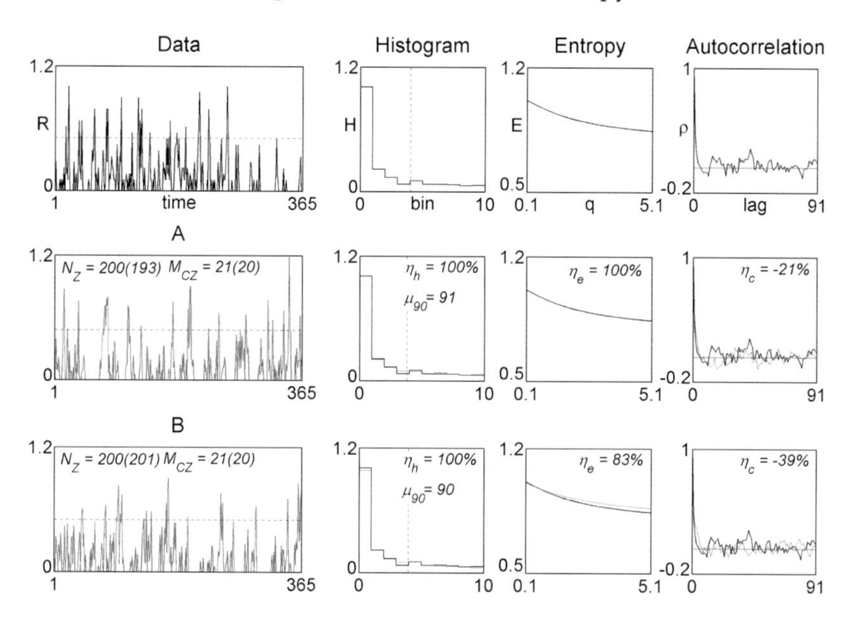

Figure 15: A rainfall set at Tinkham Creek, Washington from October 2000 to September 2001 (top) and two suitable FM simulations (bottom), based on a wire and a Cantorian representation using two maps each. While A is based on fitting the entropy, B optimizes the histogram. The two simulations also account for consecutive zeroes in the records.

comes from a Cantorian attractor generated via two maps and a subsequent threshold that seeks to fit the histogram of the data. Figure 15 shows that the simulations preserve almost perfectly the statistic included in the objective function: while simulation A matches the entropy (and also the histogram, but not optimized), simulation B very closely preserves the histogram (and does not do poorly on the entropy as implied by an η_e value of 83%). This performance – no doubt reflected by the addition of penalties into the objective function so that the number of zeros on the records and FM simulation differ at most by 10% – explains the intuitively correct feeling on the simulations. For as seen on the graphs, the total number of zeros N_Z, the maximum consecutive number of zeros M_{cz}, and the extreme μ_{90} values are all almost perfectly preserved, even if negative Nash-Sutcliffe values for properly decaying but statistically insignificant autocorrelations η_c are obtained.

5.3 Remarks

The above results show that the FM approach (and its variants) may be coupled with a heuristic optimization scheme in order to simulate a host of geophysical patterns. This is achieved encoding suitable statistics of the records rather than the records themselves, such as autocorrelation, histogram, entropy, and others. The examples herein present just a couple of possibilities, but additional sets may also be obtained by having representations based on more than two maps, by selecting various FM parameter combinations in the vicinity of an optimum, and by combining two or more attributes on an alternative objective function.

Based upon our experience, finding suitable simulations may be achieved easily and a fraction of the time required for doing encodings. As the employed statistics are much less complex than the sets, the simulations may be found (for both mildly and highly intermittent cases) iterating just two maps, hence yielding parsimonious FM representations. As it was illustrated, the simulated sets do resemble geometrically the processes under study, including their overall texture and intrinsic complexity. As such, the FM simulations may be useful to analyze the redistribution of patterns throughout the event duration and to provide alternative scenarios from which to study the intrinsic variability of a phenomenon. Certainly, the FM patterns may be used to supplement other plausible simulations found via alternative (stochastic) methods.

6. Downscaling Geophysical Sets

This section explains how the deterministic geometric FM approach may also be used to downscale geophysical sets, in particular streamflow and rainfall sets from coarse scales (e.g., weekly, monthly) to fine scales (e.g., daily). While various stochastic disaggregation techniques exist (e.g., Valencia and Schaake 1973, Koutsoyiannis 1992, 1994, Olsson 1998), they require simplified assumptions that often prevent them from adequately accounting for nonlinearities present in the records. As such, the FM method may provide an alternative approach to complement such stochastic techniques.

Given information at a coarse resolution, say accumulated every τ days, the FM method may be used to find daily values, as follows. First, accumulate the coarse records during the duration of the season in question, say a year, and then seek a suitable FM encoding of such accumulated records (as explained in Section 3). Second, given the FM parameters of such an approximation, compute the disaggregation as the output projection dy obtained at the daily resolution, just by re-computing the output histogram over the appropriate number of (increasing) bins. As the accumulated sets for fine and coarse records surely match every τ days, a close FM approximation at the coarse scale should result in a good representation at the daily scale, one that, by definition, should remain close to the overall accumulated mass. In the present study, these notions are tested for streamflow and rainfall sets using coarse scales for $\tau = 7$, 14 and 30 days.

6.1 Mildly Intermittent Records

Figure 16 shows the streamflow records at the Sacramento River for water year 2005 (minus baseflow), evaluated (from available daily values) at the aforementioned levels of aggregation (black) and the subsequent FM fits and disaggregations found via wires based on three maps and a local smoothing of 5 days (gray).

As seen on the left of the figure and as it may be expected given the previous results herein, all FM representations at the weekly, bi-weekly, and monthly resolutions produce rather faithful fits of sets and accumulated masses, with relevant search errors ϵ_{ar} and ϵ_{max} that are always below a mere 0.7 and 3.9%, respectively, and with corresponding Nash-Sutcliffe values η_d that are above a healthy 84%. As appreciated on the right hand side of the graph, the FM variants at distinct resolutions yield reasonably-looking disaggregated patterns at the daily scale, which not only closely follow the accumulated records but also preserve, even when using 30-day data, the overall texture and three main peaks of the original set.

As expected, the best disaggregated pattern, both statistically and by the naked eye, corresponds to the 7-day case. This is so, as the accumulated records at such a resolution are the closest to reality of the ones considered. Although in terms of Nash-Sutcliffe values η_d the downscaled data from 30 days are better than those from 15 days, notice that both representations

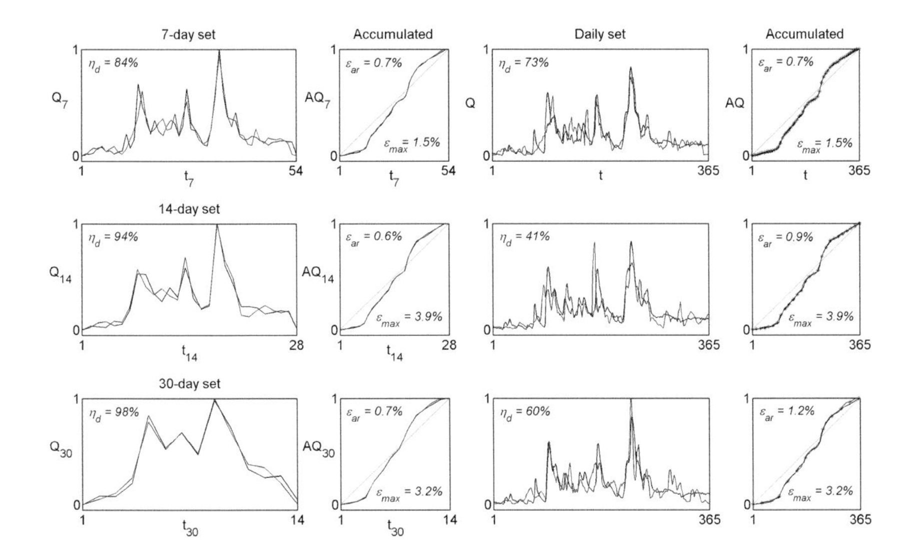

Figure 16: Sets and accumulated records associated with streamflow gathered at Sacramento River, during water year 2005. Observations (black), FM fits at coarse resolutions (gray-left); and corresponding downscalings at a daily scale (gray-right). The aggregation scales are 7, 14 and 30 days. Measurements are shown in black and FM-related information, via a wire based on three maps, are depicted in gray.

are indeed suitable as they have rather small values in $\in_{ar}$ and $\in_{max}$ that are less than 1.2 and 3.9%, respectively. As seen, all downscales are quite reasonable and they also happen to provide sensible approximations of relevant statistical information, such as autocorrelation and histogram.

6.2 Highly Intermittent Records

Figure 17 depicts rainfall records at Laikakota Bolivia for water year 1966 evaluated (from available daily values) at 7, 14 and 30 days (black) and the subsequent FM fits and rainfall disaggregations defined via wires based on three maps and a threshold (gray).

As observed on the left hand side of the figure and as it may be expected given the intrinsic complexity of rainfall sets, even though the accumulated records may be encoded with errors $\in_{ar}$ and $\in_{max}$ that are always below 3.1 and 8.7%, respectively, the corresponding encodings do not show the same degree of faithfulness as just described for streamflow. These lead, however, to the downscales shown on the right hand side, that while corresponding to increased errors $\in_{ar}$ and $\in_{max}$ below 4.3 and 11.3%, respectively, (about twice as much as what may be obtained while encoding a daily set, as in Fig. 11), do provide suitable shapes that may

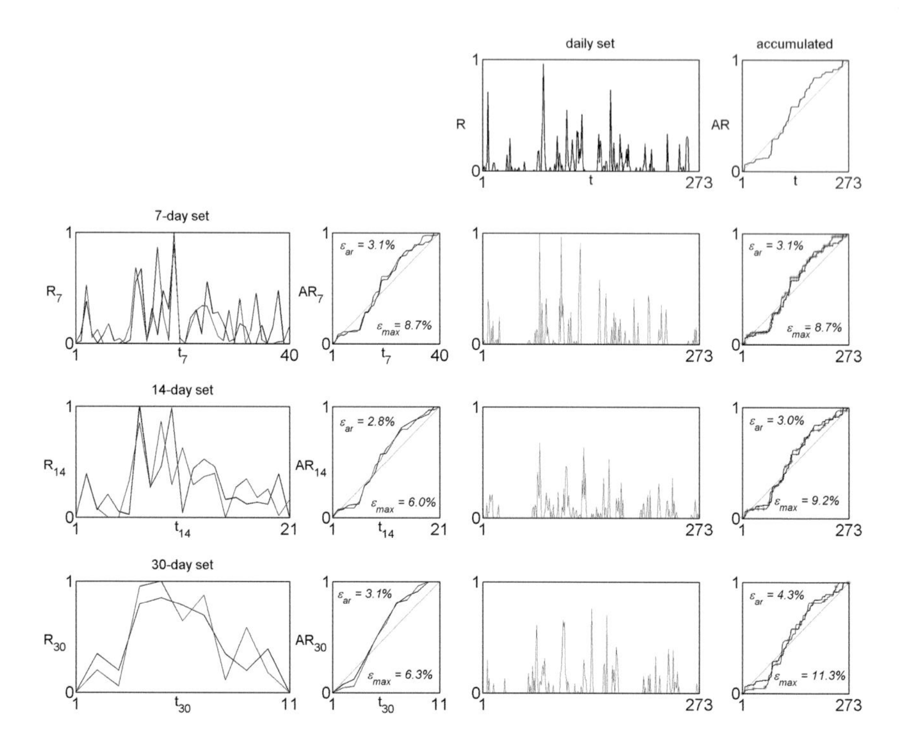

Figure 17: Sets and accumulated records associated with rainfall gathered at Laikakota, Bolivia, during water year 1966. Observations (top), FM fits at coarse resolutions (bottom-left); and corresponding downscalings at a daily scale (bottom-right). The aggregation scales are 7, 14 and 30 days. Measurements are shown in black and FM-related information, via a wire based on three maps and a threshold, are depicted in gray.

be useful in a practical setting. Certainly, the geometries of the implied downscales are sensible and they may also be used as suitably close simulations of the records, that is, representations that follow closely the distribution of rainfall throughout the year.

6.3 Remarks

The above results have clearly shown that the FM approach may be used in order to downscale streamflow and rainfall sets to the daily scale, achieving reasonable results for scales that span up to 30 days. It has been illustrated that the downscaled sets resemble the texture and implicit complexity of the involved process, as they represent suitable geometries that pass closely to the accumulated daily records. The results have established that while runoff may be downscaled with noticeable precision, the disaggregation of more complex rainfall results in only some reasonable approximations that maintain the intermittency and overall distribution of rain during the year.

The application herein shows that only 10 FM parameters may be adequate in disaggregating weekly, bi-weekly and monthly sets, with compression ratios that are as high as 37:1 for a given year. The present downscaling technique may supplement others based on stochastic methods and may clearly be used to disaggregate outputs from global circulation models (GCMs) in order to assess climate change impacts.

7. Geometric Classification and Prediction of Geophysical Records

This section is concerned with the possibility of predicting geophysical records based on the time evolution of successive FM parameters of such annual sets. Specifically, the analysis herein centers on the streamflow sets at the Sacramento River encoded before (Section 4.1.2), which shall be attempted to be forecasted based on FM-based classifications of patterns, both at the yearly and at the decadal scales. Due to space considerations and as streamflow is a relevant indicator of catchment dynamics and of potential climate change effects (e.g., Döll and Schmied 2012), the study centers on such an attribute. A similar approach, however, may also be carried out on other geophysical variables (e.g., rainfall, water temperature, etc.).

7.1 Evolution of FM Parameters for Streamflow

As previously reported, 64 years and 55 decades of successive streamflow records at the Sacramento River have been faithfully encoded via the FM approach, i.e., Figs. 5 and 7. As all such representations emanate from FM wires based on the iteration of three simple maps, the corresponding FM parameters of such sets, together with the evolving volumes of years or decades, allow visualizing (in a compressed geometric fashion) the dynamics of streamflow at the two scales.

Figures 18 and 19 present the time evolution of FM parameters for yearly and decadal encodings, respectively, as follows. As all encodings use a local smoothing parameter of 5 days (as in Fig. 1), the two sets of graphs include a total of nine FM parameters, followed by the total volume Q of a year, namely: the coordinates of the second and third interpolating points (x_1, y_1) and (x_2, y_2) (having set, without a loss of generality, the first interpolating point to (0, 0)), the vertical scaling parameters of the three maps, d_1, d_2 and d_3, and the weights that determine how the maps are iterated, p_1, p_2 and $(1 - p_1 - p_2)$% of the time. As may be appreciated, the graphs also include local means (in gray) of each parameter over a third of the domain (shown on top of each parameter frame) in order to visualize possible trends in the parameters.

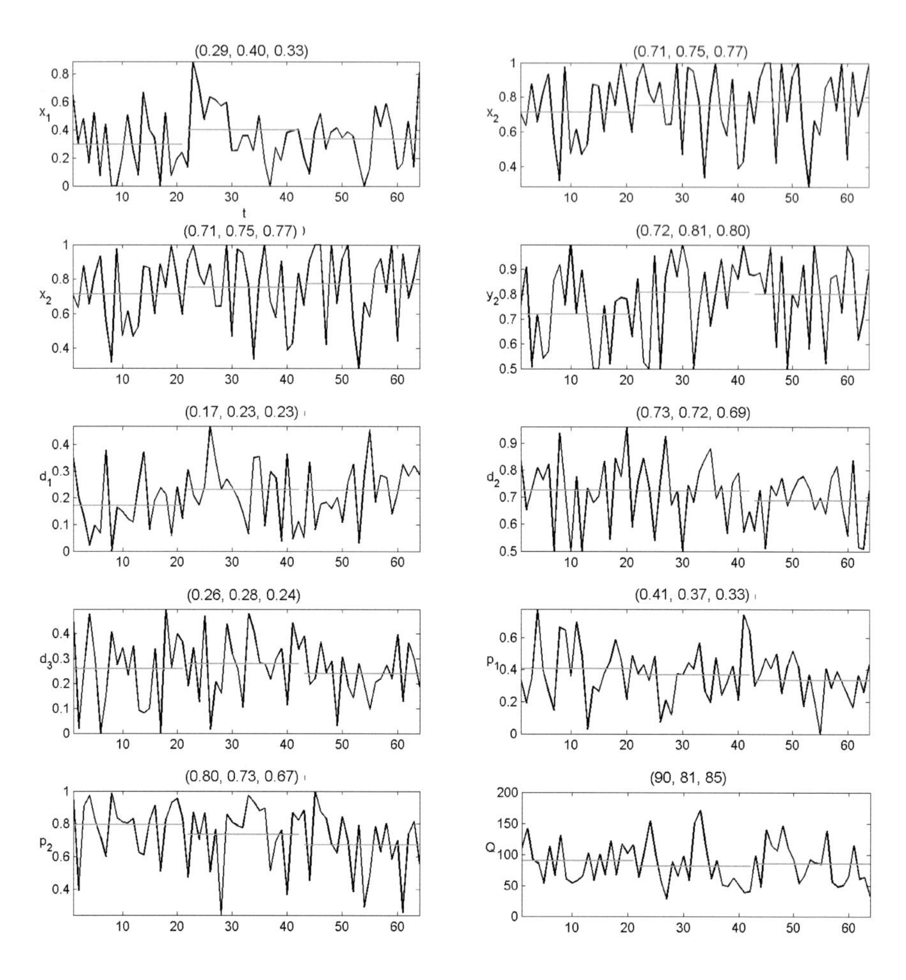

Figure 18: Time evolution of FM parameters for 64 years in the Sacramento River, corresponding to Figure 5. Gray lines represent mean trends over thirds of the time domain, with values shown above, is the total volume over a year in 100,000 cfs.

As seen while comparing the graphs, it becomes obvious that there is much more variability at the yearly scale (Fig. 18) than at the decadal scale (Fig. 19), in a manner that resembles the different shapes already reported for the spring flows at those scales in Fig. 8. Clearly, yearly FM parameters (and total flow) vary wildly and swing from high to low values and vice-versa in a manner that precludes the possibility of safely extrapolating trends into the future. Although it is clear that these variable signals do play a key role in understanding the complexity of the evolving streamflow, the geometry of the observed patterns change in a non-trivial fashion from year to year and such leads to non-specific changes every 21

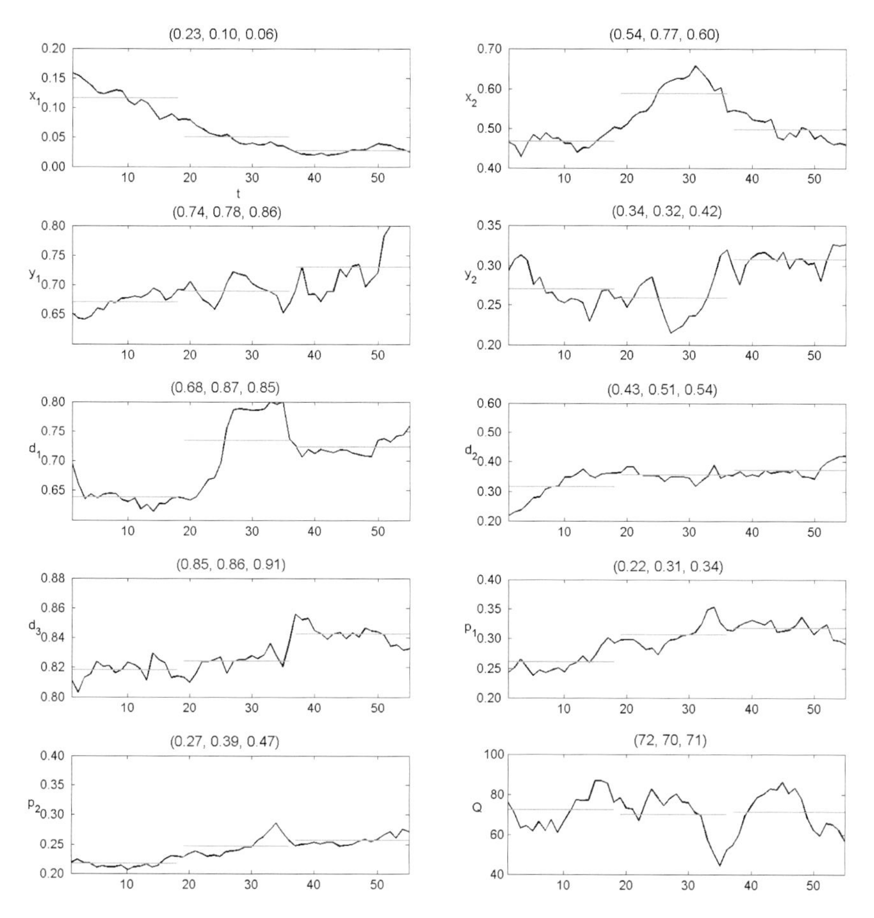

Figure 19: Time evolution of FM parameters for 55 decades in the Sacramento River, corresponding to Figure 7. Gray lines represent mean trends over thirds of the time domain, with values shown above, is the total volume over a decade in 100,000 cfs.

years (as shown in gray), which exhibit so much variability within that forecasts of individual parameters are not possible.

The evolution of FM decadal parameters and total volume in Fig. 19, on the other hand, do exhibit the expected smoothing produced by the fact that successive decadal patterns, besides being smoother, do look alike. The decadal evolutions clearly exhibit a degree of smoothness and some well-established trends when seen every 18 years (as shown by the gray horizontal lines) and such suggests that extrapolations of some FM parameters may be made into the future. Even if the flow Q remains quite variable at the decadal scale, the smaller variation in FM parameters could be used to foresee the overall geometry of decades into the future, which in turn may be translated into a yearly (disaggregated) normalized geometric prediction when decades and years share relevant correlations. Regarding plausible geometric implications of the decadal streamflow patterns to global climate change, it is interesting to note that some of the parameters in Fig. 19 do exhibit increasing and/or decreasing trends that may perhaps be related to climatic indicators.

Similar trends as those found in Figs 18 and 19 are also detected for other rivers within the United States (not reported here). While yearly FM parameter values, reflecting variable geometries, vary substantially from year to year, decadal FM parameters exhibit smoothness and some trends that allow extrapolating trends. It is anticipated that the information on such graphs may be used to study the inherent complexity of alternative locations (say by the strength of the swings in parameters) and to evaluate the presence of climatic effects at distinct regions.

7.2 Classification of Streamflow Sets via Clustering of FM Parameters

As the evolutions of FM parameters at yearly and decadal scales show ample variations (especially at the yearly scale), it becomes natural to inquire if a classification of patterns into classes may be made in order to arrive at more stable descriptions. As such, Fig. 20 shows the streamflow centroids for ten classes obtained via an unsupervised classifier, the *k*-means clustering analysis based on the Euclidean distance of FM parameters (e.g., Arthur and Sergi 2007). The shown graphs, which once again reflect the wider spread of yearly records, summarize the records only up to 1999 (and not all the records), and such are computed in an attempt to study if sensible streamflow predictions for the decade ending in year 2000 and, subsequently, for the water year 2000 may be obtained.

Having defined ten classes on yearly and decadal streamflow at the Sacramento River, Fig. 21 now shows their evolution from 1951, that is, 49 years (black) and 40 decades (gray). As expected, the classes at the yearly scale still exhibit noticeable variability, but the decadal information steadily grows in the selected classes, in a manner that suggests a stable class prediction for the decade ending in year 2000.

With this evolving information in hand, it is explained next how it may be possible to further synthesize the class evolutions in order to try to establish relations between past years and decades that may perhaps be useful in predicting as well. Following a Markovian framework, Fig. 22 summarizes the transition matrices that may be defined based on class information: from yearly to yearly, from decade to decade, and from decade to yearly sets. For instance, while the decade to decade shows simple diagonal patterns that suggest a reasonable predictability at such a scale, the yearly to yearly and decadal to yearly matrices exhibit broader distributions that nonetheless may trim away some states, say, once a prediction of the next decade is known.

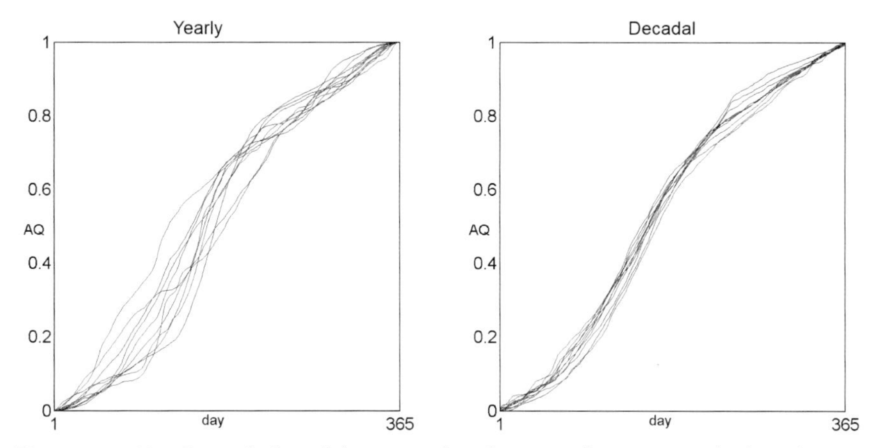

Figure 20: Yearly and decadal accumulated streamflow centroids based on a classification of FM parameters for data up to 1999 at the Sacramento River.

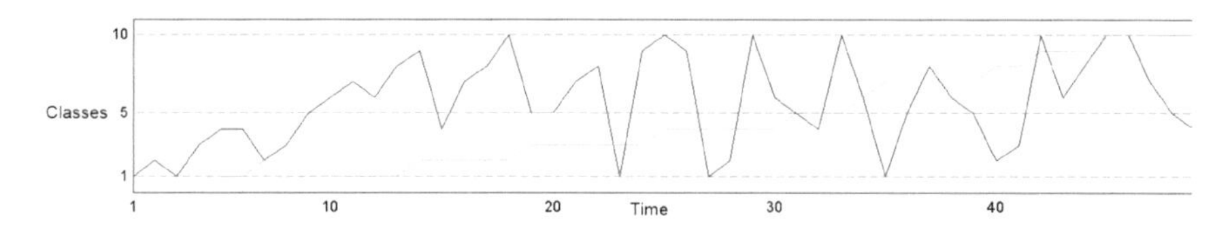

Figure 21: Yearly (black) and decadal (gray) streamflow evolution by classes for the Sacramento River up to 1999, as defined in Figure 20.

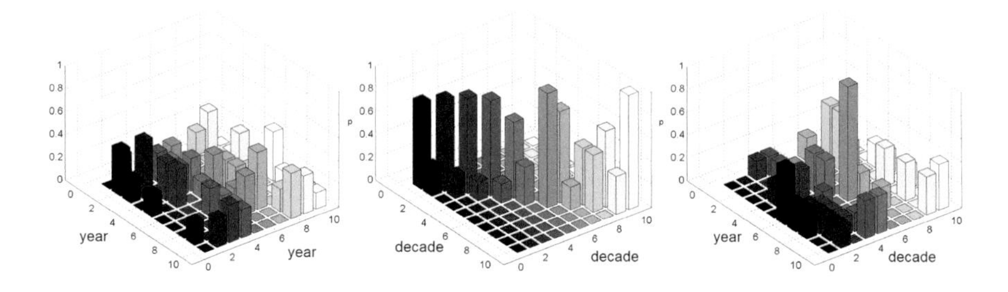

Figure 22: Transition matrices for streamflow dynamics at the Sacramento River up to 1999 corresponding to the classes in Figure 20. Graphs are read from right to left.

7.3 Predictions of Streamflow via FM Parameters and Classes

Given the synthesis of FM parameters into centroids of classes and the calculation of transition matrices up to a given year, various ideas may be used in order to define suitable predictions for the next water decade and then the next water year. As the FM parameters and classes for yearly streamflow records exhibit notorious variability for the Sacramento

River (i.e., Figs 18 and 21), predicting the next year would have to rely on forecasts of the next decade and then transfers of such via transition matrices. Some variants for predicting streamflow at the decadal and then the yearly scales are given next.

7.3.1 Decadal Predictions

Given that the evolution of decadal FM parameters and classes exhibit smoothness, it becomes logical to employ a statistical time series representation (i.e., an ARMA model) on the individual parameters and to extrapolate the class of the last available decade to the following decade.

In this spirit, Fig. 23 shows the water decade ending in year 2000 (solid black) and three plausible predictions stemming from information gathered up to year 1999: (a) a decadal set given by the FM decadal parameters extrapolated via time series models to 2000 (solid gray), (b) the set corresponding to the predicted class for year 2000 defined as the one having the largest transition probability emanating from the class of decade 1999 (dashed black), and (c) the set obtained by averaging the FM parameters used in defining the previous two sets (dashed gray). As may be seen, although the three "predictions" are less smooth than the records, their accumulated sets – depicting the timing of decadal streamflow over the year – yield rather reasonable patterns for the Sacramento River. In fact, the mean square and maximum errors in accumulated sets (ϵ_{ar} and ϵ_{max}), are, in order, (3.4, 7.2), (1.3, 2.3), and (1.7, 4.0), in percent, whose small ranges are typically encountered while repeating the analysis for other years (decades).

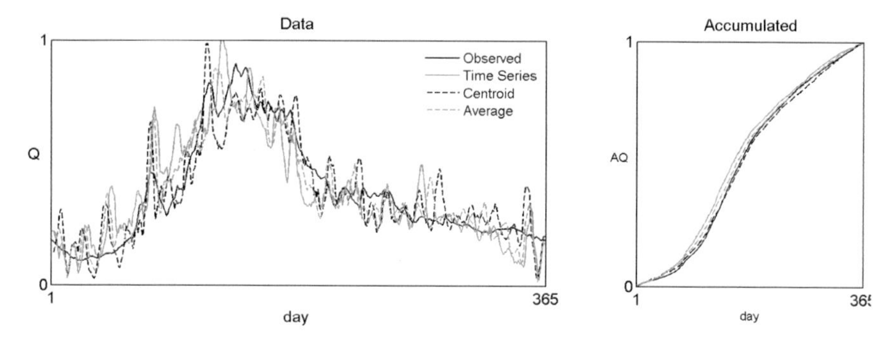

Figure 23: Plausible predictions for water decade 2000 based on FM parameter evolution and FM classes.

7.3.2 Yearly Predictions

As already mentioned, it is not possible to build meaningful time series models from the parameters in Fig. 18. However, predictions may be defined by: (a) looking at the class evolutions at the yearly scale (considering the future class that maximizes the transition probability from the current state) and (b) considering predictions at the decadal scale and transforming such into the yearly scale using the decadal to yearly transition matrix, both as reported in Fig. 22.

To continue with the example, Fig. 24 illustrates the notions for water year 2000 (solid black) and "successful" predictions based on: (a) year to year class predictions as just explained (solid gray), (b) decadal to yearly extrapolation based on the centroid-decadal prediction for decade ending in 2000 (i.e., from the best decadal prediction in Fig. 23) (dashed black), and (c) the set obtained by averaging the FM parameters used in defining the previous two sets (dashed gray). As may be seen, although the three "predictions" exhibit intrinsic variability and do not capture in detail the main peak on the yearly records, their accumulated sets – portraying the timing of streamflow over the year – yield reasonable patterns for the Sacramento River in year 2000, especially the prediction obtained by transferring the faithful predictions at the decadal scale. The mean square and maximum errors in accumulated sets for these predictions ($\in_{ar}$ and $\in_{max}$), are, in order, (5.5, 15.0), (2.0, 4.9), and (3.1, 8.8), in percent, which, although larger than the ones obtained for decadal information, hold promise that a geometric approach may produce holistic streamflow forecasts.

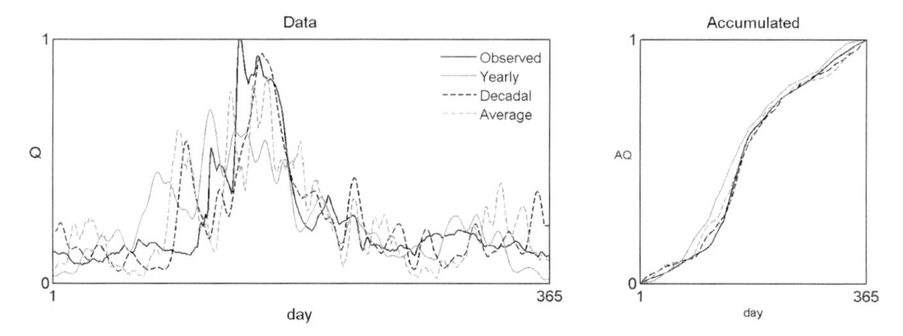

Figure 24: Plausible predictions for water year 2000 based on FM classes and transition matrices.

Even though the quality of the forecasts just presented do vary with the year considered (at the Sacramento River and elsewhere) and there are instances in which the notions provide unfaithful predictions (with maximum errors in accumulated sets greater than 30%, as reflected by the spread in Fig. 20), there is yet another idea that may be used to try to create alterative predictions. Following the same overall notions, the idea is not only to use a predicted centroid but also the years associated with such a class and then generate, in a combinatorial sense, various sets of FM parameters. These may be used to obtain a certain number of predicted "realizations" that may be analyzed to find average holistic forecasts augmented by their spreads. Results from this idea shall be reported elsewhere.

7.4 Remarks

The above results have illustrated how the FM ideas may be employed, in conjunction with further synthesis of available information, in order to define plausible future holistic scenarios of a geophysical data set at the yearly and decadal scales, and beyond. It has been shown that the FM parameters of successive streamflow sets allow us to: (a) visualize the evolution of patterns providing relevant hints about the geometric complexity of the process, (b) establish a geometric classification of sets based on clustering of FM parameters, (c) develop relations from past information into the future via transition matrices from year to year, decade to decade, and decade to year, and (d) propose a methodology for finding holistic forecasts of future years and decades.

Although surmising accurate future scenarios remains a challenge and should be tried broadly before offering widespread generalizations, it is envisioned that the ideas herein may be useful to quantify the geometry of streamflow in different rivers in distinct geographical regions, leading to a better understanding of their implicit complexity. Certainly, the same notions may be tried on sets beyond streamflow and such may naturally include observations of rainfall, water and air temperature, evaporation, and others. In regards to the water temperature and rainfall records shown before (Figs. 10 and 12), it may be said that their FM parameters exhibit ample variability (qualitatively as much as found in the yearly streamflow records), something that could be expected for the rainfall set but not so for the smoother temperature patterns. What these results stress (not shown due to space restrictions) is the fact that geometric variability from year to year is rather common in nature, which explains why it is not easy to describe and forecast geoscience phenomena.

8. Conclusions and Future Research

This chapter encapsulates the application of the deterministic fractal-multifractal FM approach to the encoding, simulation, disaggregation/downscaling, classification, and prediction of geoscience records. This work has illustrated that the FM method (and its variants), when coupled with a suitable optimization scheme, may be used to parsimoniously describe a host of patterns including rainfall events, daily rainfall sets over a year, daily streamflow sets over a year, and also daily water temperature records. Overall, mildly intermittent sets are easier to encode, as it takes, for a given set, from three to five hours of CPU on a personal computer.

Highly intermittent rainfall sets are harder to process, as their optimization process takes, for a given set, about a day of CPU time.

As hinted in Section 3, the FM parameter space corresponding to a wire or a leaf is highly complex, and even more so for Cantorian representations that also include thresholds. Although there is continuity between the FM parameters and the sets they induce (i.e. the FM graphs vary a little when a single FM parameter varies a little), multiple changes of parameters end up generating similarly looking attractors from various combinations of parameters and, hence, there is no unique optimal solution. Although the patterns shown throughout this chapter are close renderings of the target sets, the presence of alternative close solutions ought to be studied in detail. It certainly would be relevant to ascertain the dynamics of, say, the best ten solutions to a given process and to study such representations on a multi-dimensional sense trying to encounter further relationships that may help in producing improved forecasts. It is envisioned that data mining techniques and the general notion of principal components may be useful in such a research.

In regards to the plausibility of improving predictions, few ideas may be tried for further improvements. Such encompasses combining FM parameters at various resolutions beyond years and decades, using say "pentades" (aggregating the records every five years) and employing in the definition of the classes a metric that weighs the distinct parameters according to their intrinsic variability, and not equally as it has been reported herein. Certainly these notions, coupled with the usage of (ten) alternative "solutions," may yield improvements that may help elucidate climate change effects and trends.

Overall, there is much that needs to be done in order to fully study geoscience records using the FM approach. Besides trying the ideas in a variety of catchments, the following are relevant questions that represent future research. How do the FM parameters vary spatially when streamflow records of sub-catchments (upstream) are compared to those of the catchment (downstream)? How are the FM parameters of rainfall observations related to those of streamflow at the same site? Do the FM parameters vary in a systematic way when performing downscales at various scales? How are the FM parameters of various attributes related on a given site? How are the FM parameters of, say, rainfall, streamflow, and evaporation related to climate indicators? Are trends in climate relatable to discernible changes on the various attributes such that a changing hydrology may be elucidated? How are the FM parameters of distinct processes related to the underlying physics (conservation laws) of the phenomena? Are there discernible physical explanations for each one of the FM parameters?

Clearly, the scope of the FM approach is not limited to the study of geoscience records only, as, in a rather natural way, it may also be used to enhance other disciplines, such as physics, engineering, pattern recognition, general statistics, medical sciences and finance. It is envisioned that applications of the FM approach, and also for patterns in higher dimensions (Puente 2004), will appear in such fields in the future.

Symbols and Notation

$\in_{ar}$: root mean square error on accumulated records
$\in_{max}$: maximum error on accumulated records
E_q : entropy function of a data set
η_c : Nash-Sutcliffe efficiency on autocorrelation
η_d : Nash-Sutcliffe efficiency on data
η_h : Nash-Sutcliffe efficiency on histogram
η_e : Nash-Sutcliffe efficiency on entropy
Z_m : percent of zeros matched
τ_0 : lag when autocorrelation becomes zero
$\tau_{1/e}$: lag when autocorrelation becomes $1/e$
μ_{90} : mass of FM fitted histogram equivalent to 90% mass of observed histogram
M_{CZ} : maximum consecutive zero values
N_Z : number of zero values

REFERENCES

Arthur, D. and V. Sergi. 2007. K-means++: The Advantages of Careful Seeding. In: SODA '07 Proceedings of the Eighteenth Annual ACM-SIAM Symposium on Discrete Algorithms. Pp. 1027–1035.

Barnsley, M.F. 1988. Fractals Everywhere. Academic Press, San Diego.

Cortis, A., C.E. Puente and B. Sivakumar. 2009. Nonlinear extensions of a fractal–multifractal approach for environmental modeling. *Stoch. Environ. Res. Risk. Assess.* 23(7): 897–906.

Cortis, A., C.E. Puente, H.H. Huang, M.L. Maskey, B. Sivakumar and N. Obregón. 2013. A physical interpretation of the deterministic fractal-multifractal method as a realization of a generalized multiplicative cascade. *Stoch. Environ. Res. Risk Assess.* 28(6): 1421–1429.

Döll, P. and H.M. Schmied. 2012. How is the impact of climate change on river flow regimes related to the impact on mean annual runoff? A global-scale analysis. *Environ. Res. Lett.* 7(1): 014037.

Feder, J. 1988. Fractals. Plenum Press, New York.

Fernández-Martínez, J.L., E. García Gonzalo, J.P. Fernández Álvarez, H.A. Kuzma and C.O. Menéndez Pérez. 2010. PSO: A powerful algorithm to solve geophysical inverse problems – Application to a 1D-DC resistivity case. *J. Appl. Geophys.* 71(1): 13–25.

Georgakakos, K.P., A.A. Carsteanu, P.L. Sturdevant and J.A. Cramer. 1994. Observation and analysis of midwestern rain rates. *J. Appl. Meteor.* 33(12): 1433–1444.

Gupta, V.K. and E. Waymire. 1990. Multiscaling properties of spatial rainfall and river flow distributions. *J. Geophys. Res.* 95(D3): 1999–2009.

Hammel, R.D., R.J. Cooper, R.M. Slade, R.L. Haney and J.G. Arnold. 2006. Cumulative uncertainty in measured streamflow and water quality data for small watersheds. *Tran. Am. Soc. Agric. Eng.* 49(3): 689.

Huang, H.H., C.E. Puente and A. Cortis. 2012a. Geometric harnessing of precipitation records: Reexamining four storms from Iowa City. *Stoch. Environ. Res. Risk. Assess.* 27(4): 955–968.

Huang, H.H., C.E. Puente, A. Cortis and B. Sivakumar. 2012b. Closing the loop with fractal Interpolating functions for geophysical encoding. *Fractals* 20(3–4): 261–270.

Huang, H.H., C.E. Puente, A. Cortis and J.L. Fernández Martínez. 2013. An effective inversion strategy for fractal–multifractal encoding of a storm in Boston. *J. Hydrol.* 496: 205–216.

Koutsoyiannis, D. 1992. A nonlinear disaggregation method with a reduced parameter set for simulation of hydrologic series. *Water Resour. Res.* 28(12): 3175–3191.

Koutsoyiannis, D. 1994. A stochastic disaggregation method for design storm and flood synthesis. *J. Hydrol.* 156(1): 193–225.

Lanza, L.G. and E. Vuerich. 2009. The WMO field intercomparison of rain intensity gauges. *Atmos. Res.* 94(4): 534–543.

Lovejoy, S. and D. Schertzer. 2013. The Weather and Climate: Emergent Laws and Multifractal Cascades. Cambridge University Press, Cambridge.

Mandelbrot, B.B. 1982. The Fractal Geometry of Nature. HB Fenn and Company.

Mandelbrot, B.B. 1989. Multifractal measures especially for the geophysicist. pp. 1-42. In: C.H. Scholz and B.B. Mandelbrot (eds). Fractals in geophysics. Birkhanser Verlag, Basel.

Maskey, M.L., C.E. Puente and B. Sivakumar. 2015. Encoding daily rainfall records via adaptations of the fractal multifractal method. *Stoch. Environ. Res. Risk. Assess*. DOI 10.1007/s00477-015-1201-7.

Meneveau, C. and K.R. Sreenivasan. 1987. Simple multifractal cascade model for fully developed turbulence. *Phys. Rev. Lett.* 59: 1424–1427.

Nash, J. and J.V. Sutcliffe. 1970. River flow forecasting through conceptual models part I – A discussion of principles. *J. Hydrol.* 10(3): 282–290.

Obregón, N., C.E. Puente and B. Sivakumar. 2002a. A deterministic geometric representation of temporal rainfall. Sensitivity analysis for a storm in Boston. *J. Hydrol.* 269(3–4): 224–235.

Obregón, N., C.E. Puente and B. Sivakumar. 2002b. Modeling high resolution rain rates via a deterministic fractal–multifractal approach. *Fractals* 10(3): 387–394.

Olsson, J. 1998. Evaluation of a scaling cascade model for temporal rainfall disaggregation. *Hydrol. Earth Syst. Sci.* 2(1): 19–30.

Puente, C.E. 1996. A new approach to hydrologic modelling: Derived distribution revisited. *J. Hydrol.* 187: 65–80.

Puente, C.E. 2004. A universe of projections: May Plato be right? *Chaos, Solitons and Fractals* 19(2): 241–253.

Puente, C.E. and N. Obregón. 1996. A deterministic representation of temporal rainfall: Result for a storm in Boston. *Water. Resour. Res.* 32(9): 2825–2839.

Puente, C.E. and B. Sivakumar. 2007. Modeling geophysical complexity: A case for geometric determinism. *Hydrol. Earth Syst. Sci.* 11: 721–724.

Rodriguez-Iturbe, I. 1986. Scale of fluctuation of rainfall models. *Water. Resour. Res.* 22(9): 15S–37S.

Sivakumar, B. 2000. Chaos theory in hydrology: Important issues and interpretations. *J. Hydrol.* 227(1–4): 1–20.

Sivakumar, B. 2004. Chaos theory in geophysics: Past, present and future. *Chaos, Solitons and Fractals* 19(2): 441–462.

Valencia, D. and J.C. Schaake. 1973. Disaggregation processes in stochastic hydrology. *Water Resour. Res.* 9(3): 580–585.

CHAPTER

12

Use of Constructal Theory in Modeling in the Geosciences

Allen G. Hunt

Department of Physics, Wright State University, 3640 Colonel Glenn Hwy., Dayton OH 45435, United States

allen.hunt@wright.edu

1. Introduction

Regardless of their physical origin, hierarchical structures are known to allow optimal access of information, transport of mass or ideas, and thus ultimately conversion of energy, when, e.g., sources and consumers are spatially distributed. Such a realization adds impetus to understanding the dynamics of the formation of such structures. This book treats nominally the subject of "fractals" in the geosciences. In the chapter on percolation (Hunt and Yu 2017) it was pointed out that flow in disordered media will typically lead to the relevance of the fractal structures of percolation theory, but without necessarily the outcome that any durable, physical structure is produced. In this chapter, a different means, constructal theory, to create hierarchical structures is discussed in the context of optimization of heat flow, though its particular relevance in the geosciences is likely more important to the formation of drainage networks (Errera and Bejan 1998, Ledezma et al. 1997). Arguments that constructal theory leads to the appropriate dependence of the unsaturated hydraulic conductivity on a power of the saturation (Liu 2011) are not consistent with the most predictive theoretical treatments of the unsaturated hydraulic conductivity (Ghanbarian and Hunt 2012, Ghanbarian et al. 2016), which show that the unsaturated hydraulic conductivity is, rather, a function of a considerably more complex argument of the saturation, though the functional form is indeed a power law.

The fundamental principle of constructal theory was formulated thus (Bejan, 1997a, b): "For a finite-size flow system to persist in time (to live), its configuration must evolve in such a way that provides greater and greater access to the currents that flow through it." Thus, the Constructal "Law" is envisioned as an element of thermodynamics that applies to living, or at least evolving, entities. In accord with this general perspective, Bejan and Lorente (2006, 2013) suggest that the evolution of constructal theoretical perspectives, starting from a simpler, typically more nearly homogeneous model and advancing towards treatment of a more realistic, heterogeneous substrate will produce a sequence of calculations that approaches reality and simultaneously reflects evolution. In an earlier chapter (Hunt and Yu) addressing flow and transport in random porous media, however, we used percolation theoretical techniques to generate powerful predictions of soil formation and vegetation growth that could extend to 100 million year time scales. In the book "Networks on Networks: The Physics of Geobiology and Geochemistry, (Hunt and Manzoni 2016), we argued that perhaps the most important differences between "living soils" and engineered porous media, were not the sequences of feedbacks and correlations characteristic of soils, but their disorder. This perspective was already adopted by Schroedinger (1944), in the description of the building blocks of life.

Erwin Schroedinger pointed out in 1944 in his essay, "What is Life?" that the fundamental characteristic of living organisms is their reliance on aperiodic crystals, one-dimensional entities, to produce very long lifespans, and thus form a basis for persistence of information, i.e., the genetic code. He argued that it would not be possible to understand the physics of living organisms on the basis of the physics of ordered substances, which had been the fundamental purview of physics, and that new physical techniques for addressing aperiodic, or random systems would need to be developed for serious progress to be made. Pollak and Hunt (1985), Murphy (1997) and Kauffman (2007) noted that Schroedinger's perspective helped inspire the work of Watson and Crick (1953) in their search for the structure of nucleic acids, since he had predicted that this would necessarily take the form of long, one-dimensional, aperiodic molecules.

More practically speaking, constructal theory is often used either to explain complex hierarchical organization prevalent in nature or to optimize performance of engineered structures (Bejan 2000). Typical applications are to steady-state problems, but it can also be applied for transient conditions (Dan and Bejan 1998). Its most common engineering application is to problems of thermal energy removal, though Bejan (2000) also discusses many other potential problems. Indeed, among these applications is one of considerable geomorphological relevance, channel initiation and separation of lowest order channels (Montgomery and Dietrich 1988, 1989, 1992). Bejan treats this problem chiefly in terms of flow concentration and optimization, but more common geomorphological treatments are based on sediment transport. Most commonly, channel initiation is treated for steady-state conditions of precipitation and sediment transport, though it is also recognized that in more arid regions steady-state conditions are far from realistic (Bull and Kirkby 2002). While the particular example of channel formation is not treated specifically here, an example of transient heat conduction is shown to generate nearly identical results to steady-state formulations in a constructal formulation for the lowest order geometry. Such a result is already known numerically, but the present calculation is analytical in form, opening the possibility that the method can be repeated for higher order calculations more easily than numerical methods. The near coincidence of optimal architectures for steady-state and transient conditions suggests that rare pulses of precipitation may lead to similar channel networks as more regular precipitation and run-offs in more humid regions. Even if this stronger conclusion does not follow, it appears likely that for arid regions it may not be necessary to apply the fully stochastic nature of the precipitation in order to generate a realistic network.

The notion of constructal theory is generally compatible with fractal theory, but its orientation is opposite. Instead of focusing one's attention on the large spatial scales and moving to the smaller, constructal theory inverts the process. Nevertheless, both fractal and constructal approaches recognize the prominent role of hierarchical structures in optimization of fluxes of energy, mass, or even ideas. For its application, constructal theoretical methods require distinct transport coefficients in different portions of the medium corresponding to distinct transport mechanisms as a function of scale, such as gas diffusion across lung alveoli membranes and advection through blood vessels (bonded chemically to hemoglobin); or mass diffusion between structural divides and channels, with fluvial sediment transport within channels. In the case of transport in porous media a similar cross-over from diffusion to advection dominance in solute transport is noted at the scale of approximately one pore (Hunt and Manzoni 2016). However, in the geomorphological case, the sediment transport that is optimized has a crucial feedback on the network structure of the topography and its spatial scale (Montgomery and Dietrich 1992), rather than originating in the network structure.

It is important to emphasize the significant distinctions in the results of constructal theoretical calculations in two dimensions and in three (Bejan 2000). In this vein, the calculations presented here are for two-dimensional optimization, which may well work for geomorphological applications, but not for such systems as animal lungs, which are embedded in a three-dimensional object. While the optimal paths are lines in 2D, in 3D they are right circular cylinders. The three-dimensional case typically requires more serious numerical procedures to generate quantitative predictions.

One of the chief characteristics of 2D constructal theory is its focus on predicting the optimal flow paths for transporting heat or fluid, from a one-dimensional boundary to a point. An essential goal is to find the optimal separation of the finest channels, capillaries in the living organisms, first order streams in geomorphology. This type of calculation is well-documented in Bejan (2000). However, no analogous calculation can predict the separation of the critical paths (in the groundwater literature, more commonly called preferential flow paths), which carry the dominant flow. The optimization here is between the topology of percolation, and the geometry of the pore radii. This particular outcome suggests a restriction of the relevance of the quantitative approach of constructal theory to homogeneous media.

One could suggest that constructal theory, because of its "bottom up" perspective, is especially suited to the development of living organisms, which necessarily develop from the small to the large. However, in the case of drainage-basin development, the situation is complicated by several factors, including the heterogeneity of the substrate, and the history of the stream development. For example, it is less clear in which sequence streams are connected, from the top of a drainage system down (including overtopping of sills), or from the bottom up (headward erosion). Interestingly, discussions of the relative importance of the two mechanisms for stream connection are typically found within discussions of the origin of specific drainage basins, such as the Colorado River in southwestern USA.

The specific conditions for application of constructal theory are usually formulated thus:

1. Two component media,
2. Large contrasts in thermal (electrical, hydraulic) conductivity or transport coefficients,
3. Steady-state conditions,
4. Point to area (or volume) flow or the reverse.

Consider a composite medium which contains two components. One can conduct heat (or water, or gas) effectively, while the second can absorb (or dissipate) it. At the lowest level of optimization, for a two dimensional system, the procedure effectively generates the separation and width of the more highly conductive portions of the medium, while typically employing a constraint regarding the total area fraction of the more conductive part. In two dimensions, the more highly conductive region manifests itself in parallel straight lines with rectangular regions of adsorbing (or dissipating) material between. Higher levels of optimization treat the composite medium at the lowest level of the hierarchy as a new medium. Its combined transport coefficient is now larger than the original low transport component, meaning that the contrast between its transport coefficient and that of the more highly conducting portion forming the next level of hierarchy is likely to decrease. This contrast determines the number of first order channels of more highly conducting material that can join to make one second level channel that serves all the first level channels. As the contrast diminishes with increasing level in the hierarchy, so do the number of channels, until effectively that ratio is two, the inverse of bifurcation. This process continues until commensurate with the size of the system.

In the lungs, for example, diffusive exchange of gases between an animal body and its surroundings occur through the membranes of the alveoli. A large surface area favors such exchange. However, it is necessary to maximize the CO_2 expelled from the body and the O_2 absorbed. In constructal theory, the geometry of the lowest order combination of body tissues is controlled by a maximum gas exchange; if too great an area is devoted to exchange, the advective process of blood circulation is less effective. The optimization leads first to the appropriate shape of blood vessels. Although the appropriate shape in two dimensions is a straight line, in three dimensional optimizations the appropriate shape is a right circular cylinder. The optimization proceeds to higher levels by including the lower level advective and diffusive portions together as a single medium with the lower transport coefficient. At each succeeding level, the density of higher conducting channels is the required output.

1.1 Analytic Optimization of Transient Heat (or Fluid) Flow

1.1.1 General Characteristics

Our discussion here will focus on transient heat-flow from point to area in two-dimensional lowest order constructs and, as a consequence, does not follow exactly the usual format.

The fundamental strategy under steady-state conditions is to minimize the thermal resistance between the source of heat, which is normally spread out uniformly over a volume, and the sink of heat, located at a point in proximity to the volume. In two-dimensional problems, area is substituted for volume. This method leads to structures of the more highly conductive medium, which resemble fractal patterns, but which, as Bejan (2000) reminds us, are better interpreted in terms of constructing larger objects from smaller ones. Parameters that describe the geometry of the optimal structures can be derived analytically under ideal conditions (e.g., many order of magnitude contrasts in thermal conductivity), while Ledezma et al. (1997) give numerical results for real conditions with smaller conductivity contrasts that tend to the analytical results in the ideal limit. The existence of analytical solutions is useful for their role in guiding the numerical analysis.

Dan and Bejan (1998) assert that the same structures that optimize steady-state heat removal are also ideally suited to the removal of heat generated in a pulse. While these authors also produce evidence to support this contention, they do not give a general analytical approach.

An analytical approach is developed here through a change in perspective. Instead of considering how to generate a structure that is optimized for removing heat from an area, consider how to use the constructal approach to take a point heat source and spread it most rapidly through the area. Bejan's approach is to address the global optimization in an essentially piecemeal fashion from the smallest spatial scales up, in a way that a growing organism could utilize at all stages of its growth. In turning around the problem here we don't abandon the fundamental idea of constructal theory, that it is ultimately the small scale optimization that guides the larger scales. The change is adopted only for convenience of development.

At the smallest scales Bejan's structures are very simple and describe a strategy for making maximal use of a small amount of material that is valuable in removing heat. The focus here is on the basic structure, known as the zeroth order construct, and only two-dimensional conditions will be treated. But the method proposed will have wider applicability. Further, although the problem will be discussed in engineering terms it may have relevance in the study of partial differential equations on fractal as well as constructal objects.

In zeroth order constructal theory in two dimensions, Bejan considers rectangular shaped regions with a blade of highly conducting material in the middle as shown in Fig. 1. The blade extends from one side of the rectangle to the other. Heat is normally produced in the entire rectangular region and removed at the center of one side, say the left. The shape of the more highly conducting material is taken in the most elementary treatment to be rectangular, although more advanced treatments yield a position-dependent thickness. Constructal theory determines the dimensions h_0 and l_0 under the conditions that there is a fixed areal fraction of medium 1, expressed as a given ratio of d_0/h_0, and a fixed total area, A, expressed as the product, $h_0 l_0$. The steady-state calculation maximizes the heat flux for a given heat production rate. Consider instead what happens when steady-state conditions are replaced by transient conditions. Dan and Bejan (1998) state the new objective: "Determine the optimal distribution of k_p material through the given volume, such that the cooling time is minimal." The cooling time is calculated for the most disadvantaged point. The method chosen for solution of this problem is essentially the solution of the diffusion equation on a composite structure. When higher order constructs are required, the scaling up of the procedure requires, in principle, the solution of the diffusion equation on a hierarchical structure, but development of certain characteristics of the solution turns out to be sufficient.

The diffusion equation for transient heat transport reads,

$$D\Delta^2 T = \frac{\partial T}{\partial t} \tag{1}$$

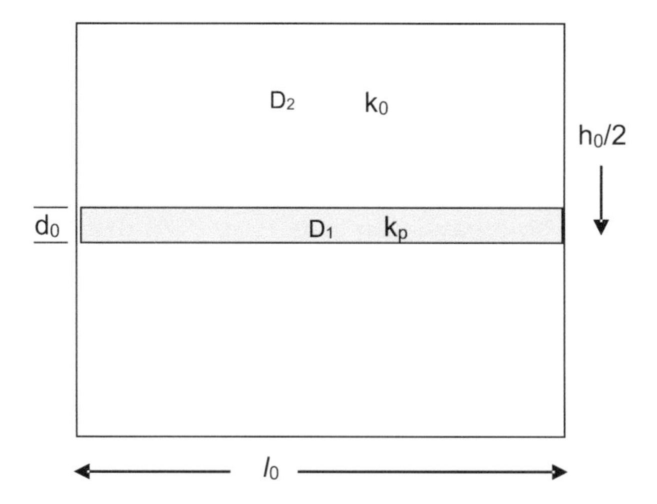

Figure 1: After Bejan, 2000, Figure 4.1. Figure shows the zeroth order two-dimensioal constructal structure and gives the definitions of the symbols used. The symbols k_0 and k_p refer to Bejan's examples from steady-state heat conduction. The symbols D_1 and D_2 refer to transient heat conduction problems and must be related to k_0 and k_p. The product $l_0 h_0 = A$ is a constant. The ratio d_0/h_0 a is also a constant, given by the ratio of the volume of material 1 to that of both material 1 and 2 together.

In this equation the diffusion constant is given by

$$D = \frac{k_t}{\rho C} \tag{2}$$

with k_t the thermal conductivity, ρ the density, and C the specific heat. Under transient conditions the relevant flow equation, Eq. (1), is always some version of the diffusion equation. Now we have to recognize that for solutions of the diffusion equation in one dimension the quantity describing mean particle transport that adds linearly is not the time, but the square root of the time. This complication must be considered since our fluid is going to flow from one medium to the other, so that we cannot represent the entire x distance as having the same diffusion constant, d. In any case for a single medium, when the distance of travel increases by a factor of β, i.e., $x \rightarrow x' = \beta x$ the time has increased from t to $t' = \beta^2 t$ and it is the square root of the increase that is proportional to β. Then, if $x^2/t = d$, $x'^2/t' = (\beta^2/\beta^2)\, x^2/t = d$ also. This means that the time to travel a distance S must be proportional to S^2. Many references can be cited to argue that the time required for diffusive transport (in one dimension) is proportional to the square of the distance, so this result is not new. But one can turn this around and say that the square root of the time elapsed is proportional to the distance of travel. The distance of travel does add. This difference is rather subtle, since the times in the transport of individual fluid molecules from place 1 to place 2 and from place 2 to place 3 add, but they are not always traveling in the same direction. For the cloud of fluid molecules represented in the diffusion equation, it is the distance of travel that is adding linearly. A more rigorous solution that was recently published in the literature on cements (Andrade et al. 1997) (but due, originally to Carslaw and Jaeger 1959 and repeated in Crank 1999) is reproduced below, but contains essentially the same physics.

This principle on the increase in total diffusion time can be applied to each of two media in series separately. Suppose that in a one-dimensional geometry a diffusion front advances across a boundary from medium 1 with d_1 to medium 2 with d_2. Then the total time, t, needed for the front to travel a distance l_1 in medium 1 and a distance l_2 in medium 2 is estimated from,

$$t^{\frac{1}{2}} = \frac{l_1}{d_1^{\frac{1}{2}}} + \frac{l_1}{d_2^{\frac{1}{2}}} \tag{3}$$

Leaving out the usual factor of $(1/2)^{1/2}$ has the effect of making this absorption time twice that of the time for the maximum flux to cross, and helping to make this calculation consistent with that of Bejan for 90% absorption. The condition on when such a calculation is accurate is that the width of the diffusion front be smaller than either l_1 or l_2.

Refer now to the results from Crank (1999). Consider diffusive flow from a medium of given width on the left and continuing into a second medium on the right (which may be, but need not be, infinite). The medium on the left has diffusion constant d_1 and the medium on the right, d_2. The width of medium 1 is l_1. The coordinate inside medium 2 is l_2, i.e., the total distance from the left boundary is $l_1 + l_2$. The boundary and initial conditions are as follows. For all $t < 0$ the concentration is everywhere zero. For $t > 0$ the left edge of the left medium is held at an arbitrary concentration. In that calculation the concentration in the second medium is a sum of a series of complementary error functions, all with arguments that have the same fundamental relationship between space (l_1 and l_2) and time t. For a delta function pulse input, the solution (by Green's function techniques) would simply be the sum of the corresponding Gaussian functions (the derivative of the complementary error function is the negative of the derivative of the error function, and that is the Gaussian). Each of the functions in the series has an argument of the form

$$\frac{\alpha_n}{2}\frac{l_1}{t^{\frac{1}{2}}d_1^{\frac{1}{2}}} + \frac{1}{2}\frac{l_1}{t^{\frac{1}{2}}d_2^{\frac{1}{2}}} \tag{4}$$

where the α_n are numerical constants (for $n = 1$, $\alpha_n = 3$, for example). Under conditions when it is possible to restrict one's attention to one or a few of the terms, then one can determine the time dependence of the position where the concentration of the diffusing material is constant by setting the above expression equal to a constant, from which one finds immediately,

$$t^{\frac{1}{2}} \propto \frac{\alpha_n l_1}{2d_1^{\frac{1}{2}}} + \frac{l_2}{2d_2^{\frac{1}{2}}} \tag{5}$$

which is very nearly the rough answer given above. The constant chosen (i.e., the α_n value) depends on the concentration level one desires. Thus, in the pulse input problem, a constant near 1 (when the argument of the lowest order Gaussian is near 1) will correspond approximately to times and positions for which the flux is a maximum.

What we need to optimize is not precisely the time, however, but the heat (or mass) flux to the corner. This can be expressed in terms of the minimum time taken for a given amount of energy (mass) to reach the most disadvantaged point.

1.1.2 Analogy to Peclet Number

In order for the thermal energy to arrive in the corner over the pathway that carries it along the center to the edge and then up the edge, it cannot diffuse off the center. The question of whether the thermal energy diffuses off the center is analogous to the problem of whether solutes will diffuse off a path that takes them through a pore of radius r by the process of advection, which is given in terms of the ratio of the advection time to the diffusion time. The reason it is given in terms of this ratio is that the diffusion equation is consistent with constant probability fluxes, or transition probabilities per unit time, and thus the actual probability is proportional to the time that the diffusion equation in a given form is applicable. This aspect of the calculation applies to our problem equally. The ratio in times for the advective flow problem yields a quantity called the (inverse of the) Peclet number (Pfannkuch 1963). In particular, the advection time is r/v and the diffusion time is r^2/d, so the desired probability is $[(r/v)/r^2/d)] = [rv/d] = P_e^{-1}$. The chief condition for the accuracy of this calculation is that the advection time be much smaller than the diffusion time ($P_e >> 1$). If the advection time is on the order of the diffusion time or smaller, the ratio can exceed 1 and cannot be interpreted in terms of a probability.

Since what we are interested in is also expressed in terms of probability fluxes, we can adapt the above argument to diffusion along the axis vs. diffusion off the axis of the blade. The fraction of the thermal energy following the path to the end of the middle blade is the ratio of the time to diffuse off to the time required to get to the end, which would be D_0^2/L_0^2, if the diffusion constants of the two media were identical. Thus the time per unit heat transport would be increased by the inverse of the fraction of heat remaining, (L_0^2/D_0^2). Here it is possible to quote Keanini (2007) who writes about random walk methods for scalar transport problems (the unbiased random walk is well-known to be consistent with diffusion in the limit of many individual steps): "Thus, consider, for example, purely diffusive transport within a slender domain in which the source term is zero and in which the aspect ratio $L_0/D_0 >> 1$ where L_0 and D_0 are the domain dimensions in the x_1 and x_2 directions, respectively. For times larger than $D_0^2/2d_1$, and solution points lying in the range $D_0 < x_1 < L_0\text{-}D_0$, most random walkers will impinge on either lateral boundary, $x_2 = 0$ or $x_2 = D_0$ (rather than exiting through the final time slice or through the end boundaries, $x_1 = 0$ and L_0). The author goes on to conclude that most of the walkers exit the lateral boundaries. The author also demonstrates *that his stochastic solution reproduces the continuum analytical solution*.

But since the diffusion constant of component 1 is much larger than in component 2, the result is slightly more complicated. Now, the loss of heat off the sides of the blade due to diffusion cannot exceed the rate at which heat is diffused into component 3. Thus the effective transverse diffusion time cannot be less than the time for diffusion into component 3.

1.1.3 Optimization

Replace $(L_0/D_0)^2$ with $(d_2/d_1)(L_0/H_0)^2$ which is proportional to $(d_2/d_1)(L_0/H_0)^2$, i.e., which assumes that only the energy that diffuses all the way to the edge of the area A at the rate of the slow component cannot make it to the end of the blade. This overestimates the transverse diffusion time and thus underestimates the loss of energy to the sides, overestimating how much energy reaches the end of the blade (and underestimating the path-parallel diffusion time). When the square root of the time along the blade is taken (to be consistent with the additive rule for the square root of time), the factors appear as $(d_2/d_1)^{1/2}(L_0/H_0)$ and one has for the total time (to the one half power),

$$t^{1/2} = \frac{H_0}{2d_2^{1/2}} + \frac{L_0}{d_1^{1/2}}\left(\frac{d_2}{d_1}\right)^{1/2}\frac{2L_0}{H_0} \tag{6}$$

Using the condition $L_0H_0 = A$ yields,

$$t^{1/2} = \frac{A}{2d_2^{1/2}L_0} + \left(\frac{d_2^{1/2}}{d_1}\right)\left(\frac{L_0^3}{A}\right) \tag{7}$$

Since t is a monotonically increasing function of $t^{1/2}$ it is sufficient to optimize $t^{1/2}$ with respect to L_0, with the result,

$$L_0 \approx \left(\frac{d_1}{d_2}\right)^{1/4} A^{1/2} \tag{8}$$

This is similar to the corresponding result obtained in the steady-state optimization, i.e., the dependence on Area A and the ratio of the thermal conductivities is the same, but there is no dependence on the ratio of $D_0/H_0 = \alpha$. We can find a more accurate value for the optimized result for L_0, however. That is accomplished by using the diffusion constant for the slower medium over the transverse length scale D_0, even within the faster medium. Furthermore, with this approximation for the present factor, there is no additional influence of the slower medium as the boundary between the two components can no longer be distinguished and heat energy (or fluid) that impinges on the boundary continues on into the other medium. Then,

$$t^{1/2} = \frac{H_0}{2d_2^{1/2}} + \frac{L_0}{d_1^{1/2}}\left(\frac{d_2}{d_1}\right)^{1/2}\frac{L_0}{D_0} = \frac{H_0}{2d_2^{1/2}} + \frac{L_0}{k_1^{1/2}}\left(\frac{d_2}{d_1}\right)^{1/2}\frac{1}{\alpha}\frac{L_0}{D_0} \tag{9}$$

Substitute again the proportionality $H_0 = A/L_0$ to eliminate H_0 as a variable. Optimization of Eq. (9) yields (where the extra factor of 3 arises from differentiating L_0^3),

$$L_0 = \left(\frac{1}{6}\right)^{1/4}\left(\frac{d_1}{d_2}\right)^{1/4} \alpha^{1/4} A^{1/2} \tag{10}$$

which is nearly identical to the steady-state result obtained above, but differs from Bejan by a factor $(1/6)^{1/4}/(1/2)^{1/2} = 0.64$. Substituting Eq. (10) into Eq. (9) and then squaring yields,

$$t = 1.04\frac{A}{d_2}\left[\frac{d_2}{d_1\alpha}\right]^{1/2} \tag{11}$$

This equation may be compared with Eq. (17) of Dan and Bejan (1998),

$$t = 1.06\frac{A}{d_2}\left[\frac{k_0}{k_p\alpha}\right]^{1/2} \tag{12}$$

after using the substitutions in their Eq. (7) and Eq. (8). If they had not made the simplification (their Eq. (24)) of assuming that the heat capacities of their two media were the same, then their Eq. (17) and our Eq. (12) would be identical, except for the 2% difference between 1.04 and 1.06. It is important that Eq. (12) is to be interpreted (Dan and Bejan 1998) as the time when 90% of the fluid (or thermal energy) has been absorbed, not a typical transport time. In the case of steady-state conditions, the factor 1.06 is replaced by 1.

Conclusions

The detailed calculation given here was for heat flow (or equivalently fluid pulses) rather than sediment transport in a drainage basin. Such a calculation for channel initiation is, at present, not possible to the same level of accuracy, since in both sediment transport by, e.g., overland flow of water, and by channel processes, the exact dependence of the flux on slope as well as fluid flux is not known. However, such calculations can be performed numerically. The extent to which important characteristics of heat flow and channel initiation are analogous cannot yet be determined precisely.

When a composite medium is constructed from individual media with greatly differing conductivities (or diffusion constants), optimization of the medium structure for transient heat pulses leads to almost the same time for the dominant transport as obtained for steady-state optimization. Note that, while the geometric characteristics of the solution scale in the same fashion as derived in Dan and Bejan (1998), the numerical coefficients are not identical, even though the total transport time is essentially the same. We have introduced an analytical treatment of such an optimization whose solution leads to a diffusion time different by only 2% from the numerical coefficient.

In landscape evolution models, it is not the composition of the medium which is variable, but the convergence and slopes of the topography, which produce a change in characteristic of sediment transport with distance from the divide. Although landscape evolution models suggest that the separations of channel heads and their distances from a landscape divide are controlled by the increase in relative importance (with increase in distance from a divide) of advective solute transport through fluvial processes when compared with diffusive transport due to a variety of mechanisms (Kirkby 1987, Willgoose et al. 1991, Howard 1994, Tucker and Bras 1998), including overland flow, the general analogy between channel formation and heat flow optimization may still be useful.

While this chapter is not intended in any sense as a wrap-up of the book, it ought to be mentioned here that the intertwined problems of landscape evolution, drainage basin formation, and channel initiation may not be amenable to a single means of solution (Whipple and Tucker 1999, 2002). Consider the case of Hack's law for the scaling of river length, L, vs. drainage basin area A, $L = A^q$ (Hack 1957, Gray 1961, Maritan et al. 1996, Rigon et al. 1998, Errera et al. 1998). Montgomery and Dietrich (1992) show that this non-Euclidean power law does not arise from changes in drainage basin shape with increasing scale, as originally assumed by Hack (1957). Maritan et al. state that, according to Gray's analysis the exponent q takes on values from 0.57 to 0.6 though Whipple and Tucker (1999) expand the range to 0.52 to 0.6, where continental scales appear to be associated with the lower range of values of Hack's law exponents, near 0.5. In Errera and Bejan (1998), values of Hack's exponent of 0.5 to 0.56 are generated. With increasing landscape scale, increasing total water flux and, presumably increased time scales, should allow for greater evolutionary capability. For example, longer channels in braided streams have longer lifetimes (Foufoula-Georgiou and Sapozhnikov 2001). Thus, one possibility is that more random elements of connection in landscape networks of varying heterogeneity may dominate river sinuosity at shorter time scales (lack of equilibration), while at longer time scales the sinuosity diminishes in the constructal optimization process. While, for a relatively homogeneous substrate or large time scales and large stream power, constructal theory may be used to guide an optimization of the channel initiation position and channel separation as well as Hack's exponent, it appears that percolation theory can best be utilized to find the stream sinuosity as a function of the heterogeneity of the substrate (Hunt 2016) (exponent values from 0.565 to 0.605), except at the largest scales. The former (latter) would indicate the necessity of incorporating feedbacks and evolution (structural heterogeneity) in a completely general formulation. In other words, the particular optimization strategy that is most successful may depend on the nature of the medium, or the aspect of the problem considered – which might suggest that a more general formulation is necessary, of which the various methods considered here are but facets, depending on one's perspective.

REFERENCES

Andrade, C., J.M. Diez and C. Alonso. 1997. Mathematical modeling of a concrete surface "skin effect" on diffusion in chloride contaminated media. *Advanced Cement Based Materials* 6: 39–44.

Bejan, A. 1997a. Constructal-theory network of conducting paths for cooling a heat generating volume. *Int. J. Heat Mass Transfer* 40: 799–816.

Bejan, A. 1997b. Advanced Engineering Thermodynamics. 2nd ed. Wiley, New York.

Bejan, A. 2000. Shape and Structure, from Engineering to Nature, Cambridge University Press.

Bejan, A. and S. Lorente. 2006. Constructal theory of generation of configuration in nature and engineering. *J. Appl. Phys.* 100; doi: 10.1063/1.2221896.

Bejan, A. and S. Lorente. 2013. Constructal law of design and evolution: Physics, biology, technology and society. *J. Appl. Phys.* 113: 151301; doi: 10.1063/1.4798429.

Bull, L. and M. Kirkby. 2002. Dryland Rivers, Hydrology and Geomorphology of Semi-arid Rivers. Wiley and Sons, Chichester, England.

Carslaw, H.S. and J.C. Jaeger. 1959. Conduction of Heat in Solids. 2nd Edition, Oxford University Press.

Crank, J. 1999. The Mathematics of Diffusion. Oxford University Press.

Dan, N. and A. Bejan. 1998. Constructal tree networks for the time-dependent discharge of a finite-size volume to one point. *J. Appl. Phys.* 84: 3042–3050.

Errera, M.R. and A. Bejan. 1998. Deterministic tree networks for river drainage basins. *Fractals* 6: 245–261.

Foufoula-Georgiou, E. and V. Sapzhnikov. 2001. Scale invariance in the morphology and evolution of braided rivers. *Math. Geol.* 33: 273–291.

Ghanbarian-Alavijeh, B. and A.G. Hunt. 2012. Unsaturated hydraulic conductivity in porous media: Percolation theory. *Geoderma*, 187: 77–84.

Ghanbarian, B., A.G. Hunt and H. Daigle. 2016. Fluid flow in porous media with rough pore-solid interface. *Water Resources Res.* DOI: 10.1002/2015WR017857.

Gray, D.M. 1961. Interrelationships of watershed characteristics. *J. Geophys. Res.* 66: 1215–1223.

Hack, J.T. 1957. Studies of longitudinal profiles in Virginia and Maryland. USGS Professional Papers 294-B, Washington DC, pp. 46–97.

Howard, A.D. 1994. A detachment-limited model of drainage basin evolution. *Water Resour. Res.* 30: 2261–2285.

Hunt, A.G. and S. Manzoni. 2016. Networks on Networks: The Physics of Geobiology and Geochemistry. Institute of Physics, Bristol UK.

Hunt, A.G. 2016. Possible explanation of the values of Hack's drainage basin, river length scaling exponent. *Non-linear Processes in Geophysics*, 23: 91–93.

Kauffman, S. 2007. Origins of Life and Evolution of the Biospheres. Springer, Berlin.

Keanini, G. 2007. Random walk methods for scalar transport problems subject to Dirichlet, Neumann and mixed boundary conditions. *Proc. R. Soc.* A463: 435–460.

Kirkby, M.J. 1987. Modelling some influences of soil erosion, landslides and valley gradient on drainage density and hollow development. *Catena Suppl.*, 10: 1–14.

Ledezma, G.A., A. Bejan and M.R. Errera. 1997. Constructal tree networks for heat transfer. *J. Appl. Phys.* 82(1): 89–100.

Liu, H.–H. 2011. A conductivity relationship for steady-state unsaturated flow processes under optimal flow conditions. *Vadose Zone J.* 10: 736.

Maritan, A., A. Rinaldo, R. Rigon, A Giacometti and I. Rodriguez-Iturbe. 1996. Scaling laws for river networks. *Phys. Rev.* E53(2): 1510–1515.

Montgomery, D. and W. Dietrich. 1988. Where do channels begin? *Nature* 336: 232–234.

Montgomery, D. and W. Dietrich. 1989. Source areas, drainage density, and channel initiation. *Water Resour. Res.* 25: 1907–1918.

Montgomery, D. and W. Dietrich. 1992. Channel initiation and the problem of landscape scale. *Science* 255: 826–830.

Murphy, 1997. What is Life? The next 50 years: Speculation on the future of biology. Cambridge University Press.

Pfannkuch, H. 1963. Contribution à l'étude des déplacements de fluides miscibles dans un milieu poreux. Contribution to the study of the displacement of miscible fluids in a porous medium. *Rev. Inst. Fr. Pét.* 2: 18.

Pollak, M. and A. Hunt. 1985. Very Slow Relaxation in Systems Lacking Translational Symmetry, with Emphasis on Disordered Insulators. *Philosophical Magazine* B52: 391 (in honor of Sir Neville Mott).

Rigon, R., I. Rodriguez-Iturbe and A. Rinaldo. 1998. Feasible optimality implies Hack's law. *Water Resour. Res.* 34(11): 3181–3189.

Schroedinger, E. 1944. What is Life? Cambridge University Press.

Tucker, G.E. and R.L. Bras. 1998. Hillslope processes, drainage density, and landscape morphology. *Water Resour. Res.* 34: 2751–2764.

Tucker, G.E. and K.X. Whipple. 2002. Topographic outcomes predicted by stream erosion models: Sensitivity analysis and intermodal comparison. *J. Geophys. Res.* 107: DOI: 10.1029/2001JB000162.

Watson, J.D. and H.C. Crick. 1953. Molecular structure of nucleic acids. *Nature* 171: 737–738.

Whipple, K.X. and G.E. Tucker. 1999. Dynamics of the stream-power river incision model: Implications for height limits of mountain ranges, landscape response timescalesm and research needs. *J. Geophys. Res.* 104: 17,661–17,674.

Willgoose, G., R.L. Bras and I. Rodriguez-Iturbe. 1991. A coupled channel network growth and hillslope evolution model. 2. Nondimensionalization and applications. *Water Resour. Res.* 27(7): 1685–1696.

Index

F

G

H

I

K

L

M

N

O

P

R

S